FRICTION STIR WELDING AND PROCESSING IV

T0188852

Friction Stir Welding and Processing IV
TMS Member Price: $58   TMS Student Member Price: $46   List Price: $83

Related Titles
- *Friction Stir Welding and Processing III*
  Edited by Kumar V. Jata, Murray W. Mahoney, Rajiv S. Mishra and Thomas J. Lienert
- *Surfaces and Interfaces in Nanostructured Materials and Trends in LIGA, Miniaturization, and Nanoscale Materials: 5th MPMD Global Innovations Symposium*
  Edited by Sharmila M. Mukhopadhyay, John Smugeresky, Sudipta Seal, Narendra B. Dahotre and Arvind Agarwal
- *Multiphase Phenomena and CFD Modeling and Simulation in Materials Processes*
  Edited by L. Nastac and B. Li

HOW TO ORDER PUBLICATIONS

For a complete listing of TMS publications, contact TMS at (800) 759-4TMS or visit the TMS Document Center at http://doc.tms.org:

- Purchase publications conveniently online.
- View complete descriptions, tables of contents and sample pages.
- Find award-winning landmark papers and reissued out-of-print titles.
- Compile customized publications that meet your unique needs.

MEMBER DISCOUNTS

TMS members receive a 30% discount on TMS publications. In addition, members receive a free subscription to the monthly technical journal *JOM* (both in print and online), discounts on meeting registrations, and additional online resources to name a few of the benefits. To begin saving immediately on TMS publications, complete a membership application when placing your order in the TMS Document Center at http://doc.tms.org or contact TMS.

Telephone: (724) 776-9000 / (800) 759-4TMS
E-mail: membership@tms.org or publications@tms.org
Web: www.tms.org

FRICTION STIR WELDING AND PROCESSING IV

Proceedings of a symposia sponsored by
the Shaping and Forming Committee of
the Materials Processing & Manufacturing Division of
TMS (The Minerals, Metals & Materials Society)

TMS 2007 Annual Meeting & Exhibition
Orlando, Florida, USA
February 25-March 1, 2007

Edited by

Rajiv S. Mishra
Murray W. Mahoney
Thomas J. Lienert
Kumar V. Jata

A Publication of

TMS

A Publication of **The Minerals, Metals & Materials Society (TMS)**
184 Thorn Hill Road
Warrendale, Pennsylvania 15086-7528
(724) 776-9000

Visit the TMS Web site at
http://www.tms.org

Statements of fact and opinion are the responsibility of the authors alone and do not imply an opinion on the part of the officers, staff, or members of TMS. TMS assumes no responsibility for the statements and opinions advanced by the contributors to its publications or by the speakers at its programs. Registered names and trademarks, etc., used in this publication, even without specific indication thereof, are not to be considered unprotected by the law.

Library of Congress Catalog Number 2006940582
ISBN Number 978-0-87339-661-5

TMS

If you are interested in purchasing a copy of this book, or if you would like to receive the latest TMS publications catalog, please telephone (724) 776-9000, ext. 270, or (800) 759-4TMS.

FRICTION STIR WELDING AND PROCESSING IV
TABLE OF CONTENTS

Friction Stir Welding and Processing IV

Session I

PREFACE

Friction stir welding was invented by TWI (formerly The Welding Institute) in Cambridge, United Kingdom, and patented in 1991. In the last 15 years, friction stir welding has seen significant growth in both technology implementation and scientific exploration. This is the fourth symposium on friction stir welding and processing held under the auspices of TMS. The interest and participation in this symposium is an indirect testimony to the growth of this field. This year, a total of 69 abstracts have been submitted, and presentations will require six full sessions. The presentations cover all aspects of friction stir welding and processing, from fundamentals to design and applications. Forty-seven manuscripts are published in the conference proceedings. Over the last six years of this reoccurring symposium, the number of presentations and manuscripts has more than doubled. This general trend is parallel to the increase in journal publications as shown in the in figure below. The journal publication data are from the Institute of Scientific Information database and do not include conference manuscripts. The manuscripts in this latest volume provide an update on the current research issues in the field of friction stir welding, and a guide for further research. After the last TMS friction stir welding and processing symposium, a major activity in the United States became the formation of the National Science Foundation Industry/University Cooperative Research Center for Friction Stir Processing at four institutions: South Dakota School of Mines and Technology, Brigham Young University, University of South Carolina and the University of Missouri-Rolla. These universities have contributed a total of 23 manuscripts to these published proceedings.

Illustration of the growth of publications on friction stir welding and processing

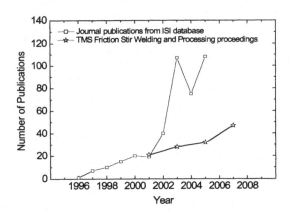

The organizers thank the Shaping and Forming Committee of the
TMS Materials Processing & Manufacturing Division for sponsoring this symposium.

Rajiv S. Mishra *(University of Missouri-Rolla, Missouri)*
Murray W. Mahoney *(formerly with Rockwell Scientific, Thousand Oaks, California)*
Thomas J. Lienert *(Los Alamos National Laboratory, New Mexico)*
Kumar V. Jata *(Air Force Research Laboratory, Ohio)*

ABOUT THE EDITORS

Rajiv S. Mishra is professor of metallurgical engineering in the materials science and engineering department at the University of Missouri-Rolla. He is also the university site director of the National Science Foundation Industry/University Cooperative Research Center for Friction Stir Processing. Dr. Mishra has authored or co-authored 150 papers in peer-reviewed journals and proceedings and is principal inventor of three U.S. patents. He has received numerous awards, including the Faculty Excellence Award from the University of Missouri-Rolla from 2001 through 2006, Young Metallurgist Award from the Indian Institute of Metals in 1993, Associate of the Indian Academy of Sciences in 1993, and Brunton Medal for the best Doctorate dissertation in the School of Materials from the University of Sheffield, United Kingdom, in 1988. Dr. Mishra also received the Firth Pre-doctoral Fellowship from the university as well as his doctorate in metallurgy in 1988.

Murray W. Mahoney is a consultant with more than 39 years experience in physical metallurgy and related disciplines. Most recently, his work has centered on developing friction stir welding and processing. This work has led to the introduction of friction stir welding to join metals considered unweldable and advance friction stir welding to higher temperature alloys. In addition, developments in friction stir processing have advanced superplasticity to very thick section structural Al alloys, enhanced room temperature formability, and improved mechanical and corrosion properties in cast Al and Cu alloys. These studies have resulted in a more complete metallurgical understanding of joining fundamentals, microstructural evolution, formability, and corrosion sensitivity. Mr. Mahoney holds a master's degree from UCLA and a bachelor's from the University of California at Berkeley, both in physical metallurgy.

Thomas J. Lienert is a technical staff member at Los Alamos National Laboratory. His current research interests include laser materials processing and friction welding, including friction stir welding. He is a principal reviewer for the *Welding Journal* and is an active reviewer for *Metallurgical and Materials Transactions*. Dr. Lienert has published over 25 papers in refereed journals and conference proceedings, and was the author of the chapter on Selection and Weldability of Aluminum Metal-Matrix Composites in the ASM Metals Handbook on Welding, Brazing and Soldering. He has also made over 60 presentations at technical conferences. Dr. Lienert has been awarded both the Charles H. Jennings Award and the McKay-Helm Award from AWS for papers published in the *Welding Journal*. He is a member of Tau Beta Pi Engineering Honors Society and Alpha Sigma Mu Metallurgy/Materials Science Honors Society. Previously, Dr. Lienert held positions at the University of South Carolina, Edison Welding Institute, and Sandia National Laboratories. He holds a doctorate and a master's degree in materials science and engineering, and a bachelor's in welding engineering, all from The Ohio State University.

For the last 25 years, **Kumar V. Jata** has worked extensively in the area of material development, processing, and structure-properties of light-weight, high-strength metallic materials for aircraft and space. His research interests include understanding of materials used in reusable thermal protection systems, metallic cryogenic tanks and aerospace structures. He has also conducted numerous structural and engine component failure analysis for the U.S. Air Force. Dr. Jata is currently the research group leader for the NDE branch at Air Force Research Laboratory (AFRL). His contributions to NDE research is in the area of sensor-material interactions and sensor development for real-time material state and damage state monitoring of space and aircraft structures. He became a Fellow of AFRL in 2006 and is the recipient of numerous technical achievement awards from the U.S. Air Force. Dr. Jata obtained his doctorate from the University of Minnesota, Minneapolis, in 1981.

FRICTION STIR WELDING AND PROCESSING IV

FRICTION STIR WELDING:
AFTER A DECADE OF DEVELOPMENT

William J. Arbegast

Director, Advanced Materials Processing and Joining Center
Director, NSF Center for Friction Stir Processing
South Dakota School of Mines & Technology;
501 E St Joseph St; Rapid City, SD, 57701, USA

Keywords: Friction Stir Welding, Friction Stir Spot Welding, Friction Stir Processing, Friction Stir Joining, Friction Stir Reaction Processing

Abstract

Friction Stir Welding (FSW) is an innovative solid state welding process invented in 1991 by TWI- The Welding Institute [1]. FSW can arguably be said to represent one of the most significant developments in joining technology over the last half century. The initial development by TWI and its industrial partners under various Group Sponsored Projects focused on single pass, complete joint penetration of arc weldable (5XXX and 6XXX) and un-weldable (2XXX and 7XXX) aluminum alloys up to 1 inch thick.

By 1995, FSW had matured to a point where it could be transitioned and implemented in the US aerospace and automotive markets. The many advantages of FSW compared to conventional arc welding have repeatedly been demonstrated with both improved joint properties and performance. Often, production costs are significantly reduced. Other times, FSW enables new product forms to be produced or skilled labor freed to perform other tasks. Research and development efforts over the last decade have resulted in improvements in FSW, and, the spin-off of a series of related technologies.

Introduction

In the 1920's and 1930's arc welding replaced rivets as the joining method for pressure vessels. Weld usage expanded through the 1940's with application to buildings, structures, and ships. By 2006, arc welding has evolved into an international industry complete with welder education and certification programs and governed by extensive specifications, design criteria and standards. A 2002 survey by the American Welding Society (AWS) estimated that US manufacturing industries spend over $34.4 billion annually on arc welding of metallic materials with an anticipated growth rate averaging 5% to 15% per year. The construction, heavy manufacturing, and light manufacturing industries make up the majority with $25 billion in annual expenditures. Industry wide repair and maintenance of welded structures are estimated to cost $4.4 billion annually. In doing so, these industries are a major consumer of energy and a producer of airborne emissions and solid waste.

Conventional arc welding of metals creates a structural joint by local melting and subsequent solidification. This normally requires the use of expensive consumables, shielding gas and filler metal. The melting of materials is energy intensive and solidifying metals are often subject to cracking, porosity, and contamination. Undesirable metallurgical changes can occur in the cast nugget due to alloying with filler metals, segregation, and thermal exposure in the heat affected zones. These may result in degraded joint strengths, extensive and costly weld repairs, and unanticipated in-service structural failures. Solid-state (non-melting) joining avoids these undesirable characteristics of arc welding.

FSW Implementation Incentives

Friction Stir Welding (FSW) is one such non-melting joining technology that has produced structural joints superior to conventional arc welds in aluminum, steel, nickel, copper and titanium alloys. FSW produces higher strength, increased fatigue life, lower distortion, less residual stress, less sensitivity to corrosion, and essentially defect free joints compared to arc welding. Since melting is not involved, shielding gases are not used during the FSW of aluminum, copper and NiAl bronze alloys while argon gas may be used during the FSW of the higher temperature ferrous and nickel alloys, mainly to protect the ceramic and refractory pin tools from oxidation. Simple argon environmental chambers and trailing shields are used during the FSW of titanium alloys to minimize interstitial pickup and contamination. Expensive consumables and filler metals are not required. An excellent state of the art review of the FSW technology is provided by Mishra and Ma [2].

FSW researchers and producers (AJT, Inc.) estimate that if 10% of the US joining market can be replaced by FSW, then 1.28×10^{13} Btu/year energy savings and 500 million pounds/ year "greenhouse" gas emission reductions can be realized. Hazardous fume emissions during the FSW of high temperature and chromium containing alloys are eliminated. Rockwell Scientific reports emission levels of Cr, Cu, Mn and Cr^{-6} (<0.03, <0.03, <0.02, and <0.01 mg/mm^3, respectively) during FSW of ferrous alloys considerably lower than those measured during GTAW (0.25, 0.11, 1.88 and 0.02 mg/mm^3, respectively).

The simplified processing, higher structural strength, increased reliability, and reduced emissions of FSW are estimated to create annual economic benefit to US industry of over $4.9 billion/year.

FSW Barriers to Implementation

The aeronautic and aerospace industries represent less than 1% ($300 million) of the total US annual welding expenditures since mechanical fastening is the joining method of choice. However, the bulk of FSW development dollars has been spent by these sectors. As a result, the broader industrial market for FSW implementation has been neglected.
As of January 2005, there were 114 FSW licenses granted by TWI (Figure 1). These were almost equally split between North America (36), Europe (37), and Asia (41) with no reported licensees in South America. Overseas, 68% of the licensees are industrial while only 36% of the North America licensees are industrial and the remaining 64% are held by government laboratories, equipment manufacturers, and academic and research institutes. This suggests that industrial implementation of the FSW process in the United States is lagging behind the overseas industries. Several over-riding issues have been identified as barriers to more extensive FSW implementation in US markets:

- Lack of industry standards and specifications
- Lack of accepted design guidelines and design allowables
- Lack of an informed workforce, and,
- The high cost of capital equipment

In 1998, the AWS D17J Subcommittee began development of a specification for Friction Stir Welding to address these issues. Acceptance and release is anticipated in the near term. In the meantime, most FSW users have developed internal specifications for application to their products. In 2002, AJT, Inc. secured American Bureau of Shipping (ABS) approval to use FSW in marine applications.

These barriers to broader market implementation in the US are also being addressed through national FSW research programs and various successful industrial implementations.

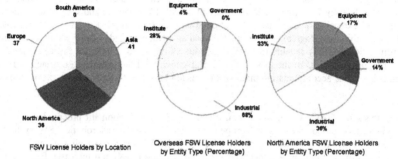

Figure 1- Demographics of FSW Licensees as of January 2005 (source: TWI)

FSW Research Programs

In 2004, the Center for Friction Stir Processing (CFSP), a National Science Foundation Industry/ University Cooperative Research Center (NSF I/UCRC), was established to bring the South Dakota School of Mines and Technology (SDSMT), University of South Carolina (USC), Brigham Young University (BYU), and the University of Missouri-Rolla (UMR) together in a collaborative FSW research program. The CFSP currently has 22 government laboratory and industrial sponsors. The center mission is to perform advanced and applied research, develop design guidelines and allowables, train scientists and engineers, and transfer the FSW technology into a broader base within the industrial sector. Current research programs at the CFSP include:

- Design allowables and analysis methodologies for FSW beam and skin stiffened panel structures
- Intelligent FSW process control algorithms
- Thermal management of titanium and aluminum FSW for property control
- Microstructural modification of aluminum and magnesium castings
- FSW of HSLA and 4340 steels
- FSW of austenitic steels and Inconel alloys
- Interactive database of FSW properties and processing parameters

The CFSP has also teamed with the Iowa State University Center for Nondestructive Evaluations (CNDE) to assess the "Effects of Defects" in aluminum alloy FSW. The probability of detection (POD) of various NDE methods are being established for the volumetric and geometric characteristic discontinuities and the relationship between flaw size and reduction in static strength and fatigue life are being determined. Statistical process control (SPC) methods are being developed based on process force and torque responses in frequency space and are being compared to the POD of the NDE methods.

5

The Edison Welding Institute Navy Joining Center (NJC) has continued to develop and demonstrate FSW technologies in thick section aluminum and titanium alloys for a variety of DOD applications. One recent technology demonstration program at the NJC used a combination of FSW, GMAW, and Hybrid Laser Welding to fabricate a large titanium structure from 0.50 inch thick Ti-6Al-4V plates (Figure 2). In this assembly, the initial corner joints were friction stir welded from the outside of the structure to establish the basic shape with the remaining structure assembled using GMAW and Hybrid Laser Welding.

Under a recently completed DARPA program, Rockwell Scientific and the NAVSEA Carderock Surface Warfare Center, in conjunction with 13 university and industrial partners, performed extensive development of friction stir processing on Al, Cu, Mn, and Fe based alloys. Within this program, MegaStir developed an advanced grade of polycrystalline cubic boron nitride (PCBN) capable of FSW of ferrous alloys up to 0.500 inch thick (Figure 3). The fracture toughness of the PCBN is sufficiently high to allow features to be machined on the tool pin thus accommodating material flow around the tool to fill the cavity in the tool's wake.

Also, this same DARPA program demonstrated the ability to friction stir process large areas on the surface of complex shaped propellers using large industrial robotic FSW systems provided by Friction Stir Link, Inc. (Figure 4). Friction stir processing eliminates near surface casting discontinuities, increases the yield strength (>2X) and increases fatigue life (>40%) compared to as-cast NiAl bronze. In addition, FSW equipment manufacturers (General Tool Corporation) are exploring alternatives to high cost multi-functional FSW equipment by developing lower cost, dedicated, single purpose systems.

A new national FSW task coordinated by Boeing is being launched under the Next Generation Manufacturing Technology Initiative (NGMTI). The NGMTI program is designed to accelerate the development and implementation of breakthrough manufacturing technologies to support the transformation of the defense industrial base and to increase the global economic competitiveness of U.S.-based manufacturing. This FSW task will bring together the DoD Tri-Services, JDMTP, DLA, FAA, NASA, and DOE with a large contingent of industrial and university partners to perform enabling and applied research to correct over-riding implementation barriers, and, to perform ManTech type demonstrations to accelerate industrial and government acceptance and implementation of Friction Stir Welding.

A second NGMTI FSW task, coordinated by Friction Stir Link, Inc. in conjunction with the University of Wisconsin, is being developed to provide a low-mass, low-power, and high-mobility robotic FSW systems. Friction Stir Link has been developing robotic FSW and Friction Stir Spot Welding (described later) for a variety of automotive and commercial applications. Integrating the FSW technology with robotics allows for flexible manufacturing approaches and reductions in production costs.

Figure 2 – The Edison Welding Institute used a combination of Friction Stir Welding, GMAW, and Hybrid Laser Welding to fabricate this demonstration article from thick section titanium plates. FSW was used to join the 0.50 inch thick plates in a corner joint configuration (arrows) to establish the basic shape of the article and GMAW and Hybrid Laser Welding was used to complete the assembly (Courtesy Edison Welding Institute)

Figure 3 –0.25 "Tapered with Flats" (left, bottom), 0.25 inch "Stepped Spiral" (left, top), and 0.500 inch "Stepped Spiral" High Temperature PCBN FSW Pin Tools (Courtesy MegaStir, Inc)

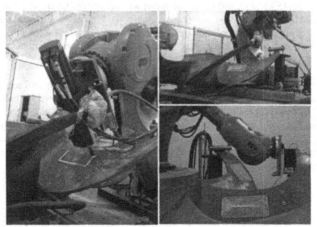

Figure 4 – Friction Stir Link, Inc., robotic FSW system processing large areas of NiAl Bronze propellers to remove near surface casting defects (Courtesy Rockwell Scientific)

Concurrent Technologies Corporation (CTC), through the Navy ManTech National Metalworking Center (NMC), has advanced the development of FSW in thick section 5083, 2195, and 2519 aluminum for ground and amphibious combat vehicles. Several large scale prototypes have been completed. The work by CTC and NMC has provided a valuable transition of the technology from sub-scale laboratory work to full-scale prototype construction – the last major step before production implementation.

FSW Process Innovations

Innovations to the FSW process are ongoing. Since 1995, over 50 US patents in FSW have been issued. Pin tool designs have evolved from those originally developed by TWI to unique designs for thick section, lap joint, high temperature, and fast travel speed joining. For example, in 2005, GKSS- GmbH reported that successful FSW at welding speeds in excess of 780 inch per minute in thin gage aluminum butt joints have been achieved.

In 1999, the NASA Marshall Space Flight Center (MSFC) and the Boeing Company developed the retractable pin tool [3] for the FSW of tapered thickness joints. MSFC is currently investigating the use of very high rotation speed (>50,000 rpm) FSW, Thermal Stir Welding (TSW) and the integration of ultrasonic energy during FSW to enable portable hand held devices.

Other researchers are also evaluating modifications to the FSW process. For example, the University of Missouri – Columbia is evaluating Electrically Enhanced Friction Stir Welding (EEFSW) where additional heat is applied by resistance heating through the pin tool. The University of Wisconsin is developing Laser Assisted Friction Stir Welding (LAFSW) of aluminum lap joints where a laser is trained ahead of the pin tool to preheat the material.

Under a collaborative research program between the Army Research Laboratory (ARL) and the SDSMT Advanced Materials Processing and Joining Center (AMP), a variety of FSW technologies are being developed, including complex curvature FSW, Friction Stir Spot Welding (FSSW), dissimilar alloy FSW, low cost fixturing and tooling, and thick plate titanium and aluminum FSW. Prototypes of advanced fuselage structures (Boeing), helicopter beams (Sikorsky), and naval gun turret weather shields (BAE Systems) have been built. The AMP Center is also developing Induction Pre-heated Friction Stir Welding (IPFSW) using an Ameritherm 20Kw remote heat station to preheat thick plate aluminum, steel, cast iron, and titanium alloys to increase travel speeds, reduce process forces, and reduce pin tool wear (Figure 5).

In 2001, the MTS Systems Corporation patented the self-reacting pin tool technology [4]. This innovation allows the FSW of tapered joints and eliminates the need for back side anvil support to react the process loads. Lockheed Martin Space Systems and the University of New Orleans National Center for Advanced Manufacturing (NCAM) have demonstrated this self-reacting pin tool on the 27 foot diameter domes of the Space Shuttle External Tank. In this application, multiple gore sections of 0.320 inch thick 2195 Al-Li were joined along a simple curvature path to create the full scale dome assembly.

*Figure 5 – MTS ISTIR 10 Friction Stir Weld System (left) with the Ameritherm 20 KW
Remote Heat Station and induction pre- heating coil (right)
(Courtesy South Dakota School of Mines)*

FSW Industrial Implementations

The Technology Readiness Level (TRL) for the FSW of aluminum alloys is high with successful industrial implementation and space flight qualification by Boeing on the 2014 aluminum propellant tanks of the Delta II and Delta IV space launch vehicles. Lockheed Martin and NASA MSFC have developed and implemented FSW on the longitudinal welds of the 2195 Al-Li liquid hydrogen and liquid oxygen barrel segments of the External Tank for the Space Shuttle (Figure 6). Lockheed Martin Missiles and Fire Control and the SDSMT have developed square box beams for mobile rocket launch system that are fabricated from thick wall "C" section extrusions joined by FSW to replace the current hollow, square tube extrusions. Airbus has announced the use of FSW in selected locations on the Airbus A350 and two new versions of the A340 (A340-500, A340-600).

In 2000, the Air Force Metals Affordability Initiative (MAI) brought together a consortium of industry and university partners to develop FSW for a variety of DOD Applications. Under Task 1, "Joining of Traditional Aluminum Assemblies" Lockheed Martin completed a development program which replaced the riveted aluminum floor structure of the C130J air transport with a FSW floor structure. Under Task 2, "Joining of Complex Aluminum Assemblies", Boeing developed a FSW cargo "slipper" pallet and implemented a FSW Cargo Ramp Toe Nail on the C17 transport. The Toe Nail is the only known friction stir welded part flying on a military aircraft. Under Task 3, "Hard Metals Joining Development", the Edison Welding Institute and General Electric developed high temperature pin tools for the FSW of steel, titanium and inconel alloys for aircraft engine applications.

Eclipse Aviation is in final FAA certification for the Eclipse 500 business class jet. First customer deliveries are scheduled for 2006. FSW lap joints are used as a rivet replacement technology to join the longitudinal and circumferential internal stiffeners to the aft fuselage section and to attach doublers at window and door cutout locations (Figure 7). The use of FSW eliminates the need for thousands of rivets and results in better quality and stronger and lighter joints at reduced assembly costs. MTS Systems Corporation designed and fabricated the custom FSW equipment and production tooling for Eclipse Aviation. This equipment permits welding complex curvatures over many sections of the fuselage, cabin, and wing structures at travel speeds in excess of 20 inches per minute (Figure 8). Because the process is faster than more conventional mechanical joining processes, the production cycle time is significantly reduced.

9

Over the last three years, the Ford Motor Company has produced several thousand Ford GT automobiles with a FSW central tunnel assembly (Figure 9). This tunnel houses and isolates the fuel tank from the interior compartment and contributes to the space frame rigidity. The top aluminum stamping is joined to two hollow aluminum extrusions along the length of the tunnel using a linear FSW lap joint. The use of FSW results in improved dimensional accuracy and a 30% increase in strength over similar GMAW welded assemblies.

The TRL for FSW of ferrous, stainless steel, nickel, copper and titanium alloys is also high with a variety of full scale demonstration programs completed. MegaStir, Inc. has developed an improved grade of the polycrystalline cubic boron nitride (PCBN) high temperature pin tools (HTPT) which has shown an acceptable service life for welding steels, nickel and copper alloys. In 2004, MegaStir, Inc. completed a prototype oil field pipeline FSW demonstration program that successfully joined 12 inch diameter x 0.25 inch wall thickness X-65 steel pipe segments using an automated internal mandrel and external FSW tooling system (Figure 10).

Chemical compatibility issues arise when welding titanium alloys with the PCBN pin tools. The University of South Carolina has shown the suitability of tungsten – rhenium (W-Re) HTPT for most titanium alloys. However, issues with pin tool wear and excessive metal adhesion still arise when welding Ti-6Al-4V. This is possibly due to reactions between the rhenium in the pin tool and the vanadium alloying elements in the titanium. Other refractory HTPT materials, such as tungsten – iridium (W-Ir), are under development at the Oak Ridge National Laboratory.

In 2005, Lockheed Martin performed FSW on 0.20 inch thick Ti-6Al-4V sheets using dispersion strengthened tungsten HTPT that alleviated the "sticking" problem and allowed for many meters of welding (Figure 11). They report that the joint efficiency ranged from 98% to 100% of base metal strength at testing temperatures ranging from -320°F to +500°F. Titanium FSW produced at the CFSP using custom designed environmental chambers and an argon atmosphere (Figure 12) showed no evidence of surface discoloration or interstitial (O, N and H) contamination.

Figure 6 – FSW process development tool at the Marshal Space Flight Center shown with a 27 foot diameter barrel segment of the 2195 Al-Li Space Shuttle External Tank LH$_2$ tank (left). Full scale LH$_2$ tank (right) at the NASA Michoud Assembly Facility in New Orleans (Courtesy NASA MSFC)

Figure 7- The Eclipse 500 business class jet is currently in final FAA certification trials (left). The internal longitudinal and circumferential aluminum stiffeners (right, top) and window and door doublers (right, bottom) are attached to the aluminum fuselage section with friction stir welded lap joints (Courtesy Eclipse Aviation).

Figure 8 – The friction stir welding equipment used to attach the stiffeners and doublers to the Eclipse 500 fuselage sections was designed and fabricated by MTS Systems Corporation. It is capable of welding a variety of component geometries through the use of interchangeable holding fixtures located beneath the multi-axis FSW head and movable gantry frame (Courtesy Eclipse Aviation).

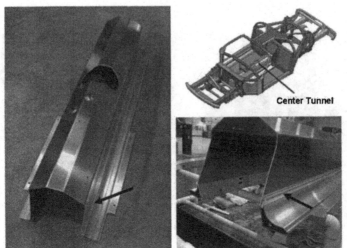

Figure 9 – The central tunnel assembly of the Ford GT is a FSW assembly made from aluminum stampings and extrusions. (Courtesy Ford Motor Company)

Figure 10 – Prototype pipe welding system showing external FSW head and internal mandrel (inset) (Courtesy MegaStir, Inc)

Figure 11 – Joining of long lengths of contamination free Ti-6Al-4V are possible with "out of chamber" Friction Stir Welding using shrouds and trailing shoe shielding gas systems (Courtesy Lockheed Martin Space Systems)

Figure 12 – Environmental chambers are used to provide an argon atmosphere and to minimize interstitial contamination in titanium FSW (Courtesy South Dakota School of Mines)

Friction Stir Spot Welding

If FSW is considered as a *"controlled path extrusion"* rather than a *"welding"* process, several spin-off technologies can be realized. Friction Stir Spot Welding (FSSW) has been in development over the last five years and has seen industrial implementation as a rivet replacement technology. Currently two variations to FSSW are being used. The "Plunge" Friction Spot Welding (PFSW) method was patented by Mazda in 2003 [5] and the "Refill" Friction Spot Welding method was patented by GKSS- GmbH in 2002 [6].

In the Mazda PFSW process, a rotating fixed pin tool similar to that used in linear FSW is plunged and retracted through the upper and lower sheets of the lap joint to locally plasticize the metal and stir the sheets together. Even though this approach leaves a pull-out hole in the center of the spot, the strength and fatigue life is sufficient to allow application at reduced production costs on the Mazda RX-8 aluminum rear door structure (Figure 13). Since 2003, Mazda has produced more than 100,000 vehicles with this PFSW rear door structure. These PFSW doors provide structural stability against side impact and impart five star roll over protection.

The GKSS RFSW is being developed at the SDSMT AMP Center under license to RIFTEC-GmbH. This process uses a rotating pin tool with a separate pin and shoulder actuation system that allows the plasticized material initially displaced by the pin to be captured under the shoulder during the first half of the cycle and subsequently re-injected into the joint during the second half of the cycle. This completely refills the joint flush to the surface (Figure 14). In addition to development as a rivet replacement technology for aerospace structures, RFSW is also being developed as a tacking method to hold and restrain parts during over-welding by linear FSW.

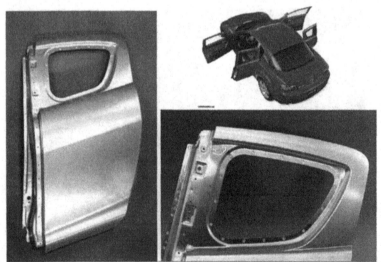

Figure 13 – Use of the Plunge Friction Spot Welding (PFSW) method on the Mazda RX-8 rear door structure provides for structural stability against side impact and five star roll over protection at reduced production costs. (Courtesy Ford Motor Company)

Figure 14 – Refill Friction Spot Welding (RFSW) using MTS ISTIR 10 system and custom designed head adapter (left). RFSW lap shear coupons (right, bottom) and metallurgical cross section of RFSW showing complete joint penetration in 0.080 inch thick 7075-T73 aluminum (right, top) (Courtesy South Dakota School of Mines)

Friction Stir Joining

Friction Stir Joining (FSJ) of thermoplastic materials uses the controlled path extrusion characteristics of the process to join ¼ inch thick sheets of polypropylene (PP), polycarbonate (PC), and high density polyethylene (HDPE) materials. Recent work at the Brigham Young University has shown joint efficiencies for these materials ranging from 83% for PC to 95% for HDPE and 98% for PP. These joint efficiencies compare favorably with other polymer joining methods such as ultrasonic, solvent, resistance, hot plate and adhesive bonding. Current work at the SDSMT AMP Center in collaboration with the Air Force Research Laboratory -Kirtland is investigating the use of FSJ to join fiber, particulate, and nanoparticle reinforced thermoplastic materials.

Friction Stir Processing

Friction Stir Processing (FSP) uses the controlled path metal working characteristics of the process to perform metallurgical processing and microstructural modification of local areas on the surface of a part. In 1997, FSP was used by Lockheed Martin to perform microstructural modification of the cast structure of 2195 Al-Li VPPA welds to remove porosity and hot short cracks. This also improved room temperature and cryogenic strength, fatigue life, and reduced the sensitivity to intersection weld cracking by crossing VPPA welds [8].

In 1998 the Department of Energy's Pacific Northwest National Laboratory (PNNL) began investigating the processing of SiC powders into the surfaces of 6061 aluminum to increase wear resistance. Initial studies showed that both SiC and Al_2O_3 could be emplaced into the surface of bulk materials to create near surface graded MMC structures. The University of Missouri-Rolla (UMR) has shown that a uniform SiC particle distribution can be achieved with appropriate tool designs and techniques, leading to significant increases in surface hardness.

In 2004, a PNNL/ SDSMT AMP Center collaborative research program investigated increasing the friction characteristics and wear resistance of heavy vehicle brake rotors by processing TiB_2 particles into the surface of Class 40 gray cast iron. This resulted in a fourfold increase in the dry abrasive wear resistance when tested per ASTM G65 (Figure 15). PNNL and Tribomaterials, LLC have performed subscale brake rotor/pad wear tests on FSP/TiB_2 cast iron rotors. These subscale brake tests have shown that FSP/TiB2 processed brake rotors have improved friction characteristics and wear resistance over baseline heavy vehicle brake friction pairs.

Friction Stir Reaction Processing (FSRP) was also investigated under this PNNL/ SDSMT FSP/TiB$_2$ program. FSRP uses the high temperatures and strain rates seen during processing to induce thermodynamically favorable "*in-situ*" chemical reactions on the surface to a depth defined by the pin tool geometry and metal flow patterns. This provides an opportunity for innovative processing methods to create new alloys on surfaces of materials and locally impart a variety of chemical, magnetic, strength, stiffness, and corrosion properties.

Studies performed at the University of Missouri-Rolla in conjunction with Rockwell Scientific have shown FSP to produce a fine grain size material and create low temperature high strain rate superplasticity in aluminum and titanium alloys. PNNL is currently investigating the application of this FSP induced superplasticity in the fabrication of large integrally stiffened structures.

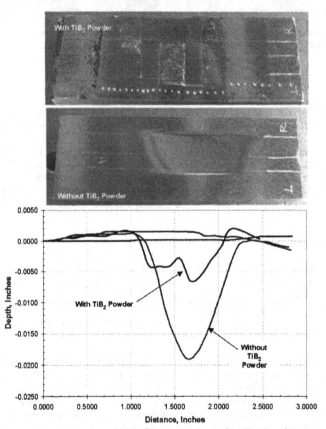

Figure 15 – Grade 40 gray cast iron ASTM G65 wear test results. Friction Stir Processing TiB$_2$ particles into the surface resulted in a fourfold increase in ASTM G65 dry abrasive wear resistance over that seen in samples without TiB$_2$ Particles
(Courtesy South Dakota School of Mines)

16

Summary

Friction Stir Welding (FSW) has matured a great deal since its introduction into the US market in 1995. The Technology Readiness Level (TRL) for aluminum alloys is high with several industrial implementations. While development efforts and property characterizations have shown that FSW can be used in ferrous, stainless, nickel, copper and titanium alloys, an industrial champion is needed.

The "metalworking" nature of the process leads to the Plunge (PFSW) and Refill (RFSW) Friction Stir Spot Welding (FSSW) methods with properties comparable to riveted and resistance spot welded joints. The use of Friction Stir Processing (FSP) to locally modify the microstructure of arc welds and castings has shown to increase strength, improve fatigue life, and remove defects. Using FSP to stir particulate materials into the surface has shown increased wear resistance and creates particulate reinforced surface layers. Friction Stir Reaction Processing (FSRP) can be used to create new materials and alloy combinations on part surfaces.

The higher strength, non-melting and environmentally friendly nature of the FSW process has shown cost reductions in a variety of applications and has enabled new product forms to be developed. Only a small percentage of the US welding and joining market has been targeted for implementation. A variety of government, industry and university collaborations are underway to accelerate the development and implementation of FSW and FSSW into these markets.

During the last decade, the defense and aerospace sectors have taken the lead in implementing FSW. Recent advances in pin tool designs and optimized processing parameters have enabled FSW and FSSW applications in the marine, ground transportation, and automotive industries. Further innovations in low cost equipment and the development of industry standards, design guidelines, and a trained workforce will enable the introduction of FSW and FSSW into the broader light manufacturing, heavy manufacturing, and construction industries during the next decade.

Acknowledgements

This paper is taken with permission in its entirety from the March 2006 issue of the Welding Journal. Contributions to that Welding Journal Article were received from Gilbert Sylva and Mike Skinner (MTS Systems Corporation), Glenn Grant (PNNL), Brent Christner (Eclipse Aviation), Doug Waldron (AJT, Inc), Jeff Ding (NASA MSFC), Tim Trapp (EWI), Tracy Nelson (BYU), Carl Sorensen (BYU), Tony Reynolds (USC), Zach Loftus (Lockheed Martin), Murray Mahoney (Rockwell Scientific), John Baumann (Boeing), Raj Talwar (Boeing), Dana Medlin (SDSMT), Anil Patnaik (SDSMT), Casey Allen (SDSMT), Rajiv Mishra (UMR), Chuck Anderson (ATI, Inc), John Hinrichs (Friction Stir Link, Inc.), Kevin Colligan (CTC Corporation), Scott Packer (MegaStir) and Tsung -Yu Pan (Ford Motor Company).

The research programs at the Center for Friction Stir Processing (CFSP) are conducted by the faculty and students at the SDSMT, BYU, UMR, and USC university sites and are funded under a grant from the National Science Foundation I/UCRC program office and the membership of the industrial and government partners. The research programs at the SDSMT Advanced Materials Processing and Joining Center are funded by the Army Research Laboratory, Air Force Research Laboratory- Kirtland, DOE Pacific Northwest National Laboratory, and the Edison Welding Institute.

References

1. W.M. Thomas et al., "Friction Stir Butt Welding", U.S. Patent No. 5,460,317
2. R.S. Mishra and Z.Y. Ma, "Friction Stir Welding and Processing", Materials Science and Engineering R50 (2005) 1–78
3. J. Ding and P. Oelgoetz, "Auto-Adjustable Pin Tool for Friction Stir Welding", US Patent 5,893,507, April 13, 1999
4. Campbell, Fullen and Skinner, "Welding Head", US Patent 6,199,745, March 13, 2001
5. Iwashita, T. et al, "Method and Apparatus for Joining", US Patent 6,601,751 B2, August, 5, 2003.
.6 C. Schilling and J. dos Santos, "Method and Device for Joining at Least Two Adjoining Work Pieces by Friction Welding", US Patent Application 2002/0179 682
7. W.J. Arbegast and P.J. Hartley, "Method of Using Friction Stir Welding to Repair Weld Defects and to Help Avoid Weld Defects in Intersecting Welds", US Patent 6,230,957, May 15, 2001

FRICTION STIR WELDING OF AN ALUMINUM COAL HOPPER RAILCAR

Casey Allen[1], Clark Oberembt[1], William Arbegast[1], Dana Medlin[2], Haven Mercer[3]

[1]Advanced Materials Processing Center;
[2]Materials and Metallurgical Engineering Department; South Dakota School of Mines & Technology;
501 E St Joseph St; Rapid City, SD 57701, USA
[3]Hutchinson Technology Incorporated
40 West Highland Park, Hutchinson MN 55350

Keywords: Railcar, Aluminum, Friction Stir Welding

Abstract

The friction Stir welding (FSW) process was used to fabricate a subscale section of an aluminum railcar designed to ship coal. Coal comprises approximately forty percent by weight of all freight shipped by rail in the United States. The primary mode of transport is via open-top hopper car. Traditional joining techniques for this type of vehicle include bolting, riveting and fusion welding. An analysis was performed to evaluate the cost effectiveness of FSW as a joining process. Key elements of the analysis included materials selection and availability, preliminary static properties, manufacturability and life-cycle cost benefits analysis. Result of this analysis demonstrated a potential cost savings of 20 percent per hopper structure while integrating with the existing rolling chassis and maintaining durability of the basic coal hopper design. The successful fabrication of a subscale prototype demonstrated overall feasibility of the concept and generated valuable data concerning joint and fixturing and tooling requirements.

Introduction

In the year 2005, freight railroads moved more than 804 million pounds of coal to electrical utilities in the United States. This tonnage comprises 42.4% of the total tons of freight originated in the country (see Figure 1 below). The total number of freight cars in service in the United States in 2005 was approximately 1.3 million with more than 35 million total carloads originated [1]. Overall it is estimated that aluminum coal cars comprise approximately 50% of the total coal railcar fleet with the majority of these used in the western region for coal that originates in Wyoming and Montana [2]. Utilization of aluminum coal cars is at or near 100% which is not the case for steel cars [3]. The aluminum coal car is 6 tons or 28% lighter than the standard 60,000 pound steel coal car [4]. This decreased weight enables additional coal to be transported with the same cost and infrastructure effectively decreasing transportation costs as can be seen in Figure 1 [5]. According to United States Department of Energy the conversion to aluminum coal cars has allowed a significant decrease in the total CO_2 emissions produced by the U.S. transportation sector [6].

19

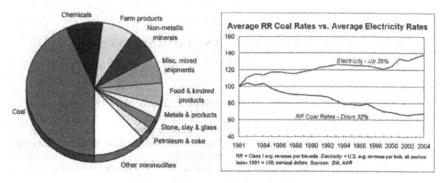

Figure 1: Relative Tonnage of Railfreight Shipped in United States [1], Left; Railroad Coal Transport Rates [5], Right

Coal transport via rail freight is accomplished using two types of freight cars; gondola and open-top hopper. The gondola car is emptied by picking the entire car and the track section it is sitting on and rolling it over to empty it. The hopper is emptied by opening doors on the under side and allowing the coal to escape. Both types of cars are used extensively in modern coal transport. The primary design feature difference is that sloped enwalls are required in the hopper design to facilitate coal flow. Both types of cars utilize a body, that holds the cargo, and trucks which incorporate the wheels and suspension. As mentioned above the majority of coal cars manufactured utilize aluminum for the car body. The focus of this work was on the open-top hopper style car body. An example of this type of car is shown in Figure 2.

Figure 2. Typical Open-top Hopper Railcar,
(Courtesy, Trinity Rail Group, LLC)

Given the large number and increasing percentage of aluminum cars utilized in coal transportation, along with the drive to decrease the cost of delivery there is an apparent opportunity to investigate the application of friction stir welding (FSW) as a joining process that could potentially decrease the total cost of manufacturing for aluminum coal car bodies.
The goals of this project were to investigate the value of implementing friction stir welding and to demonstrate that FSW can be considered a viable alternative to replace the bolted and riveted assembly of Aluminum hopper railcars and that a cost savings of 20% for joining and assembly was achievable. This work was performed by an undergraduate design group at the South Dakota School of Mines and Technology (SDSMT) and was accomplished in four tasks: conceptual design, preliminary design properties development, manufacturing demonstrator, and cost-benefits analysis.

20

Development of a design concept involved generating a body design that was modified to take advantage of the friction stir welding as a joining process while substantially maintaining the existing geometric features and component structural performance. This would ensure compatibility of the new design with existing material handling equipment and also allow direct comparison to existing production articles. Also considered were hybrid designs where the advantages of FSW could be realized while utilizing conventional joining methods where appropriate. Alternatives were considered which spanned the continuum from fully friction stir welded designs to minimally FSW designs.

Design property development focused on generating preliminary joint properties in both fatigue and static loading cases on several joint designs. The candidate joint designs and materials were selected early in the design process and some combinations may not represent those actually chosen for the optimized design.

The manufacturing demonstrator was produced to evaluate the friction stir weldability of a design that used FSW joints almost exclusively. The primary objective for this demonstration article was to implement FSW extensively so that practical approximations of cycle time, appropriate joint designs, fixturing and tooling requirements, and assembly sequence could be made. A ¼ scale model of one corner of the hopper was fabricated using full scale material thickness to approximate some of the more difficult joints and geometry on the railcar. The information obtained from this portion of the development could subsequently be fed into the cost benefit analysis.

A cost benefit analysis model was developed to characterize the cost of production for this product. The model utilized data from the tests and analyses in this project to allow a sensitivity analysis to be performed. This analysis, along with the design information was used to make a recommendation for production and design details of a friction stir welded coal hopper design. Appropriate production information from a car manufacturer was used to approximate the production and lifecycle costs for comparative purposes.

Design for Friction Stir Welding

Modifying the existing railcar design to utilize friction stir welding required an evaluation of the existing design which used bolted connections for all joints. The main structural components of the body are defined in Figure 3. This design requires the use of double thickness sheets at all joints to accommodate the bolted connections. The design as is could be substantially friction stir welded by using lap welded joints. By looking at optional joint configurations a more optimum design utilizing friction stir welding is possible.

Many joint types used in the construction of railroad hoppers are typical of those seen in a variety of existing FSW production and demonstration applications. This project evaluated material selection and joint design options with particular attention to FSW producibility, impact on manufacturing sequencing and production tooling requirements (Figure 4). The material trade ranked 6061-T6 best (cost, strength, fatigue, corrosion, and fusion weld or bolted repairability) with the second choice as 5052-H32. Note that 60601-T6 is used in the current, baseline, fastened railcar hopper. The final design was based on these trades and results in 84 total welds (9,360 in) and only 40 bolts/rivets necessary to produce a full scale railroad car hopper. The salient features of the final, friction stir welded design are compared to the original bolted design in Table 1.

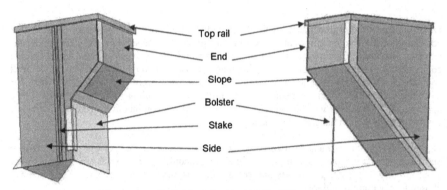

Figure 3. Structural Component Definitions for Typical Coal Hopper Car, Outside View, Left, and Inside View, Right

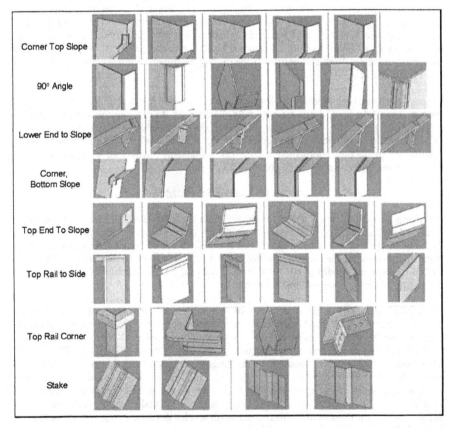

Figure 4. Rail Car Hopper FSW Joint Design Options Considered in Fabrication of ¼ Scale Corner Section Prototypes

Table 1. Comparison of Original and Friction Stir Welded Design Features

Design Feature	Original Design (Bolted)	FSW Design
Weld count	0	84
Weld inches	0	9,360
Bolt count	1200	40
Material	6061-T6	6061-T6
Stake stiffener	Hat shaped section (bolted to side)	"I" shaped section (FSW butt welded to sheet to form side)
Corner connection	Bolted (doubled thickness using angle shape)	Friction stir lap welded (doubled thickness using angle shape)

Not only does the new design eliminate the majority of the fasteners; it also eliminated the lap joints on the sheet sides. The FSW design uses butt welded "I" extrusions in line with the side sheets thus eliminating any overlapping material and saving approximately 200 pounds of material while maintaining the stiffness of the sidewall.

In addition to the largely friction stir welded design two component fabrication alternatives could reduce the cost substantially and could be implemented independently of a major design and manufacturing change:
Option 1 - Fabricate large panels with stiffeners for the side walls of any friction stir weldable design. These panels could then be integrated into the existing design using bolted connections.
Option 2 - At least one hopper car manufacturer expressed difficulty obtaining the desired material in widths that met their requirements. In this situation they are currently required to add an additional bolted joint along the length of the car in order to create the correct widths. Friction stir welding could be used to join sheets of lesser width to create larger "super-sheets".

Preliminary Material Properties

Preliminary FSW process parameter development tests were conducted on 0.188 inch to 0.160 inch dissimilar thickness 5052 aluminum sheets with joints representative of those to be expected on the full scale railroad car hopper corner section (Table 2). Metallurgical and tensile testing was conducted on these samples to determined optimum processing conditions. Note that these material combinations for each joint may differ from that used in the subscale prototype assembly. The results of this development activity were then compared with the existing fastened joint which use Huck® C50LR lock bolts.

Strength properties for the Huck® C50LR lock bolts were given by the manufacturer. Using these values and known spacing from technical drawings, the tensile strength of a bolted joint was calculated. For a bending analysis a full size overlap section was analyzed. The overlap section was idealized as a built-up beam and simply supported. The shear flow equations were applied to find the shear component acting on the bolts. Bearing failure proved to be the limiting factor in a bolted design. With these material thicknesses (0.16 inches), the bearing stress developed in the bolt hole would yield the parent material at loads about half of the rated bolts shear value. This limits the maximum line loading to 520 lb/inch for 5052-H32 and 684 lb/inch for 6061-T6 aluminum bolted structures on the current railcar design (Table 3).

23

Table 2. Material Thickness and Joint Type for FSW Process Development Testing (Note: Joints 2P and 6P were omitted from consideration for full scale design)

Joint	Weld Type	Material Size(s)*
Joint 1P**	Butt	.188 to .160
~~Joint 2P~~	~~Butt~~	~~.188 to .188~~
Joint 3P	Butt	.160 to .160
Joint 4P	Lap	.160 on .188
Joint 5P	Lap	.188 on .160
~~Joint 6P~~	~~Lap~~	~~.160 on .160~~

* .188 is 5052 H32 and .160 is 5052 H34
** "P" is for Process Development

Table 3. Baseline Rail Car Hopper Calculated Limit Line Loads

Shear Flow		Bearing Stress			
Load Distribution	1030 (z) lb/ft	Bearing area	0.1175 in^2		
Height of side (ft)	z = 9	Number of bolts	17		
Vmax =	6180 lb	Total Bearing area	1.9975 in^2		
q = 4645.3(z)					
		5052 H-32			
Spacing	0.5 ft	Fmax =	55.93 kips	Bolt Properties:	
Number of bolts	17	Strength	0.5179 kip/in	Diameter	0.625 in
Vmax	20907 lb			Shear	8.3 kips
		6061 T-6		Tension	7.67 kips
One Bolt		Fmax =	73.908 kips	Clamping	6.9 kips
t max	4009.3578 psi	Strength	0.6843 kip/in	Strength	1.383 kip/in
Vmax	1229.8235 lb				

Mechanical property testing of the 5052-H32 design allowable panels 0.165 inch to 0.165 inch butt joints and 0.165 inch – 0.187 inch lap joints using the optimized FSW parameters are shown in Table 4. Comparing the butt and lap allowable limit line load to the Huck C50LR bolt limit line load shows an increase of over 200%. Comparing these to the bearing failure limit line load for 5052-H32 in these thicknesses shows an allowable increase of over 400%. While weights calculations were not done, it is anticipated that weight will be reduced in the FSW approach due to elimination of the fasteners and potential reduction in section thickness of structural members.

Table 4. Allowable Limit Line Load for 5052-H32 FSW Butt and Lap Joints

Actual Strengths				
5052 H-32			5052 H-32	
Welds:	.165" to .165" plate - butt		Welds:	.165" to .187" plate - lap
Area	0.165 in^2		Area	0.243 in^2
			Measured	
Strength	2.937 kip/in		Strength	2.911 kip/in
Pmax =	35.244 kips		Vmax =	34.932 kips
% Difference	212.31%		% Difference	210.43%

Manufacturing Demonstrator

Although the materials trades and cost benefits analysis indicate that 6061-T6 would be the material of choice for full scale production, the ¼ scale corner section prototype was fabricated from a combination of 5052-H32, 6061-T6, and 6063-T5 materials available within the laboratory. The complexity of the joint designs required several fixturing and tooling setups. In order to efficiently develop the fabrication plan several models were made. These models progressed from CAD models of the part and critical manufacturing setups, to a foam core board model, and finally to fabrication of the final assembly. These models were used to evaluate

24

fabrication detail, assembly sequence, fit-up, clamping and fixturing, and machine interface. Throughout this process the assembly sequence was designed to minimize custom fixture fabrication and the standard welding table was utilized for all under weld anvil support, Figure 5.

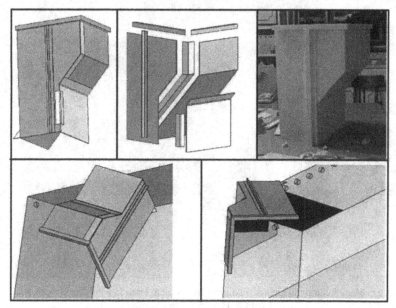

Figure 5. Top, CAD Model of ¼ Scale Demonstrator and Exploded View, Left and Center Respectively, and Bottom CAD Model of Two Weld Setups (Clamps not Shown)

Where possible, clamping was positioned within ½ inch of the joint along the length of the FSW. In fabrication of the prototype, when dissimilar alloy welds were made, the direction of travel was such that the harder 6061 material was placed on the advancing side. Using these materials, joint designs, and fixturing and tooling concepts, a ¼ scale Railroad Car Hopper Corner Section was successfully fabricated (Figure 6).

Figure 6. Manufacturing Sequence, Fixturing and Tooling Used to Fabricate ¼ Scale Railroad Coal Car Hopper Corner Section

Cost Benefits Analysis

A life cycle cost analysis was conducted to ensure meeting the >20% cost savings requirement. Group 1 variables are based on the functional needs of the full scale manufacturing process. Group 3 consists of variables may be optimized in the production process. Group 2 consists of reduced variables calculated from Groups 1 and 2. Based on these, an annual worth life cycle cost algorithm was developed (Figure 7).

Group 1:		Group 2:		Group 3:	
• L	Length of joint seam on car	• J	Joining time	• M	Material costs
• BW	Percentage of box worth	• %	Percentage of seam on car	• F	Fastener costs
• TL	Total seam length on box	• N	Number of units per increment	• E	Number of employees
• EC	Equipment costs	• V	Profits per increment	• S	Joining speed
• IC	Installation costs	• C	Initial costs	• P	Salable value
• MC	Maintenance costs	• H	Working hours per increment		
• SH	# Shifts per day	• TEC	Total employee costs		
• U	Utility costs	• R	Recurring costs		
• SV	Salvage value				
• I	Interest rate				
• LE	Life of equipment				
• EW	Employee wages				
• B	Overhead / Benefits				
• T	Analysis increment period				

$J = L/S$ $\% = L/TL$

$N = T/J$ $V = P*BW*N*\%$

$C = IC + EC$ $H = SH*2080$

$TEC = (EW*H*E) + B(EW*H*E)$ $R = ((M+F)*\%*BW) + MC + TEC + U$

Annual Worth ($$) = $-(C*((I*(1+I)^{LE})/(((1+I)^{LE})-1))) + V - R + (SV*(I/(((1+I)^{LE})-1)))$

Figure 7. Variables and Annual Net Worth Algorithm for FSW Railroad Car Hopper

The annual worth of each joint seam on a full scale design was calculated for both conventional joining techniques and for FSW. In this analysis, a FSW welding speed in excess of 28 inch per minute was required to make the 20% annual net worth benefits (Table 5).

Table 5. Calculated Annual Net Worth of Conventional and FSW Manufacturing Approaches

Joint types	Annual Worth Conventional	Annual Worth FSW	Percent Savings
Stake to sidewall	$33,653,743	$42,611,572	21.02
Top rail to sidewall	$41,715,649	$52,456,428	20.48
Slope Sheet to 90 angle	$44,837,959	$56,052,135	20.01
Top rail to endwall	$46,191,260	$57,870,192	20.18
Slope and Bolster Assy. To Side Assy.	$45,335,070	$56,367,873	19.57
Bolster to Slope Sheet	$46,336,360	$57,896,005	19.97
Bolster to 90 Degree Angle	$46,073,012	$57,184,498	19.43
	$304,143,053	$380,438,704	20.05

A sensitivity analysis was performed in order to determine which variables would affect the annual worth the most. By varying the variable values and determining the slope of the annual worth versus the varied values, the annual worth's sensitivity to each variable can be calculated. The sensitivity analysis reveals that increasing joining speed, increasing automation, decreasing fastener costs, decreasing material costs, and increasing the salable value of the product most directly affect the annual net worth (Figure 8). Of these, increasing the joining speed and degree of automation are directly related to the FSW process and the annual net worth may be additionally improved through further process optimization.

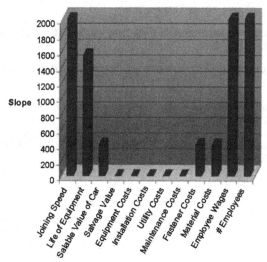

Figure 8. Sensitivity Analysis and Effects of Variables on Annual New Worth for FSW Railroad Car Hopper

Conclusions

The feasibility of implementing friction stir welding as a joining process to manufacture aluminum coal hopper railcar bodies was investigated. An overall cost saving of greater than twenty percent was targeted while requiring at least equivalent structural performance. Several designs were considered including 5xxx and 6xxx series of aluminum alloys and a large number of joint designs. Preliminary design properties were developed in order to compare to the baseline bolted design and to generate data for a cost benefit analysis. A ¼ scale corner section of the railcar was built as a manufacturing demonstrator. The demonstrator was used to demonstrate some of the practical considerations of building the full scale article with the new joining process. The cost benefits analysis determined that weld speed and level of automation were the most influential variables and that the 20% cost savings was attainable at weld speed in excess of 28 inches per minute.

Acknowledgements

This work was performed at the South Dakota School of Mines and Technology as a senior design project in the Mechanical Engineering and Materials Engineering departments. The work of students Haven Mercer, Clark Oberembt, Riley Roberdeau, Marcus Graham, and Marshall Hood was greatly appreciated along with information from Trinity Rail Group, LLC. Funding for this project was provided by the Army Research Laboratory, Cooperative Agreement: DAAD19-02-2-0011

References

1. American Association of Railroads Site, 3 November 2006,
<http//www.aar.org/PubCommon/Documents/AboutTheIndustry/Statistics.pdf>

2. Conrail Cyclopedia Site, 3 November 2006,
<http://crcyc.railfan.net/crrs/gon/gong52xproto.html>

3. Anthony Kruglinski, "A Bull Market for Coal Cars? - Railcar Manufacturing and Leasing Markets", *Railway Age*, July, 2001

4. Anthony Kruglinski, "Aluminum vs. Steel Coal Cars: Pros and Cons", *Railway Age*, October, 1992

5. American Association of Railroads Site, 3 November 2006,
<http://www.aar.org/newsroom/RailFacts.asp>

6. "Mitigating Greenhouse Gas Emissions: Voluntary Reporting" (Report DOE/EIA-0608(96), U.S Department of Energy, 1996)

DEVELOPMENT OF DESIGN CURVES FOR TENSILE STRENGTH AND FATIGUE CHARACTERISTICS OF 7075 T73 ALUMINUM FSW BUTT JOINTS

Srikanth Kandukuri[1], William J Arbegast[1], Anil K Patnaik[2] and Casey D Allen[1]

[1]Advanced Materials Processing Center
[2]Department of Civil Engineering
South Dakota School of Mines and Technology
Rapid City, SD-57701

Keywords: Pseudo Heat Index, FSW Butt Welds, Tensile Strength, Fatigue Life, Design Curves

Abstract

Aluminum 7075-T73 plates of different thicknesses were welded with varying pseudo heat index (PHI). All welds were made under position control. Earlier studies indicate that there is a relationship between the heat input into the weld process and the weld quality for friction stir welded joints. PHI is a measure of heat input into the weld process and is dependant on processing parameters like rotation speed, travel speed and forge force (or heel plunge). The joints were evaluated for their strength and for fatigue life to investigate the response of weld quality in relation to PHI. Ultimate tensile strengths were plotted against PHI with statistically based lower tolerance bounds (T90 and T99) with 95% confidence intervals using the methods mentioned in Military Handbook (MIL-HDBK-5). The study revealed that the ultimate tensile strengths of butt joints, and the corresponding fatigue life decreased with the increase of PHI.

Introduction

Aluminum 7075-T73 sheets of 0.125" thickness were welded with varying processing parameters that result in varying values of pseudo heat index (PHI). Earlier studies [1, 2] indicate that there exits a relationship between the welding temperature and corresponding FSW process parameters. The weld quality of a friction stir welded joint depends on the heat input into the weld process. The heat input into the weld process is dependant on processing parameters such as rotation speed (RPM), travel speed (IPM) and forge force (or heel plunge). PHI is defined in this project as:

$$PHI = \frac{RPM^2}{IPM \times 10,000} \qquad (1)$$

Where,
RPM is the rotation speed in revolutions per minute
IPM is the travel speed in inches per minute

In this study, friction stir welded butt joints were made with a wide range of process parameter to represent different pseudo heat index values. These joints were evaluated to determine their static tensile strength across the joint, and fatigue life. All the welds were made under position control. Relationship of weld quality, static strength and fatigue life with PHI was established. Ultimate tensile strengths were plotted against PHI with statistically based lower tolerance limits

(T_{90} and T_{99}) with 95% confidence intervals using the methods outlined in the Military Handbook MIL-HDBK-5 [3].

This paper describes the general methodology and approach followed in developing the design charts for butt welds. The procedure and analysis methods adopted in the study are described for one set of data corresponding to butt welds of 0.125 inch thick Aluminum 7075-T73 sheets. The work also includes the development of design charts for sheet thicknesses of 0.080 inch and 0.063 inch, but the design charts corresponding to these thicknesses are not presented in this paper.

Approach

Following parameters were used in the study:

Variables

> Spindle speed (RPM) – 500, 600 and 700 rotations per minute
> Travel speed (IPM) – 5, 7, 9, 11 inches per minute
> Material thickness – 0.125, 0.080, and 0.060 inch (only charts for 0.125 inch are presented in this paper)
> Joint type - butt
> Alloy type – Al 7075-T73

Pin tool, preparation of specimens, welding of panels, clamping, tilt angle and backing support were all kept constant throughout the project.

Work Sequence

The research work was organized into the following two phases:

Phase I

• Welding of Al 7075-T73 panels with suitable parameters
• Metallurgical evaluation, bend tests, and tensile testing
• Analysis of data
• Determination of relationships between ultimate tensile strength (UTS) and PHI

Phase II

• Selection of three different pseudo heat index values corresponding to hot, medium and cold welds from the UTS vs. PHI plot
• Identification of the parameters at the selected PHI value
• Welding of the panels with the three identified parameter sets
• Metallurgical evaluation, bend, tensile, and fatigue tests
• Analysis of data
• Development of S-N curves from the experimental data obtained

Weld Sequence and Marking of Panels

Phase I Weld Layout

The layout of a typical panel with four welds made with different sets of parameters is shown in Figure 1(a) for Phase I. A 48 inch aluminum panel was welded in four 10 inch long weld segments. The panel was then cut into four pieces, each 10 inch long. The 10 inch segments were marked and cut to make the required test specimens as shown in Figure 1(b).

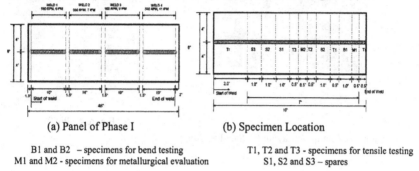

| (a) Panel of Phase I | (b) Specimen Location |

B1 and B2 – specimens for bend testing   T1, T2 and T3 - specimens for tensile testing
M1 and M2 - specimens for metallurgical evaluation   S1, S2 and S3 – spares

Figure 1. Typical Details of Weld Panels of Phase I

Phase II Weld Layout

The layout of a typical panel comprising two welds with different sets of parameters is shown in Figure 2 for Phase II. A 48 inch panel was welded in two segments of 20 inch weld for each segment. The panel was then cut into two pieces. Each 20 inch weld segment was marked and cut to make the test specimens as shown in Figure 2 (b).

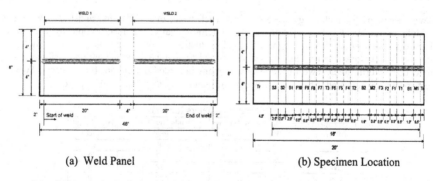

| (a) Weld Panel | (b) Specimen Location |

B1 and B2 – specimens for bend testing   T1, T2 and T3 - specimens for tensile testing
F1, F2.....F9 – specimens for fatigue testing   S1, S2 and S3 – spares
M1 and M2 - specimens for metallurgical evaluation

Figure 2. Typical Butt Weld Panels of Phase II

Typical clamping arrangement used for welding of the panels is shown in Figure 3. The pin tool geometry is shown in Figure 4. The shoulder diameter of the pin tool was 0.375 inch.

31

Figure 4. Pin Tool

Figure 3. Clamping Layout for Panel Welding

Experimental Work for Al 7075- T73 – 0.125"

Phase I

The weld parameters used for 0.125 inch thick Al7075 –T73 panels are given in Table I. Each weld length was cut to make two metallurgical specimens, three tensile specimens and two bend test specimens as shown in Figure 1(b). The rest of the weld length was saved for spare specimens.

Metallurgical Evaluation

Two specimens for metallurgy were cut from each weld, one at the end and the other at the middle of the weld. The specimens were microscopically and macroscopically evaluated. Figure 5 shows summaries of the metallurgical evaluations at two locations (end and middle) of the welds. All the welds for this series showed good rippled weld surface with no visible surface defects, or worm holes. Two of the tests specimens indicated minor root flow defects. Two of test specimens indicated differences in nugget formation and the shape of the nugget.

Table I. Weld Parameters for Phase I

Rotation Speed	Travelling speeed	Tilt Angle	Pseudo Heat Index	Weld Length
(RPM)	(IPM)	(Deg)	(PHI)	(Inch)
500	5	2	5.00	10
500	7	2	3.57	10
500	9	2	2.78	10
500	11	2	2.27	10
600	3	2	12.00	10
600	5	2	7.20	10
600	7	2	5.14	10
600	9	2	4.00	10
600	11	2	3.27	10
700	5	2	9.80	10
700	7	2	7.00	10
700	9	2	5.44	10
700	11	2	4.45	10

Weld Data Analysis

Figure 6 shows the variation of forge force recorded along the length of the weld. The welds were made under position control and therefore the forge force varied along the length of the weld. For example, for a particular weld length (as in the case of weld #14 in Figure 6), a change in forge force of about 700 lb was observed. The change in forge force over the length is believed to have caused variations in weld quality and nugget shape.

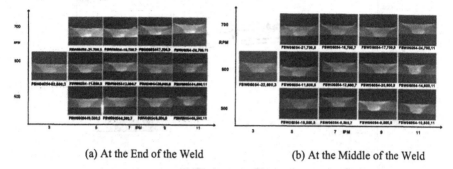

(a) At the End of the Weld   (b) At the Middle of the Weld

Figure 5. Summary of Metallurgical Evaluation (Phase I)

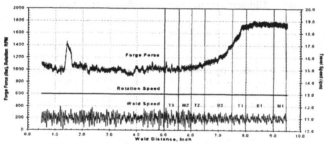

Figure 6. Forge Force over the Length of the Weld

Bend Testing

Bend test specimens were made with dimensions of 8" x 1" with uniform width. The mandrel dimensions for bending the specimens were selected from [4]. The specimens were bent to 180 degrees with the root of the weld going into tension around a mandrel with radius of 6t (six times the sheet thickness).

Two bend test specimens were tested for each weld (at the end and the middle of the weld length). All the specimens were found to pass the bend test for this data set.

Tensile Testing

Tensile testing was done on three specimens for each weld length and three specimens from the parent metal. These specimens were prepared and tested per standard procedures of ASTM E-8 and the ultimate tensile strength of each specimen was plotted against the corresponding pseudo heat index (PHI). Design curves were developed based on the test data per MIL-HDBK-5 [3] procedures. A plot of tensile strengths of butt welds for different values of pseudo heat index for a 0.125 inch thick sheet is shown in Figure 7.

33

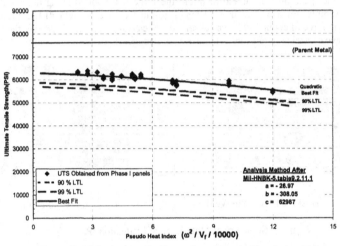

Figure 7. Ultimate Tensile Strength vs. Pseudo Heat Index Value

Statistical Design Curves

MIL-HDBK-5 suggests the following equations for developing statistical design allowable curves [3].

$$\frac{SUS}{TUS} = a + bx + cx^2 \tag{2}$$

$$T_{99} = a + bx_0 + cx_0^2 - \left(t_{0.95,\, n-3,\, \frac{2.326}{\sqrt{Q}}} \right) \sqrt{Q}\, s_y \tag{3}$$

$$T_{90} = a + bx_0 + cx_0^2 - \left(t_{0.95,\, n-3,\, \frac{1.282}{\sqrt{Q}}} \right) \sqrt{Q}\, s_y \tag{4}$$

a, b, c, Q, s_y are calculated as explained in section 9.6.3.2 of MIL-HDBK-5 [3].

Phase II

Three different pseudo heat index values (high, medium and low) were selected from the tensile strength curves for fatigue tests of Phase II from Phase I welded parameters used for 0.125 inch thick Al 7075-T73 panels. The processing parameters corresponding to these pseudo heat index values were used in Phase II welding. Typical panels were welded according the parameters shown in Table 2.

Table II. Weld Parameters for Phase II

weld no	Rpm	Ipm	PHI	Tilt Angle
				deg
06115-2	500	11	2.27	2
06115-3	700	7	7.00	2
06115-4	600	3	12.00	2

Welding

Panels were welded as shown in Figure 2(a) with the parameters shown in Table II. The panels were marked as shown in Figure 2(b). The panels were cut and the metallurgical evaluation, tensile testing, bend testing, and fatigue testing were conducted on the test specimens.

Metallurgical Evaluation

Two specimens for metallurgical evaluation were cut from each weld, one at the end and the other at the middle of the weld. Figure 8(a) and Figure 8(b) show summaries of the metallurgical evaluations at two different locations (the end and the middle of the welds).

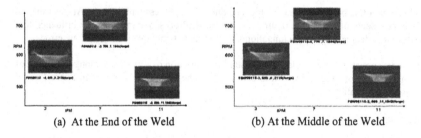

(a) At the End of the Weld   (b) At the Middle of the Weld

Figure 8. Summary of Metallurgical Evaluation (Phase II)

Figures 8(a) and (b) indicate that the welds did not have any worm holes. Root flow defect was observed on one sample. The variation of forge force in position control welds made in the panel is believed to have caused this root flow defect. Forge forces for the set of weld parameters in Phase I and Phase II are compared in Figure 9.

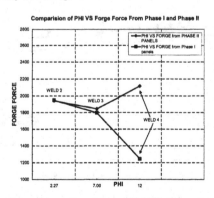

Figure 9. Comparison of Forge Forces of Phase I and Phase II Welds

35

Variations of Forge Forces Along the Length of the Weld

The variations of forge forces along the length of the welds observed in this study were attributed to the following possible causes:

1. Forge setup
2. Z-Position zero
3. Pin and tool elongation

Forge setup used in the welding of the panels in this study was consistent but not constant due to the use of a different size gauge blocks. A 0.100" gauge block was used in Phase I, but a 0.150" gauge block was used in Phase II.

The procedure for setup of zero Z-Position was consistent in this task. Tool pin tip was loaded to 10 pounds on the top of the gauge block. However material build up on the pin affected the ability to set Z-Position. It was observed that the pin length varies a few thousandths of an inch from weld to weld depending on process parameters. This variation results in errors and may have caused some difference in the plunge depth and forge force.

Pin tool elongation is believed to have happened due to the process temperature. At this time, no direct measurements of tool temperature were recorded. Process parameters may influence tool temperature to a degree where pin elongation due to process temperature may become critical.

Further studies are required to isolate the cause of the variation of forge force along the length of welds (particularly for long welds). It is believed that the buildup of material in Phase I could have caused the variation.

Tensile Testing

Tensile testing was done on three specimens from each weld and three specimens from the parent metal. The specimens were prepared and tested as explained in an earlier section of this paper. Ultimate tensile strengths of test specimens are plotted against PHI values in Figure 10.

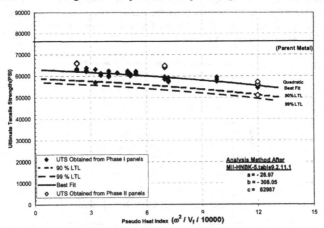

Figure 10. Comparison of Ultimate Tensile Strength Obtained from Phase I and Phase II
(0.125 inch 7075-T73 Friction Stir Welded Sheet under Position Control)

Bend Tests

Two bend test specimens were tested for each weld (one at the end and the other at the middle of the weld length). The bend specimens for weld number 4 with PHI of 12 failed due to root flow defect that was caused by high forge force. Phase II welds have root flow defect in the retreating side compared to no root flow defect for Phase I welds. Bend test specimens of Phase II failed on the retreating side as shown in Figure 11. There was also a difference between the forge forces of Phase I and Phase II welds.

Phase I Metallurgy   Phase II Metallurgy and Bend Test Results

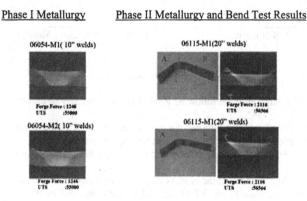

Figure 11. Comparison of Macros of Welds of Phase I and Phase II

Fatigue Tests

Six samples with varying maximum applied stresses as a percent of ultimate tensile strengths of each weld were tested. The stress ratio (which is the ratio of minimum to maximum stress, R) was maintained constant at 0.1. Number of cycles required to fail the test specimens was determined and plotted in the standard format as suggested in MIL-HDBK-5 [3].

Figure 12 shows the test data plotted in the format suggested in MIL-HDBK-5 [3] with the number of cycles on the X-axis and the maximum stress on the Y-axis. Linear best-fit relations for the data sets are shown in the Figure. The cut-off maximum stress, which is shown in the form of horizontal lines in the region of low fatigue cycles, indicates a conservative upper limit for maximum stress that can be used for fatigue life predictions for the welds. The suggested S-N curves and the maximum stress limits shown in the figure will move slightly with the addition of more test data that becomes available in future. However, the curves provide an interim basis for design.

Another approach to design for fatigue life of structures is to identify a "Safe" region in the S-N plot. A suggested "Safe" region is identified in Figure 13 which is the region below the lower bound curve. The lower and upper bounds of the test data are identified in the form of two curves. The maximum limit of stress cycles is tentatively set to 1 million cycles because the tests were "run-out" at that many cycles. The maximum stress limit represented by the horizontal line in the region of low fatigue cycles is based on judgment. The upper and lower bounds can be better defined with additional test data. In the interim, the suggested safe region is a reasonable basis for design.

37

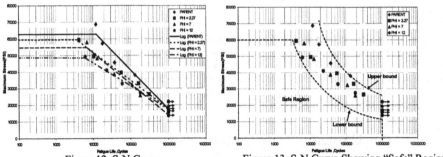

| Figure 12. S-N Curve | Figure 13. S-N Curve Showing "Safe" Region |

Conclusions

For position control butt welds, the following general conclusions may be drawn:

1. There exists a relationship between the pseudo heat index of a weld and the weld quality. As the pseudo heat index of the process increases, the ultimate tensile strength decreases. This reduction in tensile strength leads to decrease in T99 and T90 design allowables.

2. As the pseudo heat index of the process increases, the fatigue life decreases. However, the decrease in fatigue life with PHI does not appear to be significant.

3. Forge force decreases with the increase of PHI.

Acknowledgments

This work was performed under the multi-university National Science Foundation - Industry / University Co-operative Research (I/UCRC) Center for Friction Stir Processing (CFSP). The financial support at the SDSMT CFSP site provided by Army Research Laboratory, Boeing, Pacific Northwest National Laboratory, MTS Systems, BAE Systems, and Sikorsky is gratefully acknowledged. The views and conclusions contained in this paper are those of the authors and should not be interpreted as representing the official policies, either expressed or implied, of CFSP, the center members or the National Science Foundation.

References

1. W.J. Arbegast and P.J. Hartley, "Friction Stir Weld Technology Development at Lockheed Martin Michoud Systems – An Overview", *Proceedings of the Fifth International Conference on Trends in Welding Research,* Pine Mountain, GA, USA, June 1-5, 1998, p. 541.

2. R.S. Mishra, and Z.Y. Ma, "Friction Stir Welding and Processing", *Material Science and Engineering,* R50 (2005), p. 1-78.

3. Military Handbook, "Metallic Materials and Elements for Aerospace Vehicle Structures" MIL-HDBK-5H-1998.

4. Metal Handbook, "Mechanical Testing", Ninth Edition, Vol. 8, American Society for Metals, 1985.

38

Relationships Between Process Variables Related to Heat Generation in Friction Stir Welding of Aluminum

Kevin J. Colligan

Concurrent Technologies Corporation
27980 Kim Drive, Harvest, Alabama 35749, USA

Keywords: Friction stir welding

Abstract

Friction stir welding was invented in 1991 by The Welding Institute as a solid-state method of joining metals which was initially applied to the joining of aluminum alloys. Process improvements have extended the use of the technique to other metals, to thicker materials, to higher travel speeds, and to a variety of applications. In addition, detailed study of the process has led to a better understanding of how the process actually works in terms of material flow produced by different kinds of tools. However, no publications have comprehensively described the relationships between the different processing variables and how these variables affect key process conditions, such as temperature distribution, material conditions at the welding tool, etc. This paper seeks to describe the relationships between the independent process variables, such as travel speed, spindle speed, and anvil condition, and the dependent process outcomes, such as the spindle torque, heat generation, and workpiece temperature. A conceptual model is proposed that relates the main processing variables by showing the influence of physical effects on heat generation in the workpiece. Case studies are used to confirm and explore the relationships expressed in the conceptual model.

Introduction

It is important for users of friction stir welding to have a conceptual understanding of how the process works. Such an understanding is helpful in deciding how to change process conditions to achieve desired effects, such as for improving joint strength, for eliminating certain common weld defects, and for transferring a known welding procedure to new conditions. Computer based models and empirical studies of the stir welding process have been developed by many researchers to examine heat transfer, metallurgical evolution, and material flow. This work has been are very useful for building, piece by piece, an understanding different physical effects in friction stir welding. A general examination of the relationships between variables can now be used to develop a conceptual model of the process. In spite of the apparent simplicity of the process, the interaction between process variables, which sometimes compete to produce counterintuitive results, complicates developing a conceptual understanding of the effects of the process variables. The goal of this paper is to propose a conceptual model that relates the different process variables with key process conditions in a way that will allow the practitioner of FSW to develop a general understanding of the process.

Conceptual Model Formulation

A conceptual model is proposed that relates the main process variables, structured around the interrelation of the main categories of physical effects that come into play in the generation and distribution of heat in friction stir welding. These physical effects include metallurgical effects, mechanical effects, and heat transfer effects. As shown in Figure 1, mechanical effects are further subdivided to distinguish between effects related to the generation of torque and effects related to the generation of heat. At this level the model is very simple and it offers an intuitive way of relating the physical effects: workpiece material properties affect the generation of torque, torque affects the generation of heat, which in turn drives heat transfer, which affects material properties in advance of the welding tool. The inherent stability of the process is reflected in the closed-loop nature of the relationships between these effects. It should be noted that the present model focuses on heat generation, and does not address the effect of process variables on the lateral and axial forces generated by the welding tool, nor does the present model address the effect of process variables on defect formation.

The four categories of physical effects can be expanded to show each of the process variables and how they interact to influence the workpiece flow stress, torque, heat generation, and temperature, as shown in Figure 2.

First, the flow stress of the workpiece at the welding tool pin is the result of the thermal history, the amount of strain (work hardening), and the strain rate. Thermal history effects can further be subdivided into slow processes that depend on the temperature distribution in advance of the welding tool, which result in metallurgical degradation of the workpiece, and rapid, adiabatic heating that takes place as the material is deformed by the pin. The relatively slow metallurgical degradation of the workpiece can be thought of as pre-conditioning that the workpiece experiences as it heats from room temperature to the temperature just in front of the pin. This pre-conditioning has the effect of eliminating any tempering in the workpiece [1], so that alloys of identical composition behave essentially the same in welding. Softening from adiabatic heating is the last bit of softening that occurs from the rapid temperature rise that occurs from plastic work caused by the welding tool pin.

The flow stress is also generally influenced by strain and strain rate. For aluminum alloys, strain hardening is a very weak effect at the temperatures that are reached near the welding tool, and may therefore be ignored when considering the interaction of variables. Strain rate effects however may not be insignificant and should be considered. The influence of strain rate on flow stress is difficult to analyze since detailed knowledge of the plastic deformation process is required in order to know the strain rate at a given location. However, for the purpose of constructing this conceptual model detailed knowledge of the plastic deformation is not required, only what physical effects and variables influence the process, so it is sufficient to assert that the strain rate is a function of the peripheral velocity of the pin and the apparent vertical velocity of the threads.

Continuing in Figure 2, flow stress leads to the development of spindle torque as the workpiece material resists plastic deformation. Torque is simply the force exerted at the surface of the pin multiplied by the pin radius. The tangential force at the surface of the pin may be from gross plastic deformation or from sliding friction, and it can be related to the flow stress (the resistance to plastic work), the forward travel per feature per revolution (how much material the reentrant features on the tool surface are capturing on each revolution, equivalent to chip-load in machining), and the length of the pin. The travel per feature per revolution is further dependent on the spindle speed, the travel speed, and the number of reentrant features. If the welding tool being used doesn't have reentrant features, the travel per feature per revolution can be omitted from consideration.

When torque is applied through rotation at a given speed, heat is generated in the workpiece. The torque can be used to calculate the power (energy per unit time) delivered to the workpiece simply by multiplying the torque by the spindle speed. The relationship between the two-dimensional temperature distribution some distance from the heat source and the power and velocity of a moving heat source was formulated by Rosenthal [1] as,

$$t - t_0 = \frac{Q_p}{2\pi\kappa} e^{-\lambda v \xi} \frac{K_0(\lambda v r)}{g},$$ (1)

where,

t = temperature in the plate,

t_0 = initial temperature in the plate,

Q_p = weld power,

κ = thermal conductivity of the plate,

λ = 1/(2 * thermal diffusivity),

v = welding speed,

ξ = distance from the weld center in the direction of welding = $x - vs$,

s = time,

x = position along the joint,

r = distance from the weld center to point (ξ,y)

K_0 = modified Bessel function of the second kind, 0th order, and

g = plate thickness.

This function gives a means for calculating the temperature as a function of position in an infinite plate with a moving heat source. However, for the purpose of the conceptual model the Rosenthal equation is more than is required, since a precise solution is not needed. Instead, a functional relationship between the temperature distribution, welding speed, and power is needed so that one can predict the general effect of the welding speed and power on the temperature distribution. The following functional relationship is proposed:

$$t - t_0 \propto \frac{Q_p}{v}$$ (2)

Note that the exponential term and the Bessel function term in (1) are both a function of welding speed. If the product of these two dimensionless terms is plotted as a function of the welding speed, as shown in Figure 3, we see that the curve has a trend that is similar to 1/v. Although the two functions are certainly not equal to each other, they both trend downward with velocity. For the purpose of a conceptual model, the specific energy, or energy input per unit length of weld, $\frac{Q_p}{v}$, sufficiently describes the variation in the temperature distribution with welding speed and weld power, but is much simpler than the Rosenthal solution, or other similar, analytic solutions, and is therefore a more convenient indicator of the temperature distribution. The specific energy can be calculated and displayed in real time on any welding machine that has a means of measuring spindle torque, spindle speed, and travel speed. The effect of plate thickness and workpiece thermal conductivity, included in equation (1) will be accounted for elsewhere in the conceptual model.

The variation in maximum workpiece temperature near the welding tool with respect to welding speed is not well represented by the specific energy in experimental results. Instead, the weld power is proposed as an indicator of maximum temperature, which is demonstrated in the case studies to follow. The welding speed certainly has an effect on the maximum temperature, but this effect influences the maximum temperature through change in the torque that accompanies the change in welding speed.

Returning to discussion of Figure 2, the temperature distribution is the result of the balance between heat generation and heat loss to the environment. Important variables in the determination of the temperature distribution are the thermal diffusivity of the workpiece and anvil (how easily does the heat dissipate), the size of the workpiece and anvil, and the workpiece surface convection characteristics. The effect of thermal conductivity (directly proportional to the diffusivity) and plate thickness from the Rosenthal temperature solution, equation (1), above, is included here in the conceptual model. The maximum temperature is not linked to the flow stress in Figure 2, since this effect is already captured in the adiabatic heating effect, discussed above.

Conceptual modeling of the relationships between the variables can be further enhanced by showing whether the relationship is a direct relationship or an inverse relationship, as shown in Figure 4. In this diagram the relationship is coded as a plus sign for a direct relationship and a minus sign for an inverse relationship. As will be seen in the case studies that follow, the relationships between variables can be used to help understand how changing welding conditions will influence the key features of flow stress, torque, heat generation, and workpiece temperature. These case studies were gathered from existing weld data and analyzed in terms of the variable relationships model shown in Figure 4, both for the purpose of explaining what was observed in the testing and for confirming the formulation of the model.

Two published papers were used in the case studies that follow [3, 4]. These were supplemented by data from unpublished work performed by the author. Aluminum alloy 5083, a non-heat treatable, solution strengthened alloy, was used in the published studies and aluminum-lithium alloy 2195 was used in the unpublished work. The nominal alloy compositions are shown in Table 1. In the first study 25.4-mm 5083-H131 aluminum plate was welded using different spindle speeds, travel speeds, and using pins with different numbers of features (in this case, the reentrant features were flats cut into the surface of a threaded pin) in order to isolate the effects of travel per feature per revolution, spindle speed and travel speed. The welds performed are listed in Table 2, and the tool design used is summarized in Table 3. In the second study welds were made in 25.4-mm 5083-H116 with embedded thermocouples to measure the mid-plane temperature of the workpiece as the welding tool approached and passed. Welds were made at different travel speeds, with different shoulder diameters, and with different anvil conditions (mica insulated vs. non-insulated). A summary of the welding conditions is presented in Table 4. The tool design was identical to the tool used in the Colligan study, summarized in Table 3, except that a pin with four flats was used in all of the welds made.

Case Study 1 – Changing Travel Speed

Three sets of findings are used here to examine the effect of changing travel speed on heat generation in friction stir welding of aluminum. In the first study, by Colligan [3], welds were made with increasing travel speed while maintaining constant spindle speed and travel per feature per revolution. The spindle torque was measured and used to calculate the weld power and weld specific energy (heat input per unit length of weld), as shown in Figure 5. As can be seen in the figure, when the travel speed was increased the torque and weld power were seen to increase and the weld specific energy was observed to decrease. From the study by Xu [4] in welds that had a mica-insulated anvil, Figure 6 shows that increasing travel speed has the effect of decreasing the temperature away from the welding tool and increasing the maximum temperature recorded on the thermocouple. The torque, specific energy, and weld power measured in the Xu study are plotted in Figure 7. The trend of this data is similar to that observed in the Colligan study. The slight differences between the two studies are presumably due to the fact that in the Xu study an insulated anvil was used, and no attempt was made to hold the travel per flat per revolution constant as the travel speed increased. Finally, in an unpublished experiment a weld in 0.725-inch thick 2195 plate was started at 127 mm/min travel speed, and then the travel speed was increased to 152 mm/min. It was observed that the torque and weld power increased, but it took a full 75-mm of travel before the full effect was observed, as shown in Figure 8.

The observations from these studies can be explained by looking at the influence of increasing travel speed on the key physical effects, as shown in Figure 9. In this figure upward

42

influence on a variable is represented by a heavy black line, downward influence on a variable is represented by a heavy gray line, a neutral influence is represented by a heavy dotted line, and an unaffected influence is left as a thin black line. The five points below correlate with the numbers in the figure.

1. First, it is known from experiment that torque goes up with increasing travel speed, as pointed out in Figure 5 and in Figure 7. This in turn increases the weld power, since the spindle speed is unchanged, and exerts an upward influence on the specific energy. However, the increase in travel speed also exerts a downward influence on the specific energy. In the Colligan and Xu studies the competition between these variables was dominated by the travel speed, since the specific energy was seen to decrease with increasing travel speed. This effect was also observed by Reynolds [1].

2. The increase in weld power is reflected in the increased maximum temperature observed in the Xu study, shown in Figure 6.

3. The decreased specific energy resulted in a lower temperature distribution away from the weld zone, since the workpiece and anvil heat loss characteristics were not changed.

4. The reduced temperature distribution has the effect of increasing the flow stress of material arriving at the pin, since there is less metallurgical degradation from precipitate coarsening and dissolution and possibly less softening from recovery.

5. The increased flow stress results in increased torque, in agreement with 1, above.

In the Colligan study the influence of changing travel speed on the travel per flat per revolution was neutralized by changing the number of pin flats. In the Xu study this was not done, so increasing the travel speed should have increased the travel per flat per revolution, exerting an additional upward influence on the torque. So in the Xu study the effects of flow stress and travel per flat per revolution combine to increase the torque.

In the case of the weld in 0.725-in 2195-T8 it was observed that it took about 75-mm of travel for an increase in travel speed to have the full effect on changing the torque. This result is in agreement with Figure 9, since the increase in torque is the result of the reduced temperature away from the weld zone, which results in reduced metallurgical degradation, which takes time to develop as the tool travels.

Case Study 2 – Changing Spindle Speed

In the Colligan study welds were also performed to evaluate the effect of increasing spindle speed on the torque, specific energy, and weld power while keeping the travel per feature per revolution and the travel speed constant, as shown in Figure 10. It was observed that increasing the spindle speed dramatically reduced the spindle torque. The weld power and specific energy increased significantly at first, but reached a limiting value. In an unpublished experiment by the author, a weld in 25.4-mm 2195-T8P4 was started at 170 rev/min, then the spindle speed was increased to 200 rev/min while keeping the travel speed constant, as shown in Figure 11. The spindle torque was seen to immediately drop in response to the increased spindle speed, and it took just two data samples (about 2 seconds) for the torque change to reach its full effect. These observations can be examined within the context of the conceptual model, shown in Figure 12.

1. It was observed in Figure 10 that increasing the spindle speed resulted in decreased torque, which, according to the conceptual model, would exert a downward influence on the power and specific energy.

2. The increase in spindle speed also exerts an upward influence on the power and specific energy, and it was observed that initially these variables increased, but then reached a limiting value of about 10 kW for the power and about 5 kJ/mm for the specific energy. The

increasing power and specific energy would be expected to have the effect of increasing the maximum temperature and the overall temperature distribution as long as the power and specific energy are increasing, but no such test data is available to the author at this time to confirm this prediction in the model. However, a study by Record, et al. [7], it was concluded that spindle speed was the most significant factor in determining welding tool temperature, supporting the prediction that increasing spindle speed leads to increased power and specific energy, and increasing workpiece temperature.

3. Since no change was made to the heat loss characteristics, one would expect the increasing spindle speed to initially increase the temperature distribution, but only to a limited value, since the specific energy reaches a limited value. This would be expected to increase metallurgical degradation, decreasing the flow stress to a point, but above 333 rev/min (weld PE6 in Table 2 and Figure 10) there should be no additional increase in temperature or in metallurgical softening.

4. Increasing the spindle speed has the effect of increasing the thermal softening from adiabatic heating, exerting an additional downward influence on the flow stress at the tool. This effect is counteracted, to some unknown extent, by the presumed increase in strain rate. However, if it is known that the torque goes down with increasing spindle speed (Figure 10), even beyond the point at which no additional far-field temperature distribution increase is expected, it must be the case that the adiabatic heating effect is stronger than the effect of increasing strain rate, resulting in a continued decreased flow stress with increasing spindle speed.

5. Since the effect of travel per flat per revolution was neutralized in the Colligan study by changing the number of flats on the pin, the decreased flow stress results in decreased torque, which is in agreement with 1, above.

There is further evidence that the decrease in flow stress must not be entirely due to an elevated temperature distribution (and the resultant metallurgical degradation) with increasing spindle speed. In the unpublished experiment in 25.4-mm 2195 aluminum-lithium it was observed that increasing the spindle speed resulted in an immediate change in the spindle torque. This would not be the case if the decrease in flow stress were the result of metallurgical degradation, which takes time to develop.

The strain rate sensitivity, which one would expect to increase flow stress with increasing spindle speed, apparently doesn't come into play in FSW of aluminum, presumably because the plastic deformation is not isothermal. In the absence of rate sensitivity, increasing spindle speed reduces the flow stress through adiabatic heating and thermal softening, resulting in a decrease in torque.

It is possible that the weld power and specific energy are limited by the instantaneous maximum temperature in workpiece material being worked by the pin approaching the melting point of the workpiece, which has the effect of placing an upper bound on the weld power and specific energy. Welding conditions at very high spindle speeds have not been extensively explored since the welds produced at high spindle speeds usually have gross defects in the stir zone, presumably due to the overheating and very low flow stress.

Case Study 3 – Changing Anvil Insulation Condition

In the Xu study tests were also carried out to examine the effect of different anvil conditions on the temperature field in friction stir welds. A solid anvil and an anvil insulated by mica were both used with identical weld settings, and thermocouples were used at mid-plane in the plate of 25.4-mm 5083 aluminum. The mid-plane temperature as a function of position for the two welds is shown in Figure 13. The data showed that insulating the anvil had the effect of raising the far-field temperature distribution but lowering the mean maximum temperature. The torque, specific energy, and weld power for these two welds are tabulated in Table 4. This result is instructive since it shows the effect of changing the far-field temperature distribution without changing the independent process variables. The observed effects can be examined within the context of the conceptual model, shown in Figure 14.

1. Adding anvil insulation had the effect of increasing the temperature distribution in advance of the tool, since heat loss to the anvil is reduced. At the same time the maximum temperature decreased.

2. Increased far-field temperature has the effect of increasing metallurgical degradation, reducing the flow stress of the workpiece as it arrives at the pin. No other process variables were changed, so the change in far-field temperature has a direct effect on flow stress.

3. The reduced flow stress results in decreased torque, as shown in the table.

4. The reduced torque results in decreased weld power and specific energy.

5. The reduced weld power results in decreased maximum temperature, which is in agreement with 1, above. The reduced specific energy did not result in reducing the temperature distribution, since this effect was overpowered by the change in anvil condition noted in 1.

The increased temperature distribution in advance of the tool has the effect of increasing the softening the workpiece material by metallurgical degradation effects, which decreased the flow stress, the torque, the weld power, and the peak temperature. It is speculated that this same situation could be achieved by reducing the workpiece size (such as by welding very narrow plates), by reducing the anvil size, or by making bobbin or self-reacted welds (where the anvil is eliminated). This also highlights the critical nature of workpiece and anvil size effects, which becomes an important consideration when optimizing an application in the laboratory environment on a sub-size workpiece for subsequent transition to production of larger components.

It was noted in 5 that the competition between reduced specific energy (tending to lower the temperature distribution) and reduced heat loss (tending to increase the temperature distribution) was won by the reduced heat loss. In the work by Reynolds [1] a similar competition existed when he noted that aluminum alloy 6061 had the highest specific energy of all the alloys tested in identical conditions. He speculated that this was due to 6061 having the highest thermal conductivity of the alloys tested. In the conceptual model proposed here, increased workpiece thermal conductivity leads to increased heat loss, decreased temperature distribution, increased flow stress, increased torque, and increased specific energy, which is in competition with the increased thermal conductivity. Apparently the effect of changing the heat loss characteristics is a stronger effect than the change in specific energy.

Conclusions

A conceptual model is proposed to relate the process variables and physical effects in friction stir welding of aluminum and its alloys. This model provides a framework for examination of the effect of changing process conditions, so that a better intuitive understanding of the friction stir welding process can be developed. Three case studies were examined within the context of the conceptual model to test the model against experimental observations. The case studies were found to generally be consistent with the proposed model, and in some situations the case studies showed how competition between variables resulted in different outcomes. Perhaps as more case studies are examined, experience will dictate how to assign weights to different effects as an aid to predicting the result of competition between different variables.

45

References

1. Reynolds, A.P., and Tang, W., "Alloy, Tool Geometry, and Process Parameter Effects on Friction Stir Weld Energies and Resultant FSW Joint Properties," *Friction Stir Welding and Processing*, ed. Jata, K.V., Mahoney, M.W., Mishra, R.S., Semiatin, S.L., and Field, D.P., TMS, 2001, pp. 15–23.
2. Rosenthal, D., "Mathematical Theory of Heat Distribution During Welding and Cutting," *Welding Journal*, 1941, 20(5), 220s–243s.
3. Colligan, K.J., Xu, J., and Pickens, J.R., "Welding Tool and Process Parameter Effects in Friction Stir Welding of Aluminum Alloys," *Friction Stir Welding and Processing II*, ed. Jata, K.V., Mahoney, M.W., Mishra, R.S., Semiatin, S.L., and Lienert, T., TMS, 2003.
4. Xu, J., Vaze, S.P., Ritter, R.J., Colligan, K.J., and Pickens, J.R., "Experimental and Numerical Study of Thermal Process in Friction Stir Welding," International Conference on Joining of Advanced and Specialty Materials VI, ASM, 2003.
5. Department of Defense, Military Standard, "Detail Specification: Armor Plate, Aluminum Alloy, Weldable 5083 and 5456," MIL-DTL-46027J, September 4, 1998.
6. Brown, W. F., Jr., Aerospace Materials Handbook, Nonferrous Alloys – AlWT 2195, p.6.
7. Record, J.H., Covington, J.L., Nelson, T.W., Sorensen, C.D., and Webb, B.W., "Fundamental Characterization of Friction Stir Welding," Proceedings of the 5th International Symposium on Friction Stir Welding, September, 2004, Metz, France.

Table 1. Specified chemical composition ranges for aluminum alloys 5083 and 2195.

Alloy	Chemical Compositions, wt. %												Impurities, %	
	Cu	Mn	Mg	Fe	Si	Zn	Ti	Cr	Zr	V	Ag	Al	Each	Tot.
5083 [5]	0-0.10	0.40-1.00	4.0-4.9	0-0.40	0-0.40	0-0.25	0-0.15	0.05-0.25	-	-	-	Bal.	0.05	0.15
2195 [6]	3.70-4.30	0-0.25	0.25-0.80	0-0.15	0-0.12	0-0.25	0-0.10	-	0.08-0.16	-	0.25-0.60	Bal.	0.05	0.15

Table 2. Summary of welding conditions, welds in 25.4-mm 5083-H131 [1].

Weld Identification	Spindle Speed rev/min	Travel Speed mm/s (in/min)	Number of Pin Features (flats)	Travel per Feature per Rev. mm/feature-rev (in/feature-rev)
PE1	250	2.12 (5.0)	4	0.127 (0.0050)
PE2	250	2.12 (5.0)	3	0.170 (0.0067)
PE3	250	2.12 (5.0)	2	0.254 (0.0100)
PE4	250	1.59 (3.7)	3	0.127 (0.0050)
PE5	250	1.06 (2.5)	2	0.127 (0.0050)
PE6	333	2.12 (5.0)	3	0.127 (0.0050)
PE7	500	2.12 (5.0)	2	0.127 (0.0050)
PE8	250	2.12 (5.0)	0	0.508[1] (0.0200)

[1] mm/rev

Table 3. Tool design for welds in 1-in 5083-H131 [3].

Shoulder Diameter, mm (in)	42 (1.65)
Scroll Depth, mm (in)	1.3 (0.05)
Scroll Pitch, mm (in)	3.3 (0.13)
Scroll Width, mm (in)	2.5 (0.10)
Nominal Pin Length, mm (in)	25.2 (0.992)
Pin Maximum Diameter, mm (in)	15.2 (0.598)
Thread Pitch, mm (in)	1.41 (0.056)
Number of Features (flats)	0, 2, 3 and 4

Table 4. Summary of welding conditions, welds in 25.4-mm 5083-H116 [4].

Plate Number	Spindle Speed rev/min	Travel Speed mm/s (in/min)	Insulating Conditions	Torque N-m (ft-lbs)	Specific Energy kJ/mm (BTU/in)	Power kW
Plate 04	250	1.69 (4.0)	With mica	295 (218)	4.1 (110)	7.8
Plate 02	250	2.12 (5.0)	With mica	326 (240)	3.6 (97)	8.5
Plate 05	250	2.54 (6.0)	With mica	358 (264)	3.3 (89)	9.4
Plate 08	250	2.12 (5.0)	no mica	369 (272)	4.1 (110)	9.7

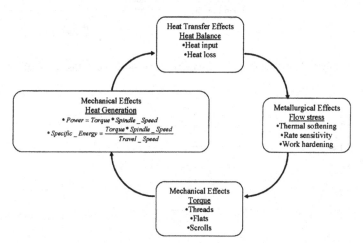

Figure 1. Relationships between the main groups of physical effects.

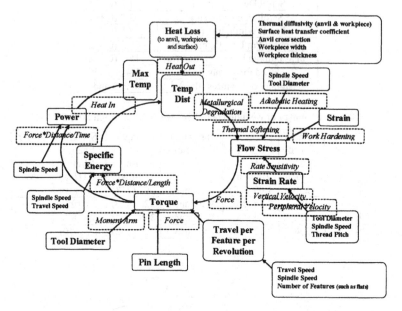

Figure 2. Relationships between variables with physical effects overlaid.

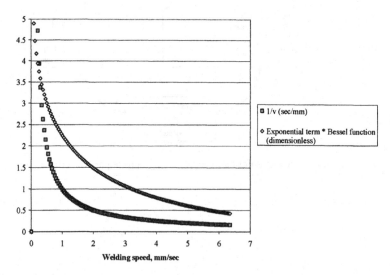

Figure 3. Comparison between 1/v and the product of the exponential term and the Bessel function from Rosenthal's two-dimensional temperature solution.

48

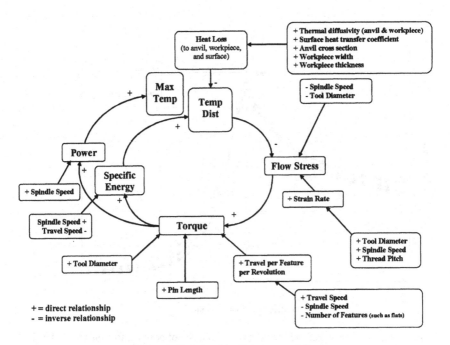

Figure 4. Relationships between variables.

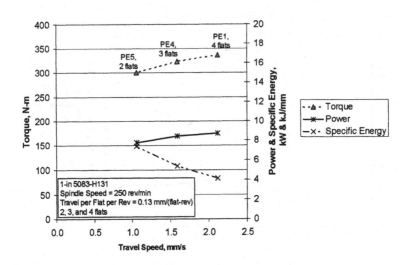

Figure 5. Torque, Power, and Specific Energy as a function of travel speed, weld in 25.4-mm 5083-H131 aluminum plate [3].

49

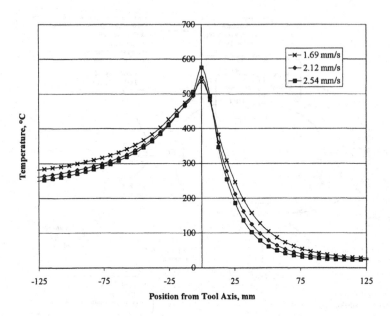

Figure 6. Temperature at plate mid-thickness as a function of position, friction stir weld in 25.4-mm 5083-H116 plate [4].

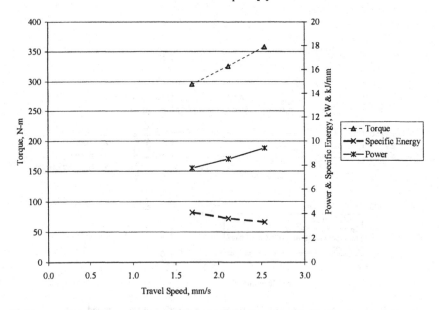

Figure 7. Torque, power, and specific energy as a function of welding speed, friction stir welds in 5083-H116 aluminum plate [4].

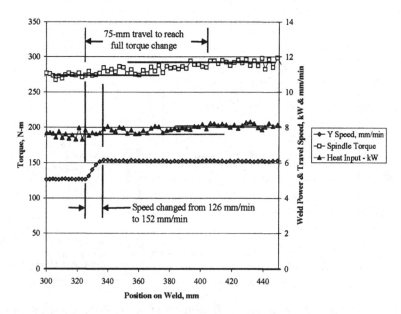

Figure 8. Torque, weld power, and travel speed as a function of position on the weld from FSW of 25.4-mm 5083-H116 aluminum plate, showing sudden increase in travel speed.

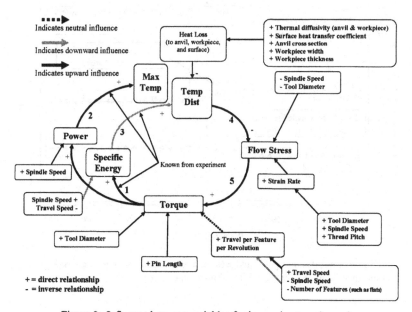

Figure 9. Influence between variables for increasing travel speed.

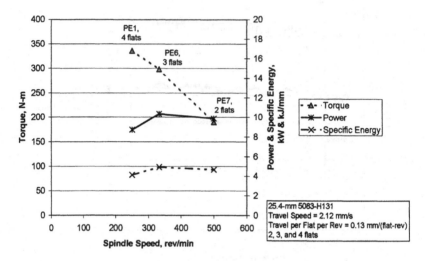

Figure 10. Torque, weld power, and specific energy as a function of spindle speed, friction stir welding of 25.4-mm 5083-H131 aluminum plate.

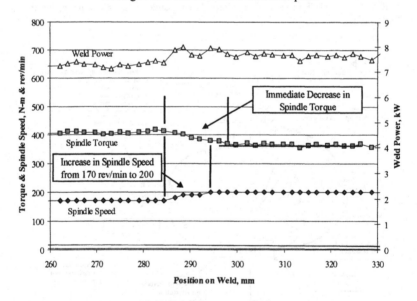

Figure 11. Torque, spindle speed, and weld power as a function of position on weld from friction stir weld in 25.4-mm 2195-T8, showing the effect of a change in spindle speed.

52

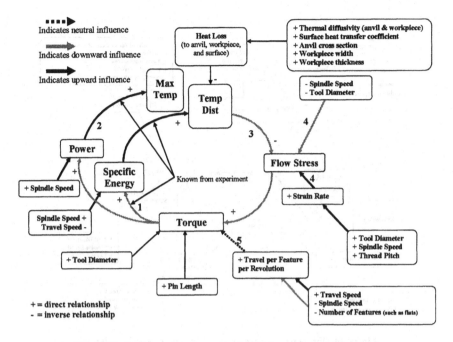

Figure 12. Influence between variables for increasing spindle speed.

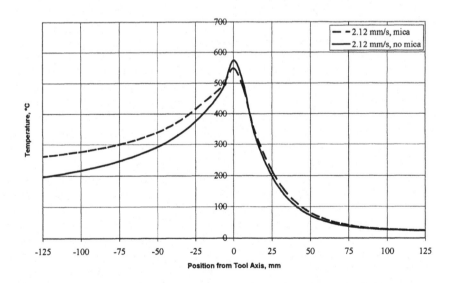

Figure 13. Plate mid-plane temperature as a function of time for friction stir welding of 25.4-mm 5083-H131 aluminum plate, showing the effect of anvil insulation.

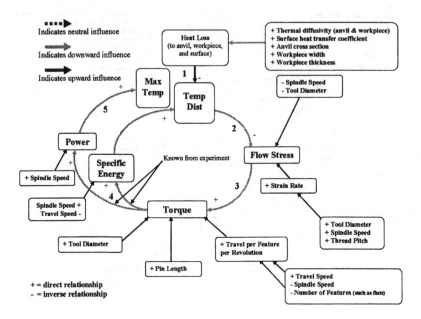

Figure 14. Influence between variables for adding anvil insulation.

54

Analysis of Self-Reacting Friction Stir Welds in a 2024-T351 Alloy

Tomasz Neumann[1], Rudolf Zettler[1], Pedro Vilaça[2], Jorge F. dos Santos[1], Luisa Quintino[2]

[1] GKSS-Forschungszentrum, Institute for Materials Research, Geesthacht, Germany

[2] Universidade Técnica de Lisboa, Instituto Superior Técnico, Portugal

Keywords: Classical friction stir welding, bobbin tool welding, self-reacting welding, tool design.

Abstract

In this study the GKSS force controlled self-reacting unit has been employed to produce welds in a 4mm thick 2024-T351 aluminium alloy plates. Three pins geometries were investigated for their suitability in joining the workpieces. Processing temperatures were measured in both sides of the welded joint at a depth of 1mm from the top and bottom surfaces of the plates. Microhardness testing indicated that considerable thermal softening occurs in each side of the joint. A comparison of welding forces, welding temperatures and microhardness measurements performed in the same alloy for similar FSW parameters, produced using standard FSW technique indicates that the self-reacting welds have higher processing temperatures with slightly lower welding forces for both in the weld travel and transverse to the welding direction. Temperature development in self-reacting FSW has been analysed using the *i*STIR code enabling to assess heat losses and heat power dissipated during welding.

Introduction

Friction Stir Welding (FSW) is a relatively recent joining technique developed and patented by TWI Ltd., Cambridge, England [1]. The thermo mechanical joining process takes place with each workpiece remaining in a solid-state i.e. there is no bulk melting in the region of the joint as would occur for conventional fusion welding processes. Rather the workpieces are joined by the interaction of a rotating and non-consumable tool, which is plunged and traversed along the joint line. Friction and deformation produce the heat necessary to cause the material in direct contact with the tool to soften and flow [2-4]. Thermally softened material from both sides of the joint then combine to produce what some researchers have termed a layered or onion ring structure in the region of the weld nugget [5-6]. Unlike fusion welding there is no evidence of an as cast microstructure across the stir zone. Rather the plastic deformation that takes place within the nugget results in recrystalisation of the parent material grains. The outcome is a much finer and equiaxed microstructure than previously existed for the base material. Surrounding the weld nugget is a thermo-mechanically affected zone (TMAZ), which is encapsulated, by a heat affected zone (HAZ) before one comes back into contact with the parent material.

Although numerous FSW patents exist pertaining to welding machines and welding tools there are essentially two main welding configurations; classical or otherwise referred to as single-sided where the pin and shoulder are applied from one direction only and self-reacting where both sides of the joint are processed simultaneously using a FSW tool having two shoulders. The bobbin tool investigated in this study is a self-reacting tool for which no

55

backing or supporting anvil is required [7]. In comparison to fixed gap bobbin tools where the distance between the interfacing tool shoulders cannot be adjusted during welding the self-reacting tool can apply or negate axial loading of the workpieces during processing.

All bobbin tools have two rotating shoulders, one above and one below the workpiece. Connecting both shoulders is a pin that penetrates entirely the workpiece thickness. Both shoulders when in contact with the workpiece provide for frictional heating but just as importantly prevent plasticized material from being expelled from the immediate weld zone. Unlike the classical FSW process the lower shoulder of a bobbin tool replaces the backing element. Significant processing flexibility can therefore be gained through the absence of a backing element, especially in the instance where hollow profiles are to be joined and where space restricts the introduction of any such backing element. In addition to increased flexibility the ability to friction stir weld simultaneously a workpiece from both sides has the potential to eliminate root flaw formation. Although several defect types are known to exist for FSW, such as volumetric i.e. voids or lack of penetration defects the occurrence of flaws is generally much lower when compared to those produced using conventional fusion welding technology. Since bobbin tool welds are devoid of a weld root it is not expected that such defects should occur and since the formation of root flaws are detrimental to the joint properties the possibility of avoiding such a defect can be viewed as a major process advantage.

Current available information concerning the bobbin tool friction stir welding can be found in the literature [8-11]. The MTS System Corporation demonstrated that 2195 and 2219 aluminium alloys in thickness range from 4mm up to 25mm could be successfully joined using self-reacting FSW [8, 9]. This includes dissimilar alloy welds (2219/2195) and tapered thickness (8mm to 16,5mm over 610mm length) configurations. AdAPT welding head was also proved to perform self – reacting butt welds on 4mm to 30mm thick aluminium alloys (2219, 2519, 2195, 5083, 6061, 2024 and 7020). Acceptable weld quality was obtained with various pin tool configurations, demonstrating the robustness of the process. The performance of the welds exceeded that of typical fusion welds and met or exceeded that of traditional FSW.

Bobbin tool welding of thin (1.8mm, 2mm, and 3mm) aluminium alloy 6061-T6 at traversing speeds grater than 1m/min has been demonstrated using MTS Process Development System Jr located at Lockheed – Martin [10].

Marie at al. demonstrated feasibility of bobbin tool FSW on various aluminium alloys (2024-T3, 2219, 6056-T4, 7449 TAF) in a thickness range 4mm to 15mm using ESAB's SuperStir FSW machine [11]. The work shows that bobbin tool FSW weld ability seems to be better for medium and thick sections than for thin materials, especially for 2000 series (tool failure). Authors conclude that bobbin tool technique is not suitable for increasing the travel speed but appears as a potential way to weld thicker sections, with no limitation from huge down forces.

The feasibility of using the formulation of heat flow in conventional FSW via the analytical code iSTIR in the establishment of correlations between FSW parameters and the thermal efficiency has already been demonstrated [12,13]. This computational model is an experimental/analytical procedure based on an inverse engineering approach. The complete formulation supporting the iSTIR code is discussed in detail in [13, 14]. In summary iSTIR results are generic solutions of the differential equation for heat conduction in a solid body, formulated for a point heat source. This source is considered to be concentrated at the mid-thickness simulating the typical location of the centre of the nugget and travelling with constant linear velocity, for 3D flow in bodies with limited width with surface losses. A simplified form of the differential equation of heat conduction, gives rise to Eq. (1), where: k (W/Km) is the thermal conductivity and A (W/m^3) is the heat rate field p/volume unit, generated (+) or dissipated (-) internally or at the surface. The temperature field, T (K), is expressed by Eq. (2), where: $e^{-vw/2\alpha}$ is the shape function for the temperature distribution, T_0 is initial or reference temperature and φ (K) is the axissymmetric temperature field, to be determined, by substituting Eq. (2) into Eq. (1) and solving according to the specific boundary conditions.

$$-\frac{v}{\alpha}\frac{\partial T}{\partial w} = \frac{\partial^2 T}{\partial w^2} + \frac{\partial^2 T}{\partial y^2} + \frac{\partial^2 T}{\partial z^2} + \frac{A}{k} \qquad (1)$$

$$T = T_0 + e^{\frac{vw}{2\alpha}}\varphi \qquad (2)$$

Experimental Programme

A schematic of the GKSS bobbin tool can be found in Figure 1. Axial force is applied onto both side of the workpiece via the actuation of a hydraulically operated piston contained within the body of the unit. Connected to the piston is the tool pin, which in turn is connected to the lower tool shoulder. By controlling the pressure on either side i.e. above or below the piston a defined force can be applied to the workpieces.

Figure 1. GKSS bobbin tool.

Figure 2. GKSS bobbin tool weld set up.

The GKSS bobbin tool was designed as a portable unit that could be handled by any system capable of accepting and actuating a rotating spindle. In this investigation all welds were performed using a Tricept TR805 robot at the GKSS-Forschungszentrum, Geesthacht, Germany. Weld set-up, Figure 2, consisted of the GKSS bobbin tool being mounted directly to the Tricept 805 spindle. To ensure optimal rigidity of the system only the Z-axis of the robot was actuated during welding trials. Weld traverse was achieved using a single axis programmable motorised table. A Kistler type force measurement table was mounted on top of the traversing table. The 4mm thick 2024-T351 aluminium alloy plates measuring 495 x 160 were firmly clamped to a frame mounted on a table supported by a load cells system, as seen in Figure 3. Table 1 presents the mechanical properties of the base material used in this study.

Figure 3. Workpiece clamping system.

Actuation and disengagement of the axial force during FSW was achieved using two hydraulic pumps. To determine the accuracy of the axial force placed on the workpieces a manometer was inserted between pump and spindle. The manometer had a range between zero and 10bar at increments of 0.2bar.

Based on the experience gained from FSW the 2024-T351 alloy using the classical approach [3] three tool pin profiles were designed and manufactured for welding with the GKSS bobbin tool. The first pin design possessed a simple cylindrical profile (Pin C). The second pin profile was identical to the first but had two counter rotating threads (Pin CT). The third pin profile was identical to the second but had three flats machined into the portion of the pin in the vicinity of the threads, (Pin CT3F). A schematic of pins CT and CT3F can be found in Figure 4. Each pin had a diameter of 5mm and was used in conjunction with 13mm scroll shoulders. Table II presents the process parameters investigated in this study and a summary indication of the obtained joint quality.

Figure 4. Bobbin tool welding pins CT (left) and CT3F (right).

Table I. Mechanical properties of the 2024-T351 base material

	Hardness Vickers	Ultimate Tensile Strength	Tensile Yield Strength	Elongation at Break
ASM: 2024-T351	137	478 MPa	333 MPa	20 %
Tested: 2024-T351	142	469 MPa	324 MPa	18.7 %

Table II Weld parameters investigated for each welding tool pin

Weld Number	Pin Tool	Rotational speed at start [rpm]	Rotational speed during welding [rpm]	Traverse speed [mm/min]	Comments
041010 BTW1	Pin C	600	450	100	Good weld
041010 BTW2	Pin CT	600	450	100	Good weld
041010 BTW3	Pin CT	600	600	100	Non bonding defect at the weld centre line
041010 BTW4	Pin CT3F	450	-	100	Pin broken at weld start
041010 BTW5	Pin CT3F	600	450	100	Good weld

FSW process temperatures were measured for both sides and for three depths within the workpieces, in the locations displayed in Figure 5. K-type thermocouples measuring 0.5mm in diameter were placed into 0.6mm diameter holes. Temperatures were recorded over a weld length of 70mm at the middle zone of the weld length ensuring that a steady state thermal condition had been achieved. Data was recorded at 20Hz.

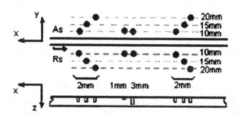

Figure 5. Top view of the thermocouples arrangement and side view representing the hole depths.

In addition to temperature measurements welding forces were also recorded by a Kistler Type 9366B force measurement table coupled to an 8 Channel Type 5019/5017 Amplifier. The multi-component kit consisted of 4 load cells allowing for a range up to 40kN. Each element contains a preloaded force sensor. The outputs of the four force measuring elements are connected in the summing box so that the orthogonal forces Fx, Fy, Fz as well as 6-component force and moment measurements Fx, Fy, Fz, Mx, My, Mz are possible. Data was recorded at 10Hz.

After welding two bend specimens were wet cut from each weld in order to determine bend properties for the top and bottom surfaces of the joint. The specimens were ground transverse to the weld direction prior to bending in order to eliminate any possible notch effect due to the initial cutting of the specimens from the plate.

Further specimens were removed from each weld for hardness testing purposes. Hardness profiles were obtained through the thickness of each weld; at 1mm below the top surface, at mid plate thickness and 3mm below the top surface of the workpieces. All specimens were ground and polished prior to testing where a load of 0.2kg and an indentation distance of 0.5mm spacing was used.

Results and Discussion

Process loads. As a result of two shoulders making contact with the workpieces during FSW it was anticipated that frictional forces would promote a high transverse force causing the bobbin tool to deflect sideways away from its weld direction path. This fear however did not eventuate. Weld forces in the direction of weld travel (-Fx) and transverse to the weld travel direction (Fy), Figure 6 were relatively small and ranged between 800 to 900N during welding.

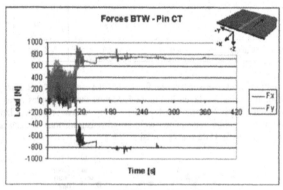

Figure 6. Process in-plane loads during bobbin tool FSW.

Comparing welds having an equivalent weld pitch i.e. ratio between the rotation and travel speed as produced in the same 4mm thick 2024-T351 alloy, Figure 6 for the bobbin tool and Figure 7 for the classical friction stir weld performed using the Tricept 805 robot, it can be seen that the bobbin tool weld has the lower processing forces. Note in both cases the welds were performed using tools having a scroll shoulder. Although no bobbin tool welds were directly performed with the robot traversing along the joint line between the workpieces the fact that the measured side and traversing forces were so low indicates the possibility of producing these same welds solely with the aid of the robot. One area of concern however prior to weld traverse is the initial onset of the axial force. Figure 6 clearly demonstrates large surges in force as can be seen for the first 60 seconds of measurement due to the shoulders engaging the workpiece.

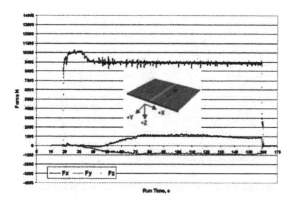

Figure 7. Classical FSW process loads.

When bringing the two shoulders of the bobbin tool into contact with the workpiece it was found necessary to gradually build up the axial force over time. Even with minimal loading a fierce vibration would often occur until the base material under the shoulders was sufficiently plasticised. During the investigated welding trials it was found that the necessary axial force for FSW with the bobbin tool (i.e. force applied between the shoulders) varied between 1000N and 1400N. This value (1000N) is almost one tenth that of the axial load required to produce the same weld using the classical friction stir weld procedure.

Process. Although weld plates measured 495mm in length the investigated bobbin tool welds produced in this study had a weld length measuring 350mm. The reason for this shorter weld length was due to the fact that this was the maximum traverse distance of the motorised table.

Not only was it necessary to gradually apply axial loading to the shoulders as they engaged the workpieces. It was also necessary to determine the minimum rotation speed. This speed was found to be 600rpm. Welding trials were conducted where the start up rotation speed was lesser, e.g., 450rpm. This rotation speed only led to tool breakage and subsequently the start up rotation speed was increased to 600rpm.

Among the welds performed the following three welds were selected for further analysis (Table 1): BTW1, BTW2, BTW5 produced using tool pins C, CT and CT3F, respectively. All welds were performed having a constant weld traverse of 100mm/min and a tool rotation speed of 450rpm. Visual examination of the three welds demonstrated that bonding of the workpieces had occurred. Surface appearance of the top and bottom bobbin tool friction stir welded plate surfaces can be found in Figures 8-10.

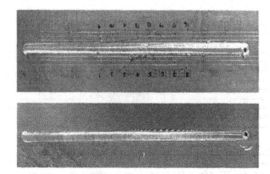

Figure 8. Top face above and bottom face below for bobbin tool weld "Pin C".

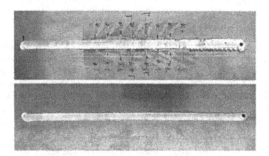

Figure 9. Top face above and bottom face below for bobbin tool weld "Pin CT".

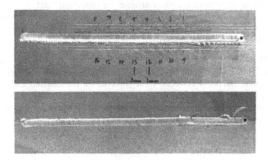

Figure 10. Top face above and bottom face below for bobbin tool weld "Pin CT3F".

In all cases significant sections of the weld length appeared defect free and of high quality, the exception being the last 80mm of each weld. More often a separation both vertical and horizontal of the workpieces was observed to have occurred. This separation caused significant flash formation and led in the case of the weld produced using Pin CT3F, Figure 10 having an open running void in the underside of the workpieces. Although this weld was more rigidly clamped than both the welds produced using pins C and CT (this was undertaken so as to avoid sideways separation of the workpieces) it was found that the high sideways clamping force attributed to vertical separation i.e. upwards bowing of one of the two workpieces and this led to defect formation.

Macrostructure and bend tests results Weld macrographs were taken from mid weld length where it was found through temperature measurement that the process conditions had achieved a steady thermal state.

Figures 11-13 present the three welds as viewed in the direction of weld travel i.e. from behind the welding tool pin. Welds were performed with the bobbin tool rotating in a clockwise direction. Hence the advancing side of the weld is on the left and the retreating side is on the right of each macrograph.

In the case of the bobbin tool weld produced using pin C, Figure 11, it can be seen that the weld nugget is for the most part symmetric in shape with a distinct boundary appearing between the thermo-mechanically affected zone (TMAZ) and heat affected zone (HAZ) in the advancing side of the weld. At mid thickness of the workpieces a small tunnel defect was seen to have occurred in the weld nugget. This defect was more prominent for the advancing side but could also be found in the retreating side of the weld nugget, Figure 11.

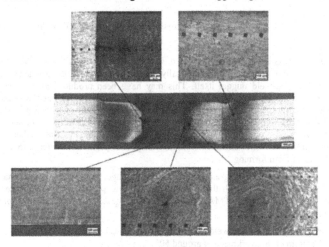

Figure 11. Macrograph and microstructures of the "Pin C" weld.

Two bend test samples were cut out at mid weld distance of the weld produced using pin C. The results indicated a bend angle of up to 25° for the top face and a root bend angle up to 58° could be achieved from this weld. The bend tests have been performed up to the first visible failure in the weld.

In the case of the weld produced using the pin CT, Figure 12, it was observed that the weld was once again almost symmetric in shape. The shape of the weld nugget however was clearly quite different when compared to the weld produced using pin C, Figure 11.

63

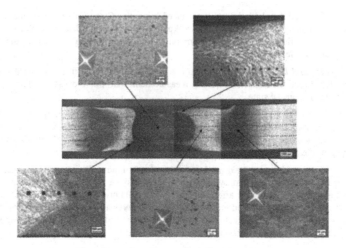

Figure 12. Macrograph and microstructures of the "Pin CT" weld.

In Figure 12 it is possible to observe that not only was the weld nugget bulbous but there also existed a clearly definable thermo-mechanically affected zone which divided both shoulder flow zones from the weld nugget itself. This may have been produced as a result of the counter rotating threads creating a compressive material flow towards the mid thickness of the workpieces. Micrographs taken in and around the weld nugget, shown in Figure 12, also confirmed that there were no volumetric defects in the mid thickness of the workpiece as was the case for the weld produced using pin C, emphasized in Figure 11. This further supports the assumption that there was sufficient vertical transport of plasticised material to avoid any volumetric defect from forming.

As can be seen in Figure 12 the grains within the weld nugget resulted much finer than those of the base material and as with the weld produced using pin C, where in Figure 11 the microstructure contained in the TMAZ and the HAZ are visibly distinguishable from that of the parent material.

Bending tests conducted on the weld performed using pin CT achieved bend angles for both the top and bottom of the workpiece of around 50°.

Unlike the welds performed using pins C and CT the weld produced using pin CT3F, Figure 13, were not symmetric in shape. The upper portion of the weld nugget (inclusive here is the shoulder induced flow zone) was distinctly larger than that of the lower portion of the weld.

64

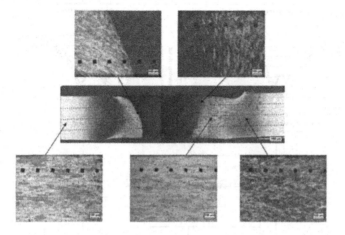

Figure 13. Macrograph and microstructures of the "Pin CT3F" weld.

It is thought that the shape of the stir zone, in Figure 13, may have been influenced by not having had the counter rotating pin threads central with that of the mid thickness of the workpieces. Examination of the pin after welding indicated much more of the upper threaded portion of the pin had been in contact with the workpieces in comparison to the lower threads.

Although pin CT3F had three flats machined into the surface of the threaded portion of the pin no volumetric defects have been formed. Bend samples cut out of the pin CT3F weld were again tested and demonstrated a top face bend angle up to 28° and root bend angle up to 37°.

Hardness Three hardness profiles were performed as indicated in the experimental program. The results from this investigation are summarised in Figure 14.

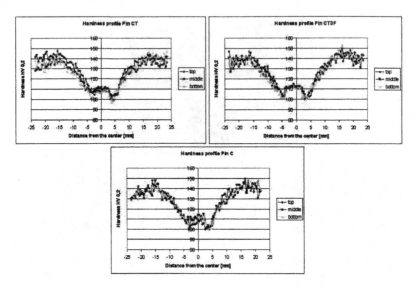

Figure 14. HV0.2 profiles of the GKSS bobbin tool welds.

Hardness Vickers (HV0.2) tests performed on the 2024-T351 base material produced an average of 137HV over the ten indentations. In the case of the bobbin tool welds, which had a measurement length of 23mm either side from the centre of the stir zone hardness values varied between 100HV and 150HV indicating that thermal softening had occurred. Measurements taken 30mm from the center of each weld again only indicated a maximum hardness of 134 HV. Based on these results it can be concluded that the HAZ is spread much further into the base material compared to equivalent classical friction stir welds, shown in Figure 15. Here the results indicate that there is not only an increase in hardness comparable to values obtained for the base material after only 8mm from the weld centre, but also that there appears to be a distinctly different hardness regions within the through thickness of the workpiece material.

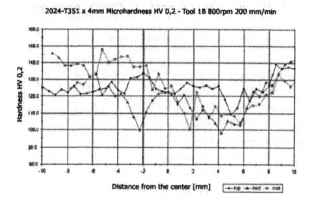

Figure 15. Hardness profiles of classical friction stir welds in 2024-T351.

Thermal Cycle: FSW process temperatures were nominally measured 10mm, 15mm and 20mm from the either side of the joint line 1mm from the top surface, mid workpiece thickness and 3mm below the top surface as portrayed in Figure 5. During welding it has been observed that the welding plane forces were relatively small however big enough to push the bobbin tool out of the joint line up to 1mm into either the advancing or the retreating side for the performed welds. Using the general conduction theory expressed by the Fourier Law thermal cycles for 10mm distance have been calculated based on the actual measurements and thermocouple location after welding. Figure 16 shows the calculated temperature profile measured 2mm bellow the surface for weld Pin CT.

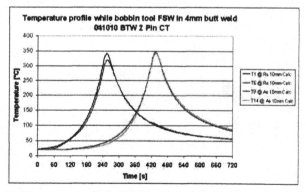

Figure 16. Temperature profile in GKSS bobbin tool friction stir welds in 2024-T351.

Temperature profiles for the bobbin tool welds regardless of the pin used were very similar and indicated a peak processing temperature of around 350°C, for about 10mm from the weld joint line. Friction stir welds performed for comparable processing parameters using the classical FSW approach indicated a peak processing temperature of around 282°C again measured 10mm from the weld joint line and 1mm below the workpiece surface, Figure 17.

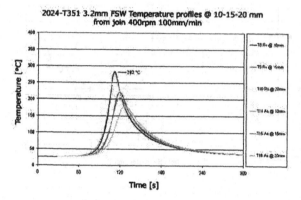

Figure 17. Temperature profile in classical friction stir weld in 2024-T351.

Not only are the processing temperatures higher for the bobbin tool welds, but heating and cooling rates are much inferior for these welds in comparison to the classical friction stir weld. This fact emphasises the small heat dissipation during the bobbin tool welds of aluminium alloys.

When one compares peak processing temperatures over the measurement area for the bobbin tool welds it can be seen that slight differences occur dependant on the tool pin form used to produce the weld, Figure 18. In all the investigated welds it appeared that the retreating side of the weld was the hotter of the two sides. Figure 18, also confirms that the different shaped pins, pin C, CT and CT3F have only a marginal effect on temperature profiles.

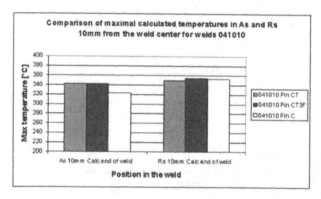

Figure 18. Peak temperature for different GKSS bobbin tool pins configuration about 10mm far from the center of weld bead into both sides of the weld bead.

It can also be observed that the temperature as measured 3mm below the surface of the workpieces is approximately 15°C higher than that measured 1mm below the workpiece surface, Figure 19. The reason for this may be due to the inability of the bottom shoulder to dissipate heat as effectively as the top half of the welding tool.

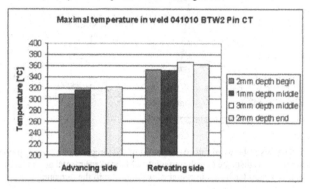

Figure 19. Peak temperatures at different depths for the GKSS bobbin tool weld Pin CT about 10mm far from the centre of weld bead into the retreating side.

Modelling The application of the analytical code *i*STIR to the thermal data resultant from the GKSS bobbin tool trials enabled to determine the heat power dissipated by conduction through the material thickness. One relevant issue is to investigate the differences of the heat power dissipated between the GKSS bobbin tool welds and the classical FSW approach where the bottom surface of the plates is in intimate (continuous and under high pressure) contact with a rigid metallic anvil, resulting in additional losses by conduction between the plates and the anvil.

The heat power dissipated corresponds to the best fit between the computational analytical and the experimentally measured temperature developments at both sides of the weld, iterating on the equivalent surface losses coefficient which stands for the overall heat power losses, i.e., by superficial mechanisms to the environment (by convection and radiation mechanisms) and by conduction to the tool and the clamping system.

*i*STIR code was applied to the available data considering a similar friction stir welded joint with 3D heat flux in a limited thickness and width domain, including heat losses at the surfaces. The thermo-physical data considered and the final value obtained for the equivalent surface losses coefficient are present in Table III. The very small value obtained for the equivalent losses coefficient, $2W/m^{2o}C$ (almost null), should be emphasized.

Table III Thermo-physical properties considered for the AA2024-T351

Initial Temperature	Thermal Conductivity	Equivalent Surface Losses
$20°C$	$140 W/m\cdot°C$	$2W/m^2\cdot°C$

As a sample of the results obtained, Figure 20 shows the fitting between the computational and the experimental temperature curves at the retreating side of the welded plates. The value obtained for the heat power dissipated was 205w. The error for the maximum temperature estimated is about 8% in excess which is within the error of the measured temperatures and the fitting of the heating and cooling phases is considered acceptable although heating and cooling rates are over estimated.

69

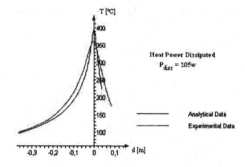

Figure 20. Comparison between the development of the temperature (2mm depth i.e. plate mid-thickness) versus distance (parallel to joint line) for experimental and modelling data, about 10mm far from the center of weld bead into the retreating side.

*i*STIR code was previously applied to the classical FSW approach for the same material and thickness by Vilaça et al. [12,13]. FSW trials under hot conditions, i.e., rotation speed of 800rpm, travel speed of 100mm/min and a vertical downward force of 11kN, resulted in values of heat power dissipated of 319w and equivalent losses coefficient of 130w/m^2°C.

Comparing the GKSS bobbin tool welds with the classical FSW, the heat power dissipated is about 67% and the equivalent losses coefficient is about 1.5%. Note that even though the maximum temperature due to heat flow into the plates (with the source of heat located on the mechanical processing of the plates by the tool during weld) resulting from the bobbin tool is higher than in the classical approach, the heat power is inferior because the losses are very low. The thermal efficiency of the bobbin tool welds will be established after measurement of torque during welds are available. As soon as torque data is available, it will be possible to determine the mechanical power delivered by the tools into the parts being welded.

Thus it is possible to state that from all the mechanical power delivered by the tools, into the parts being welded, the heat power dissipated from bobbin tool is about 2/3 of the classical tool, which is not a significant difference, but the most relevant is definitely the surface losses which are almost inexistent for the GKSS bobbin tool welds.

Conclusions

The experimental programme thus conducted allows for the following conclusions to be drawn:

1. The GKSS bobbin tool is capable of successfully producing butt welds in 4mm thick 2024 T351 aluminium alloy.

2. Tool pins CT and CT3F have produced satisfactory results when FSW the 2024 T351 alloy with a bobbin tool.

3. Results of preliminary mechanical tests demonstrate that considerable thermal softening of the workpieces occurs as a result of bobbin tool welding.

4. In comparison to the classical FSW technique it appears that a greater heat input and an inability of bobbin tool welds to dissipate heat due to the absence of a backing bar

70

or anvil are the main reasons for the increased thermal softening experienced for these welds.

5. Using *i*STIR programme it has been found that in the classical FSW technique the superficial heat losses, e.g., via backing bar and clamping, is about 65 times higher as in the bobbin tool process.

6. Although higher temperatures are developed in the plates during FSW, with the bobbin tool, *i*STIR code indicated that the heat power dissipated is 2/3 lower than in the classical friction stir welds.

7. Axial force when FSW can be considerably reduced when using the bobbin tool. Side and traversing forces however appear similar to those produced using the classical FSW approach under comparable welding parameters.

8. Plate clamping is an issue which must be further investigated when FSW using the bobbin tool. High side clamping forces have shown to lead to plate bowing i.e. vertical separation during FSW. This leads to excessive flash and defect formation.

References

1. Thomas W. M., Nicholas E. D., Needham J. C., Murch M. G., Temple – Smith P. and Dawes C. J. (TWI): Improvements relating to friction welding. EP 0 615 480 B1.

2. Zettler R, Lomolino S, dos Santos J, Donath T, Beckmann F, Lippman T and Lohwasser D. A Study on Material Flow in FSW of AA2024-T351 and AA6056-T4 Alloys. 5th FSW Symposium, Metz, France, 2004.

3. Li Y., Murr L. E., McClure J. C.: Flow visualisation and residual microstructures associated with the friction stir welding of 2024 aluminium to 6061 aluminium. Materials Science and Engineering A 271 (1-2): 213-223. 1999.

4. v. Strombeck A., dos Santos J. F., Torster F., Kocak M. (GKSS): Bruchmechanische Untersuchungen an Reibrührgeschweißten Aluminiumlegierungen, Werkstoffwoche 1998, 12. – 15. 10. 1998 München.

5. Kallee S. W. (TWI): FSW: Process and applications. GKSS/TWI Workshop, Reibrüschweißen, GKSS-Forschungszentrum, Geesthacht, 3. Mai 1999.

6. Backlund J., Norlin A., Anderson A.: Friction Stir Welding – Weld properties and manufacturing techniques. Preprints of Inalco '98, 16 April 1998, Cambridge, UK.

7. v. Strombeck A., dos Santos J. F. (GKSS): Vorrichtung zum Verbinden von Werkstücken nach der Methode des Reibrührschweißens. Deutsches Patentschrift DE 199 57 136 C1.

8. Skinner M. J., Edwards R. L. (MTS Systems Corporation); Adams G. (Lockheed Martin / Michoud Operations); Li Z. (Edison Welding Institute): Improvements to the FSW Process using the Self – Reacting Technology. 4th International Symposium on Friction Stir Welding 14 – 16 May 2003, Park City, Utah, USA.

9. Skinner M. J., Edwards R. L. (MTS Systems Corporation): Improvements to the FSW Process using the Self – Reacting Technology. International Conference on Processing & Manufacturing of Advanced Materials, July 7-11, 2003, Leganés, Madrid, Spain.

10. Gil S., Edwards R. (MTS Systems Corporation); Sassa T. (Nippon Sharyo, LTD.): A feasibility study for Self Reacting Pin Tool Welding of thin section aluminium.

5th International Friction Stir Welding Symposium 14 – 16 September 2004, Metz, France.

11. Marie F., Allehaux D., Esmiller B. (EADS Corporate Research Center France): Development of the bobbin tool technique on various aluminium alloys. 5th International Symposium on Friction Stir Welding 14 – 16 September 2004, Metz, France.

12. Vilaça P., Quintino L., dos Santos J. F., Zettler R., Sheikhi S.: Quality assessment of friction stir welding joints via an analytical thermal model, iSTIR, accepted for publication in Materials Science & Engineering A, 25 September 2006.

13. Vilaça P., Fundamentos do Processo de Soldadura por Fricção Linear – Análise Experimental e Modelação Analítica, PhD thesis, IST, Technical University of Lisbon, Portugal, September 2003.

14. Vilaça P., Quintino L., dos Santos J. F.: iSTIR–Analytical thermal model for friction stir welding, Journal of Materials Processing Technology, pp.452-465, Volume 169, Issue 3, 1 December 2005.

DEVELOPMENT OF A TORQUE-BASED WELD POWER MODEL FOR FRICTION STIR WELDING

Jefferson W. Pew[1], Tracy W. Nelson[2] and Carl D. Sorensen[2]

[1]Formerly with Brigham Young University, now with Exxon Mobil

[2]Brigham Young University
Department of Mechanical Engineering
435 CTB
Provo, UT 84602

Keywords: Friction Stir Welding, Process Understanding, Process Development, Process Fundamentals, Heat Input, Torque, Weld Parameters

Abstract

The basis for fusion weld computational models is welding heat input. Heat input is fairly well understood for most arc welding processes. However, this traditional approach is not as straightforward for Friction Stir Welding (FSW). To date, there is no definitive relationship to quantify the heat input for FSW. An important step to establish a heat input model is to explore effects that FSW process parameters have on the weld power.

This study details the relationship between weld parameters and torque for aluminum alloy 2024. A quantitative weld power and heat input model is created from the torque input. Heat input model shows that decreasing the spindle speed or increasing the feed rate significantly decreases the heat input at low feed rates. At high feed rates, feed rate and spindle speed have little effect on the heat input.

In addition, this study outlines and validates the use of variable spindle speed tests for determining torque over a broad range of parameters. The variable spindle speed tests are found to be a significant improvement over previous methods of determining torque as they enable the torque to be modeled over a broad range of parameters with a minimal number of welds. The methods described in this study can be easily used to develop torque models for different alloys and materials.

Introduction

Despite extensive research into the fundamentals of Friction Stir Welding (FSW), a definitive model for weld heat input has not been determined. Similar to traditional arc welding processes, selection of heat input is essential for both modeling FSW and attaining desired weld properties. In traditional welds, the weld power can be easily set by adjusting current and voltage. In contrast, there are no weld variables in FSW whereby power can be directly controlled. In FSW, weld power is a function of many variables including, 1) spindle speed, 2) travel speed, 3) tool geometry, and 4) tool depth. In addition, the power requirement is strongly affected by the material being joined. There is no well-established relationship between these weld variables and weld power. Currently, FSW parameters must be established through trial and error. This approach may not result in the optimum heat input and is costly both in terms of time and money.

Some preliminary research [1-3] has shown the necessary power requirement is likely a function of various parameters that include (but may not be limited to) the material type, feed rate, spindle speed, tool geometry, and tool depth.

The purpose of obtaining a weld power model is to develop a heat input model. To clarify, heat input is a measure of energy/length. Weld power is a measure of energy/time. Although the heat input can be calculated by dividing the weld power by the feed rate, many researchers have attempted to measure the heat input directly.

Numerous papers in the literature have attempted to characterize or model heat input during FSW. Some of the earliest research attempted to measure heat input through thermocouple placement [4-12], heat affected zone (HAZ) measurements [13], or a combination of these two methods [14-16]. Others calculate heat input from the power or torque input to the weld [17-25]. These approaches have deficiencies because the torque input is not constant for all weld parameters and tooling, or from one material to another.

We will detail the effects of weld parameters (feed rate, spindle speed, and tool depth), on the weld power during FSW for one tool geometry in 2024 Al. The results of this study can be easily applied to new alloys and different tool designs to rapidly create weld power models. The results can also be used to establish a weld heat input model for a given alloy.

Experimental Approach

The weld power can be calculated from spindle torque through the relationship shown in Equation 1.

$$Power = \Omega M \hspace{6cm} \text{eqn. 1}$$

In this equation, Ω represents the spindle speed and M represents the torque. Although the FSW machine used for this study can record the power output of the motor, this power output includes motor and drive train losses. For this reason, calculating the weld power from the torque directly from the spindle is a more accurate estimation of the power into the weld.

The FSW tool used for all the tests was a Convex Scrolled Shoulder Step Spiral (CS4) tool made from heat-treated H13 tool steel (see Figure 1). As with all CS4 tools, no head tilt was required. The CS4 tool had a pin length of 6.35 mm (0.25 in.) and a shoulder diameter of 25.4 mm (1.0 in.). Only one tool was used for all tests. The tool penetration depth was gauged by a dial indicator attached to the tool holder.

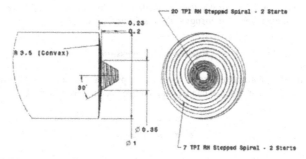

Figure 1. Geometry of FSW tool used in all welds.

Variable Spindle Speed Validation

Variable spindle speed tests may be a viable method to calculate weld power for any given material [3]. In a variable spindle speed test, the feed rate and tool depth are held constant throughout the weld while the spindle speed is varied continuously from one value to another. The use of these tests was first reported by Reynolds [24, 26] for determining forces in FSW over a broad range of parameters.

Material

The alloy explored in this study was Al 2024-T3 and was selected because it has a relatively small FSW window. This meant that a set of tests could be defined covering all weld parameters. This alloy was also selected because of its common usage and the considerable interest based on the FSW literature.

Test plates, 13cm (5in) wide by 16.5cm (5.4ft) long, were sheared from larger plates. The plate thickness was 9.5mm (0.375in.). The plate thickness was deliberately kept greater than the pin length to avoid interactions between the tool and anvil. The authors realize this may not represent actual full penetration welding conditions, but the objective was to develop a method and model.

Preliminary work was performed to determine the parameters whereby the 2024-T3 Al could be successfully welded. Weld parameters included tool depth, feed rate, and spindle speed. The weld parameters evaluated for 2024-T3 Al are shown in Table 1.

Table 1. Weld parameters evaluated for 2024-T3 Al.

Alloy	Feed Rates (mm/min)	Spindle Speed Range (RPM)	Tool Depth (mm.)
2024	51, 102, 152	175 – 350	5.2, 5.3, 5.5

Several measures were taken to ensure consistency between welds. First, the spindle speed ramp rate was held constant for all welds. Second, an extra minute of weld time was added to the start of each weld where the spindle speed was held constant. The dwell time allowed the weld depth to be adjusted appropriately prior to data acquisition. Third, after each weld, the anvil was cooled to room temperature using dry ice and methanol. Cooling was used to prevent the anvil from preheating the next weld and changing the resulting forces.

Results and Discussion

All welds exhibited a distinctive torque curve. A representative torque curve for 2024 Al is shown in Figure 2. The curve shape can be characterized accurately as a third-order polynomial.

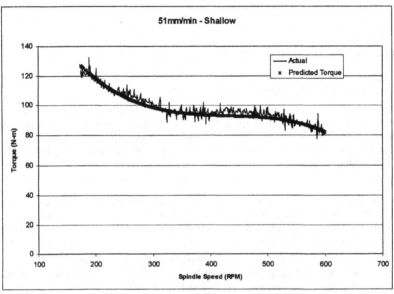

Figure 2. Graph of torque versus spindle speed in friction stir welded 2024 Al. Weld parameters were 51mm/min at a shallow depth.

To create a model of torque as a function of parameters, a regression analysis was performed on the each data set in MINITAB (a statistical software package). The independent variables considered for the model included "Feed Rate", "Depth", and "Spindle Speed" as well as the four interaction terms "(Feed Rate)*Depth", "(Feed Rate)*(Spindle Speed)", "Depth*(Spindle Speed)", and "Depth*(Feed Rate)*(Spindle Speed)". Two additional variables, "(Spindle Speed)^2" and "(Spindle Speed)^3", were also included together with their interaction terms with "Feed Rate" and "Depth". These higher order variables of Spindle Speed were added because, as mentioned previously, the data were described best by third-order polynomials. This accounts for a total of 13 possible independent variables in each model.

Backward elimination was used to find a combination of parameters resulting in the best fit without extra terms. Backward elimination is a statistical tool for removing excess terms from a model. This allows the model to be more accurate since excess variables simply may be describing noise in the data.

Once a model had been determined for the entire set of data, an R^2 value was calculated for each individual weld. To check the accuracy of the model, extra welds were made with new parameters. Again, a R^2 value was calculated to determine the goodness-of-fit of the model to these extra runs. The overall R^2-adjusted values for alloy 2024 Al were 97.0%.

Weld power (watts) was then calculated using equation 2. In this equation, Ω represents the spindle speed and M represents the predicted torque (N·m). The spindle speed is multiplied by 2π and divided by 60 to convert from revolutions/minute to radians/second.

$$WeldPower = \frac{(2\pi)\Omega M}{60}$$ eqn. 2

From the weld power, a predicted heat input can be calculated by dividing the weld power by the travel speed (v) using Equation 3. This is only an estimate of the heat input since there may be losses that depend on the input parameters. For example, the rate of heat loss, through radiation or by conduction to the anvil and tool, may change based on the parameters. Graphs can be plotted to show how these factors changed as a function of depth.

$$H.I. = \frac{P.I.}{v}$$ eqn. 3

A predicted heat input model can be created from the weld power model using Equation 3. Using this model, predicted heat inputs are shown in Figure 3 for alloy 2024 for all three tool depths investigated. These graphs show that the heat input decreases both with decreasing spindle speed and increasing feed rate. The heat input shifts towards higher values as the tool depth increases, but the overall shape of the surface changes very little.

The rate of change in heat input decreases rapidly as the feed rate is increased. For example, (in Figure 3a) at 215 rpm, the change in heat input from 25 mm/min to 76 mm/min (1 ipm to 3 ipm) is nearly 4000 J/mm while the difference between 76 mm/min to 177 mm/min (4 ipm to 7 ipm) is less than 1000 J/mm. Little benefit in heat input is gained by increasing the feed rate above 177 mm/min (7ipm) except to make faster welds. If a further decrease in the heat input is required above this feed rate, then it would be more advantageous to decrease either the tool depth or the spindle speed.

Also, RPM has a much greater affect on heat input at lower travel speeds that at higher travel speeds. Over the range of 177 to 576 rpm, the change in heat input at 177 mm/min (7 ipm) is only about 1000 J/mm. In contrast, at 25 mm/min (1 ipm) the change in heat input over the same range of RPM is approximately six times greater.

This new approach to determine FSW heat input is much more accurate than other reported means of determining heat input. Fewer welds are required to develop a full understanding of the complex relationship between weld parameters, tooling, and material. In addition, a high degree of accuracy can be achieved. These results provide a foundation for future work in processes and tooling development.

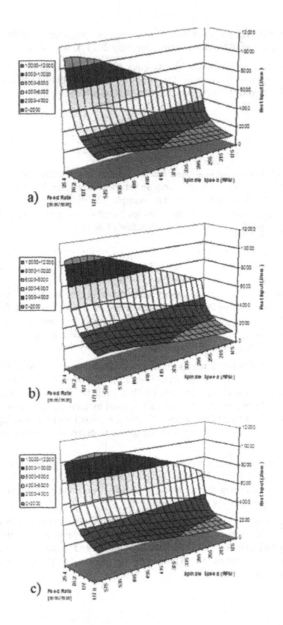

Figure 3. Heat input predictions using the model for alloy 2024 at tool penetration depths of a) 5.2mm, b) 5.35mm, and c) 5.5mm

Conclusions

1. The variable spindle speed method was utilized to determine the weld power. This new method not only captures more information than previous methods, but is also both easier and requires fewer welds than previous methods.

2. Empirical models were developed that accurately describe weld power as a function of weld parameters for aluminum alloy 2024. The same method can likely be applied to any alloy.

3. The empirical models developed have an accuracy of 93% or greater in predicting the weld power.

Acknowledgments

The authors acknowledge financial support for this work from the Office of Naval Research, contract No. N00014-03-1-0792, Dr. Julie Christodoulou, Program Manager.

References

1. Record, J., *Statistical Investigation of Friction Stir Processing Parameter Relationship*, Brigham Young University, Provo, UT (2005).

2. Record, J., J. Covington, B.W. Webb, and T.W. Nelson, *Proceedings of the Fifth International Symposium on Friction Stir Welding*, Metz, FR, September 2004.

3. Pew, J., J. Record, and T.W. Nelson, "Development of a heat input model for friction stir welding", *7th International Conference on Trends in Welding Research*, Pine Mountain, GA, May 2005.

4. Russell, M.J., and H.R. Shercliff, "Analytical Modeling of Friction Stir Welding", INALCO '98, TWI, Cambridge, U.K., 1998.

5. McClure, J.C., W. Tang, L.E. Murr, X. Guo, Z. Feng, and J.E. Gould, "A Thermal Model of Friction Stir Welding", *Proceedings of the 5th International Conference on Trends in Welding Research*, Pine Mountain, Georgia, June 1998.

6. Chao, Y.J. and X. Qi, "Thermal and Thermo-Mechanical Modeling of Friction Stir Welding of Aluminum Alloy 6061-T6", *Journal of Materials Processing and Manufacturing Science*, 7, pp. 215-233 (1998).

7. Song, M. and R. Kovacevic, "Thermal Modeling of Friction Stir Welding in a Moving Coordinate System and Its Validation", *International Journal of Machine Tools and Manufacture*, 43, pp. 605-615 (2003).

8. Colegrove, P., "Three Dimensional Flow and Thermal Modeling of the Friction Stir Welding Process", *Proceedings of the 2nd International Symposium on Friction Stir Welding*, 26-28 June 2000, Gothenburg, Sweden.

9. Schmidt, H.N.B. and J. Hattel, "Heat Source Models in Simulation of Heat Flow in Friction Stir Welding", *International Journal of Offshore and Polar Engineering*, 14, pp. 296-304 (2004).

10. Tang, W., X. Guo, J.C. McClure and L.E. Murr, "Heat Input and Temperature Distribution in Friction Stir Welding", *Journal of Materials Processing and Manufacturing Science*, 7, pp. 163-172 (1998).

11. Chao, Yuh J., X. Qi, W. Tang, "Heat Transfer in Friction Stir Welding – Experimental and Numerical Studies", *Journal of Manufacturing Science and Engineering*, vol. 125, pp. 138-145 (2003).

12. Song, M. and R. Kovacevic, "Numerical and Experimental Study of the Heat Transfer Process in Friction Stir Welding", *Proceedings of the Instn. of Mechanical Engineer*, Part B: *Journal of Engineering Manufacture*, 217, pp. 73-85 (2003).

13. Midling, O.T. and G. Rørvik, "Effect of Tool Shoulder Material on Heat Input During Friction Stir Welding", *Proceedings of the 1st International Symposium on Friction Stir Welding*, 14-16 June 1999, Thousand Oaks, CA, USA.

14. Gould, J.E. and Z. Feng, "Heat Flow Model for Friction Stir Welding of Aluminum Alloys", *Journal of Materials Processing and Manufacturing Science*, 7, pp. 185-194 (1998).

15. Frigaard, Ø., Ø. Grong, B. Bjørneklett, and O.T. Midling, "Modelling of the Thermal and Microstructure Fields during Friction Stir Welding of Aluminum Alloys", *Proceedings of the 1st International Symposium on Friction Stir Welding*, 14-16 June 1999, Thousand Oaks, CA, USA.

16. Song, M., R. Kovacevic, J. Ouyang, and Mike Valant, "A Detailed Three-Dimensional Transient Heat Transfer Model For Friction Stir Welding", *6th International Trends in Welding Research Conference Proceedings*, 15-19 April 2002, Pine Mountain, GA, pp. 212-217.

17. Linder, K., Z. Khandkar, J. Khan, W. Tang, and A.P. Reynolds, "Rationalization of hardness distribution in alloy 7050 friction stir welds based on weld energy, weld power, and time/temperature history", *Proceedings of the 4th International Symposium on Friction Stir Welding*, 14-16 May 2003, Park City, UT, USA

18. Zahedul, M., H. Khandkar, J.A. Khan, and Anthony P. Reynolds, *6th International Trends in Welding Research Conference Proceedings*, 15-19 April 2002, Pine Mountain, GA, pp. 218-223.

19. Khandkar, M.Z.H., J.A. Khan, and A.P. Reynolds, "Prediction of Temperature Distribution and Thermal History During Friction Stir Welding: Input Torque Based Model", *Science and Technology of Welding and Joining*, 8, pp. 165-174 (2003).

20. Schmidt, H., J. Hattel and J. Wert, "An Analytical Model for the Heat Generation in Friction Stir Welding", *Modelling and Simulation in Material Science and Engineering*, 12, pp. 143-157 (2004).

21. Leinart, T.J., W.L. Stellway, Jr., and L.R. Lehman, "Heat Inputs, Peak Temperatures and Process Efficiencies for FSW", *Proceedings of the 4th International Symposium on Friction Stir Welding*, 14-16 May 2003, Park City, UT, USA.

22. Colegrove, P.A., and H.R. Shercliff, "Experimental and numerical analysis of aluminium alloy 7075-T7351 friction stir welds", *Science and Technology of Welding and Joining*, vol. 8, no. 5, pp. 360-368 (2003).

23. Reynolds, A.P., W. Tang, Z. Khandkar, J.A. Khan, and K. Lindner, "Relationships between weld parameters, hardness distribution and temperature history in alloy 7050 friction stir welds", *Science and Technology of Welding and Joining*, vol. 10, no. 2, pp. 190-199 (2005).

24. Reynolds, A.P., Z. Khandkar, T. Long, W. Tang, and J. Khan, "Utility of Relatively Simple Models for Understanding Process Parameter Effects on Friction Stir Welding", *Materials Science Forum*, Vols. 426-432, pp. 2959-2694 (2003).

25. Reynolds, A.P., and W. Tang, "Alloy, Tool Geometry, and Process Parameter Effects on Friction Stir Weld Energies and Resultant FSW Joint Properties", *Proceedings of Symposium on Friction Stir Welding and Processing*, 4-8 November 2001, Indianapolis, IN, USA, pp. 115-122.

26. Reynolds, A.P., ONR Workshop on Friction Stir Welding, February 2004.

Eulerian Simulation of Friction Stir Welding in 7075 Aluminum

[1]S. Guerdoux, [1]L. Fourment, [2]M. Miles, [2]T. Nelson

[1]Ecole des Mines de Paris, CEMEF
1 Rue Claude Daunesse, 06904 Sophia-Antipolis, France
[2]Brigham Young University
265 CTB, Provo, UT 84602, USA

Keywords: Friction Stir Welding, Modeling, Eulerian

Abstract

An Eulerian simulation of friction stir welding was developed, where heat from friction is modeled using the Norton friction law. The Norton law has been used to model hot forging operations, where the shear stress at the interface between the tool and the workpiece is a function of the local flow stress of the material, the relative sliding velocity at the interface, and a friction coefficient α_f. Different levels of friction coefficient were used to determine temperature profiles at steady state conditions. Temperature predictions were in the range of 500-600°C under the tool shoulder and around the pin, where a level of friction of $\alpha_f = 0.2$ appears to provide the most reasonable results compared to experiment.

Introduction

Most modeling of friction stir welding has thus far been done using an Eulerian formulation, where the material is modeled as a viscous fluid and where the heat from material deformation and from friction has been estimated using analytical thermal models. Some early models employed the Rosenthal equation [1] to estimate the thermal profile in the part. Other approaches have used analytical heat sources to model the friction stir welding tool as a moving heat source [2]. This is useful for computing residual stresses in the part, but a Lagrangian formulation was used and the material flow was not computed [3]. A third group of simulations follow the Computational Fluid Dynamics (CFD) approach, in which the material is modeled as a viscous fluid, but where the heat contribution from friction is not explicitly included in the computation [4, 5]. This paper presents a simulation approach where the heat from deformation and friction, imparted by the friction stir welding tool, is computed in a thermo-mechanical steady state model. The friction law employed is viscoplastic, where the shear stress at the interface is dependent on the local flow stress of the material, the relative sliding velocity, and a friction coefficient. By varying the friction coefficient, while holding all other parameters constant, the effect of friction on the heating of material around the tool was evaluated.

Modeling Approach

Mechanical equations
The finite element discretization of the part is based on an enhanced (P1+/P1) 4-node tetrahedron element. A mixed velocity-pressure formulation is written for any virtual velocity field v* and

any virtual pressure field p*. The weak form of the momentum equation and weak form of incompressibility of the plastic deformation are written as:

$$
\begin{cases}
\forall v^* & \int_\Omega s : \dot{\varepsilon}^* dV - \int_\Omega p \nabla \cdot v^* dV - \int_{\partial\Omega} T \cdot v^* dS \\[2mm]
& - \int_\Omega \rho g \cdot v^* dV + \int_\Omega \rho \gamma \cdot v^* dV = 0 \qquad (1) \\[2mm]
\forall p^* & \int_\Omega p^* (\operatorname{tr}\dot{\varepsilon} - \alpha T) dV = 0
\end{cases}
$$

where s is the stress deviator, $\dot{\varepsilon}$ is the strain rate, p is the pressure field, T is the surface traction, ρ is the density, and α is the linear expansion coefficient.

Elasto-plastic and elasto-viscoplastic constitutive models are available to simulate both elastic and plastic deformation during warm and cold forming, together with elastic springback and residual stress calculations after forming. The 3D isotropic viscoplastic Norton-Hoff law is used:

$$
s = 2K\left(\sqrt{3}\dot{\bar{\varepsilon}}\right)^{m-1}\dot{\varepsilon} \qquad (2)
$$

where $\dot{\bar{\varepsilon}}$ is the effective strain rate:

$$
\dot{\bar{\varepsilon}} = \sqrt{\frac{2}{3}\dot{\varepsilon} : \dot{\varepsilon}} \qquad (3)
$$

K is the material consistency, a function of temperature T, and equivalent strain $\bar{\varepsilon}$:

$$
K = K_0 (\varepsilon_0 + \bar{\varepsilon})^n e^{\frac{\beta}{T}} \qquad (4)
$$

Another available constitutive equation is the Hansel-Spittel law defined as follows:

$$
\sigma_f = A e^{m_1 T} \bar{\varepsilon}^{m_2} \dot{\bar{\varepsilon}}^{m_3} e^{\frac{m_4}{\bar{\varepsilon}}} (1 + \bar{\varepsilon})^{m_5 T} e^{m_7 \bar{\varepsilon}} \dot{\bar{\varepsilon}}^{m_8 T} \qquad (5)
$$

For the simulations presented in this paper the values of K and m have been tabulated in order to match two different Hansel-Spittel constitutive laws (for hot and cold temperatures). The tabulated values are shown in Table 1.

Table 1. Tabulated values of K and m as a function of temperature.

Temperature T (°C)	0	250	350	400	460	480	500	525	550	600
material consistency K	200	120	110	100	60	40	20	12	5	0.5
strain rate sensitivity m	0.01	0.05	0.07	0.09	0.098	0.1	0.13	0.3	0.6	0.9

Contact and friction conditions at the interface between part and tools are extremely complicated and modeled by Tresca, Coulomb or Norton friction laws. The last one is defined as a non linear relation between the shear stress $\tau = \sigma_n - (\sigma_n . n) n$ and the tangential velocity difference between the tool and the part:

$$\tau = -\alpha_f K |v_g|^{p-1} v_g \qquad (6)$$

where α_f is the friction coefficient and p is the sensitivity to the sliding velocity. The unilateral contact condition is applied to these surfaces by means of a nodal penalty formulation. The Norton friction law was used for the simulations presented in this paper.

As an Eulerian description is utilized, the grid velocity is null. A specific technique has been implemented in order to deal with convective terms, as shown in the following equation:

$$\frac{d_g \varphi}{dt} = \dot{\varphi} + (v_{msh} - v_{mat}) \cdot \nabla \varphi \qquad (7)$$

Thermal Coupling
The integral formulation is written for any test function w:

$$\int_\Omega \rho c \frac{dT}{dt} w dV + \int_\Omega k \nabla T \cdot \nabla w dV = \int_\Omega \dot{q}_v w \, dV + \int_{\partial\Omega} \phi \cdot w dS \qquad (8)$$

where ϕ is the boundary heat flux, and where the material density ρ, the heat capacity c, and the conductivity k are functions of temperature. The heat dissipation $\dot{q}_v$ is given by:

$$\dot{q}_v = f \overline{\sigma} \dot{\overline{\varepsilon}} \qquad (9)$$

where $\overline{\sigma} = \sqrt{\frac{3}{2} s : s}$ is the equivalent stress. For a Norton-Hoff viscoplastic material the heat generation rate from material deformation is presented as follows:

$$\dot{q}_v = f K (\sqrt{3} \dot{\overline{\varepsilon}})^{m+1} \qquad (10)$$

where the factor f takes into account the fraction of energy which is converted into heat, taken as 0.9 in this paper. Prescribed outward heat fluxes or prescribed temperatures are also taken into account as convection and radiation boundary conditions which are expressed in combination:

$$-\lambda \nabla T \cdot n = h_{cr} (T + T_{ext}) \qquad (11)$$

$$h_{cr} = h_c + \varepsilon_r \sigma_r (T^2 + T_{ext}^2)(T + T_{ext}) \qquad (12)$$

Conduction with the tool and heat from friction must also be taken into account:

$$-k \frac{\partial T}{\partial n} = -h_{cd} (T + T_{tool}) + \frac{b}{b + b_{tool}} \tau . v_g \qquad (13)$$

In the particular case of the Norton friction law we have:

$$\tau . v_g = \alpha_f K |v_g|^p \qquad (14)$$

85

Results and Discussion

Three simulations were done for aluminum alloy 7075, with different levels of friction coefficient α_f: 0.2, 0.3, and 0.4. Each simulation took approximately 3 days in real time to run to completion on a PC with a dual core, 3.3GHz processor. In the model the tool was maintained at 19°C on the shank, to simulate a cooled tool holder, while the workpiece started at 25°C and was supported by an aluminum backing plate with the same initial temperature. The tool speed was 650 rpm, while the feed was 3 mm/sec. Temperature maps for three different friction coefficients are shown in figure 1 at steady-state conditions (after 60 seconds of welding), where the rate sensitivity parameter was held constant at p=0.125 for all three cases.

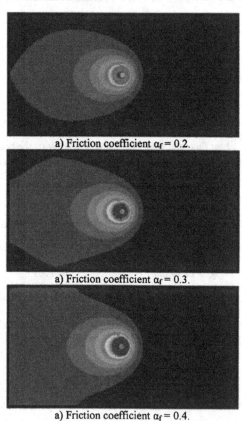

a) Friction coefficient α_f = 0.2.

a) Friction coefficient α_f = 0.3.

a) Friction coefficient α_f = 0.4.

Figure 1. Top view of temperature maps of the 7075 aluminum plate at steady state conditions, using various coefficients of friction. All other parameters were the same for these simulations. The effect of increasing the level of friction is significant in the generation of heat during the welding process. A significant portion of the area under the shoulder and around the pin is greater than 550°C for α_f = 0.4, while for α_f = 0.2 most of this same region is between 500°C and 550°C.

86

The law used to model friction in this case (Norton law: $\tau = -\alpha_f K |v_g|^{p-1} v_g$) is a function of the flow stress of the alloy, the relative sliding velocity at the interface between the tool and the workpiece, and the friction coefficient α_f. The law is also rate sensitive, as governed by the parameter p, although this parameter has not been varied for these results. Side views of the weld are shown in figure 2.

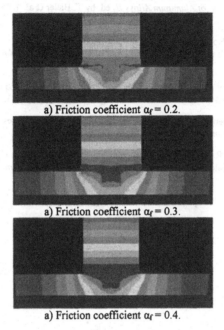

a) Friction coefficient $\alpha_f = 0.2$.

a) Friction coefficient $\alpha_f = 0.3$.

a) Friction coefficient $\alpha_f = 0.4$.

Figure 2. Side view of temperature maps of the 7075 aluminum plate at steady state conditions. Heat generation from increasing friction is significant, where all other simulation parameters were held constant. A friction coefficient of 0.2, shown in a), provides the most reasonable temperature profile, with the highest temperatures located under the shoulder and ranging between 500°C and 550°C.

The temperatures prediction for steady state friction stir welding of AA 7075 appears to be most reasonable for $\alpha_f = 0.2$, where temperatures under a small region under the tool shoulder exceed 550°C, but where most temperatures around the tool are between 500°C and 550°C. We don't have a direct comparison for AA 7075, but these temperatures are consistent with experimental data for FSW of AA 6061 [6]. This result will be compared in future work to results obtained using other friction laws, including the Tresca and Coulomb laws. The Norton law used in the present work is derived from a physical interpretation of friction which we think matches more closely what actually happens at the interface in friction stir welding than other laws. The basic mechanism for transmitting a shear stress in the Norton model is through a boundary layer of viscoplastic material. This boundary layer has the same mechanical properties as the workpiece material, which can soften with increased temperatures and which can harden or be rate sensitive based on the local strain and strain rate. In contrast, the Coulomb law is more appropriate for sliding friction at room temperature, while the Tresca law has been used for modeling hot working processes, but is not rate sensitive. The relative importance of heat generated by friction versus that produced by material deformation needs to be studied and modeled properly in order for simulation predictions to be useful. Therefore, we plan further study of the Norton law, versus other alternatives, and will evaluate our modeling efforts with additional experimental data.

References

1. Z. Feng, J.E. Gould and T.J. Lienert, "A Heat Flow Model for Friction Stir Welding of Aluminum Alloys", *Hot Deformation of Aluminum Alloys II,* ed. by T. Bieler et al, TMS, Warrendale, PA, 1998, pp.149-158.

2. C.M. Chen and R. Kovacevic, "Finite element modeling of friction stir welding-thermal and thermomechanical analysis", *International Journal of Machine Tools & Manufacture*, 43 (13), 2003, pp.1319-1326.

3. Y.J. Chao, X. Qi and W. Tang, "Heat Transfer in Friction Stir Welding: Experimental and Numerical Studies", *Journal of Manufacturing Science and Engineering,* 125 (1), 2003, p. 138-145.

4. P. Ulysse, "Three-dimensional modeling of the friction stir-welding process", *International Journal of Machine Tools & Manufacture*, 42 (14), 2002, p.1549-1557.

5. P.A. Colegrove and H.R. Shercliff, "Experimental and Numerical Analysis of Aluminum Alloy 7075-T7351 Friction Stir Welds", *Science and Technology of Welding and Joining*, 8, Maney Publishing, 2003, p. 360-368.

6. S. Darby and T. Nelson: "Unpublished research", August 2006.

TRANSPORT PHENOMENA IN FRICTION STIR WELDING

L. St-Georges[1], V. Dassylva-Raymond[2], L.I. Kiss[2], A.L. Perron[2]

[1]REMAC Industrial Innovators; 2953 boul. du Royaume; Jonquière, QC, G7X 7V3, Canada
[2]Université du Québec à Chicoutimi; 555 boul. de l'Université ; Chicoutimi, QC, G7H 2B1, Canada

Keywords: friction stir welding, thermo-fluid model, welding parameters, velocity field, temperature distribution

Abstract

The friction stir welding is a solid-state process where intense plastic deformations and temperature fluctuations are produced by a rotating tool. In this process, the plastic deformations and the temperature field have a significant impact on the microstructural modifications observed in the welded zone and consequently, on the final mechanical properties of the welded joint. To understand the effect of the tool rotation on the temperature and the velocity field, a three dimensional thermo fluid model has been developed and is presented for AA 6061-T6. The global shape of the velocity field is found to be similar to a Rankine combined vortex. The contribution of this nearly perfect potential vortex to the flow vorticity is negligible; the vorticity basically results from the translational movement of the tool. For the particles located near the weld axis, trajectories with complete revolutions around the tool and with only deviations around the tool are observed. For this latter case, the particles go from one side of the weld, pass around the tool and pursue their trajectory on the other side of the weld axis. Finally, the effect of the rotation of the tool on the temperature field is found to be significant on the temperature history of the welded joint. The tool rotation has a predominant effect when the rotational speed is relatively high compared to the traverse speed.

Introduction

In friction stir welding, the temperature elevation is generated by the friction between the tool and the material and by the plastic deformations within the material. Intense plastic deformations are observed at elevated temperature and relevant microstructure modifications are obtained in the welded zone. Since the beginning of the friction stir welding, studies have been conducted to understand the effects of the welding parameters, to optimize the tools used and to determine better combination of welding parameters [1-3].

Mathematical models have been developed to understand the effects of the welding parameters. Thermal 2D and 3D models [4-8], thermomechanical [9-12] and viscoplastic models [13] have been used to understand and to reproduce the physical phenomena observed. The slip-stick motion at the tool/material interface has been described and models have been developed to produce a better representation of this interface [14, 15]. Based on friction stir welding experiments, the presence of a liquid film at the interface and a dominant sliding contact has been suggested in the pin-workpiece region [16].

89

To obtain optimal welds, it is of practical importance to have a good understanding of the material flow pattern and its effects on the weld characteristics. The material flow during the friction stir welding is dependant of the tool geometry, of the process parameters and of the nature of the material to be welded. To visualize the flow pattern during the friction stir welding, different approaches have been used. Markers and dissimilar materials have been employed as tracer techniques and the final position of these tracers have been observed, without specific indication on the path followed by the material [17, 18]. Because of its importance on the final microstructure, the temperature history of the welded joints has been the subject of many investigations. Various experimental tests and numerical calculations have been performed on this subject [19-25].

Although complex models have been developed and numerous experimental tests conducted, the utility of relatively simple numerical models has been shown [26]. Despite all the investigations realised, only little information is still available on the relative effect of the material flow produced by the rotation of the tool on the temperature field and consequently on the temperature history. The goal of the investigation presented here is to extract the effect of the tool rotation on the temperature field and more globally, on the transport phenomena observed in the welded material.

Mathematical Model

The friction stir welding is a solid-state process where a specially designed rotating tool is used. To realise a weld, the rotating tool is inserted into the material and it moves along the weld axis with a given traverse speed. The mechanical energy used to realise the weld is then transformed into heat and plastic deformation inside the material. The portion of energy related to the plastic deformation is assumed to be low, typically less than 10% of the total input.

In the formulation used here, the mechanical energy provided to the tool is assumed to be entirely transformed into heat within the material and is expressed as

$$\dot{Q}_{Total} = T\omega + Fv \tag{1}$$

where $\dot{Q}_{Total}$ is the total rate of heat transfer, T and F are respectively the torque and the traverse force applied on the tool, ω and v are the imposed rotational and traverse speeds. For AA6061-T6, in typical welding conditions, the mechanical energy provided by the linear motion of the tool is low, less than 5%. For simplification, this term is neglected. The total rate of heat transfer is thus calculated using the torque and the rotational speed data measured during experiments.

In our model, the total rate of heat transfer is then transferred into the shoulder and the probe volume as a volumetric heat source, as illustrated in Figure 1. To limit the dispersion of the heat into the tool, the lateral and the upper surfaces of the shoulder have been isolated. In reality, these surfaces are not isolated but the thermal conductivity of the tool is low in comparison with the thermal conductivity of the aluminium (about 18 to 20 times). Consequently, the resultant portion of the heat transfer through the tool is negligibly low.

The model allows the use of temperature dependent properties, heat transfer and material flow. The temperature field is obtained by the solution of the Kirchhoff-Stefan equation in a fixed coordinate system:

$$\frac{\partial T}{\partial t} + \vec{V}\,\overrightarrow{grad}\,T = \alpha \nabla^2 T \tag{4}$$

while the velocity field $\vec{V} = \vec{V}(x,y,z,t)$ is obtained by the solution of the continuity and Navier Stokes equations.

90

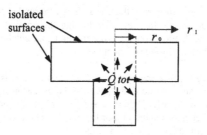

Figure 1. Heat transfer in the welding tool

Material Behavior

An essential feature of the model is how the fluid dynamic behavior of the material is described. The material is assumed to be a fluid; its physical properties are adjusted to reproduce AA6061-T6 material behaviour (see Table I). For the dynamic viscosity, a hypothetic fluid which behaves as a viscose liquid at elevated temperatures and as a solid at low temperature is used.

Table I. Mechanical and Thermal Properties of AA6061-T6

Property	Value
Limit of elasticity	276 Mpa
Maximum shear stress	207 Mpa
Thermal conductivity	167 W/m-K
Density	2700kg/m³
Solidus temperature	582°C
Liquidus temperature	652 °C

Boundary Conditions

Heat Transfer. Convective and radiative boundary conditions are applied on the lateral and top surfaces of the plate. As the speed of the tool is relatively low, a natural convection is assumed in the simulation:

$$\dot{q}_{convection} = h(T_s - T_\infty) \tag{2}$$

The heat transfer coefficient h is calculated at an ambient temperature of 22°C, T_s is the temperature of the material surface and T_∞ is the air temperature.

During the welding operation, the temperature of the workpiece surface may become higher than 500°C. According to this elevated temperature, the heat transfer by radiation is non-negligible and has to be considered. The radiative flux along the surface of the material is described by the relationship

$$\dot{q}_{rad} = \varepsilon \sigma (T_s^4 - T_\infty^4) \tag{3}$$

where ε is the emissivity. For an aluminium plate in the range of temperature considered here ε can be estimated as 0,03.

For the bottom surface of the welded plate, different boundary conditions are used. This surface is in contact with an anvil made of steel or stainless steel. To evaluate the rate of the heat transfer through this surface during the welding process, simulations were run with and without a possible heat transfer through the anvil, assuming a perfect contact between the components (no thermal resistance). Qualitative comparisons were then made and no relevant differences in the temperature field were observed. Moreover, there is always a thermal contact resistance between the plates, so the real impact of the heat transfer through the anvil does not have a relevant impact on the temperature field. Based on these observations, an adiabatic plane is used as the boundary condition for the bottom surface of the material.

Kinematic Condition. During the welding operation, the material flow is principally constricted under the shoulder and no visual plastic deformation is observed on the upper surface of the welded plate. To simulate this material flow, a slipping condition is imposed between the shoulder and the upper surface of the material. For the pin, a sticking condition is imposed at the pin-workpiece interface to reproduce the rotation of the flow.

Model Geometry

The geometrical model used in the simulation is composed of a rectangular flat plate and of a cylindrical tool. This model is presented in Figure 2. The global referential system is located on the central axis of the tool, on the upper surface of the material. A computational fluid dynamic code (CosmosFlow) is used for the simulation.

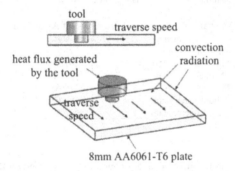

Figure 2. Geometric model

Numerical Simulations

Prior to realize the numerical simulations, experiments were performed on a fully instrumented MTS I-Stir PDS welding equipment, to identify realistic welding conditions [27]. The tool geometry and the vertical loads imposed in the simulation are representative of the ones measured in normal welding conditions.

To understand and to visualize the effects of the tool rotation on the temperature field and to see the material flow in various welding conditions, four different test cases have been run. In the different simulations realized, different ratios of rotational / traverse speeds have been used, as presented in Table II. Each case has been run and repeated with a zero rotational speed to see the effect of this parameter on the velocity and on the temperature field.

Table II. Welding Parameters Used in the Simulation

Welding parameters	Case 1	Case 2	Case 3	Case 4
Rotational speed, ω (rpm)	300	600	900	1200
Traverse speed, v (mm/min)	500	750	1250	1250
Rotational / traverse speed, k (rot. / mm)	0,6	0,8	0,72	0,96
Pin radius (mm)	5,5	5,5	5,5	5,5
Pin length (mm)	7	7	7	7
Shoulder radius (mm)	11,5	11,5	11,5	11,5

Numerical Results

Velocity Field

The velocity field obtained at 4mm below the upper surface of the material in the transverse direction of the weld axis for the case 1 and 2 is presented in Figure 3. In these graphics, the tangential velocity is expressed as a function of the transverse direction (perpendicular to the weld axis). Moreover, the velocity field is presented for a rotating and a non-rotating tool ($k = 0$). The advancing and the retreating side appear respectively on the right and on the left side of the graphics. As illustrated, the velocity field has a global behaviour similar to the one of a Rankine combined vortex. In its central portion (not shown), the tangential velocity can be expressed as the solid body motion of the pin tool and increases linearly with the radius. For the flow around the pin tool, the tangential velocity declines hyperbolically with the radius. In contrast to the flow obtained for a non-rotating tool, the tangential velocity obtained on the advancing side for a rotating tool is reversed compared to the one obtained with a non-rotating tool (opposite direction).

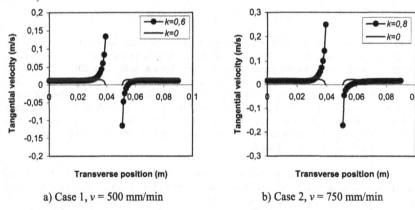

a) Case 1, $v = 500$ mm/min   b) Case 2, $v = 750$ mm/min

Figure 3. Transverse velocity field

To see the effect of the tool rotation on the material flow, the path followed by particles located at different radius around the tool have been followed and plotted (see Figure 4). As illustrated, the particles distant from the tool axis pass around the pin tool and follow a deviated trajectory. When the distance between the lateral surface of the tool and the particles is reduced, some particles begin to describe a complete revolution. As the relative distance decreases further, more and more particles describe a complete revolution and in some cases, several complete revolutions are defined.

93

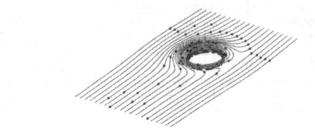

Figure 4. Global particles motion

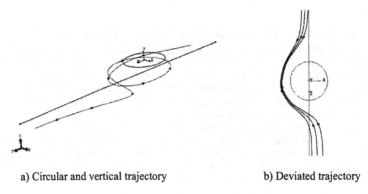

a) Circular and vertical trajectory   b) Deviated trajectory

Figure 5. Types of particles trajectory

The circular motion around the tool is accompanied by a vertical displacement, as illustrated in Figure 5a by the trajectory of a single particle. This trajectory is typical of the ones followed by the particles near the tool. Among these particles, others have a completely different trajectory. Some others particles, also located near the weld axis, describe only a deviated trajectory and pass around the tool, as presented in Figure 5b. These particles initially located on the left side of the weld move around the tool and pursue their trajectory on the right side of the welded joint. Although the mixing mechanism needed to realize a sound joint in friction stir welding is not well known, the trajectory followed by these particles have a significant impact on the mixing mechanism observed in the welded zone.

For the case of a purely cylindrical tool, the vertical material flow is produced by the internal characteristics of the velocity field itself, and could be reduced or increased by the use of special features on the pin tool (fillets, grooves, etc.). A global picture of the transverse material flow, as obtained for the case 2, is presented in Figure 6. As depicted by the particle paths obtained, close to the lateral surface of the tool, the velocity is oriented vertically, towards the upper surface of the material. The vertical component of the velocity field decreases for higher radius and its orientation changes gradually to become inverted at a specific location.

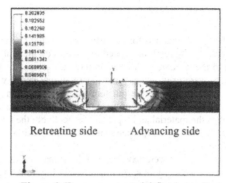

Retreating side   Advancing side

Figure 6. Transverse material flow, case 2

Beside the particle trajectories and velocity fields, in Figure 7 the distribution of the vorticity is also shown. Although there is no any well-defined criteria about the quality of a friction stir welded joint available, we think that vorticity and shearing provide at least a partial insight into the mechanism of mixing in the welding zone. It is interesting to note, that in the frame of our fluid dynamic and rheological model, the vorticity basically results from the translational movement of the tool. As it was mentioned earlier, the rotational movement results in a nearly perfect potential vortex so its contribution to the vorticity is negligible. Naturally this does not mean that the tool rotation does not play an important role in the mixing of the two plates together, as the particle trajectories clearly show it.

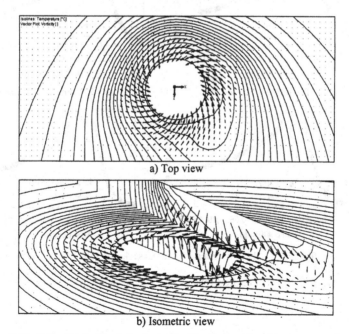

a) Top view

b) Isometric view

Figure 7. Vorticity distribution in horizontal and vertical sections/planes across the plates

<u>Temperature Field</u>

The temperature field has been obtained for the different welding conditions described in Table II. In order to see the effect of the rotational speed of the tool on the temperature field, four other conditions were added to the welding conditions described before. In these new conditions, the same welding parameters were used, except that the rotational speed was imposed to be zero. The effect of the flow rotation around the pin tool on the temperature profile is presented in Figures 8 and 9. The temperature profiles presented in these graphics have been obtained at 4mm below the upper surface of the material and represent respectively the temperature distribution in the weld axis and in the transverse axis (perpendicularly to the weld axis).

The temperature distribution in the weld axis has a major impact on the final microstructure and mechanical properties of the welded joints. As illustrated in Figure 8, the rotation of the tool has a predominant effect when the rotational speed is relatively high compared to the traverse speed. As the rotational/traverse speed ratio decreases, the difference between the temperature fields obtained for the rotating and the non-rotating tool decreases. The same trend is observed in the temperature field obtained in the transverse axis (see Figure 9). Furthermore, the fluid temperature observed on the retreating side for the rotating tool is lower than the one observed for a non-rotating tool. This trend is reverse on the advancing side.

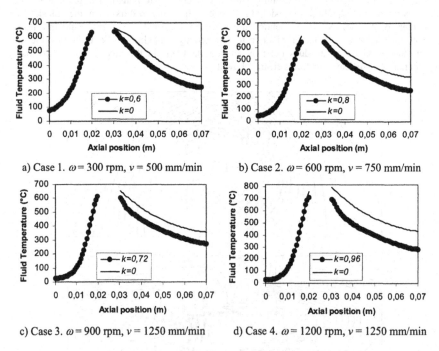

a) Case 1. $\omega = 300$ rpm, $v = 500$ mm/min

b) Case 2. $\omega = 600$ rpm, $v = 750$ mm/min

c) Case 3. $\omega = 900$ rpm, $v = 1250$ mm/min

d) Case 4. $\omega = 1200$ rpm, $v = 1250$ mm/min

Figure 8. Effect of the flow rotation on the temperature field, weld axis

96

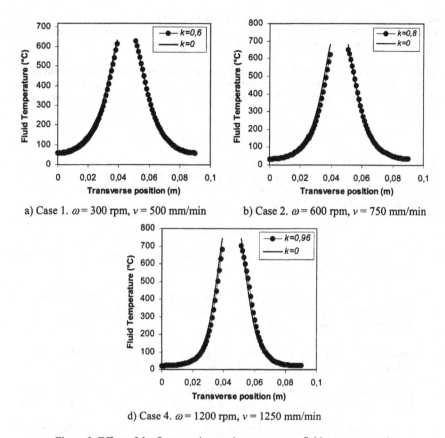

a) Case 1. $\omega = 300$ rpm, $v = 500$ mm/min

b) Case 2. $\omega = 600$ rpm, $v = 750$ mm/min

d) Case 4. $\omega = 1200$ rpm, $v = 1250$ mm/min

Figure 9. Effect of the flow rotation on the temperature field, transverse axis

Conclusions

The rotation of the tool used in friction stir welding influences the material flow. As the rotational/traverse speed ratio (k) increases, the flow becomes more complex. For the particles located near the weld axis, complete revolutions around the pin tool can be observed. These revolutions are accompanied by vertical displacements. The vertical displacements of the particles during the welding can be explained by the velocity field obtained around the tool. For a cylindrical tool, close to the lateral surface of the pin tool, the velocity field has a major vertical component oriented toward the upper surface of the material. Among the particles located near the weld axis, other particles move around the tool and go from one side of the weld to the other and contribute significantly to the mixing mechanism in the welded zone.

The tangential velocity observed in the velocity field is similar to a Rankine combined vortex. For the central portion of the velocity field, the tangential velocity can be expressed as the solid body motion of the pin tool and increases linearly with the radius. For the material around the pin tool, the tangential velocity declines hyperbolically. The contribution of this nearly perfect potential vortex to the flow vorticity is negligible. The vorticity basically results from the translational movement of the tool.

Although there is no any well-defined criteria about the quality of a friction stir welded joint available, the trajectory followed by the particles, the vorticity and the shearing provide at least a partial insight into the mechanism of mixing in the welding zone.

Finally, the effect of the rotation of the tool on the temperature field is found to be non-negligible (typically 20-45%) for rotation/traverse speed ratios between 0.6 and 0.96. For speed ratios in this range, the rotation of the tool has a significant effect on the temperature history of the welded joint.

Acknowledgements

The authors would like to address acknowledgements to the Aluminium Technology Centre (National Research Council of Canada) for the use of their MTS I-Stir PDS equipment. The authors also want to address acknowledgements to the technicians Mario Ouellet and Marc–André Allard for their useful support during the experiments.

References

1. R.S. Mishra and Z.Y.Ma, "Friction Stir Welding and Processing", *Materials Science and Engineering*, R50 (2005), 1-78.

2. Y.G. Kim, H. Fujii, T. Tsumura, T. Komazaki and K. Nakata, "Three Defect Type in Friction Stir Welding of Aluminum Die Casting Alloy", *Materials Science and Engineering*, A415 (2006), 250-254.

3. Dubourg, L. and P. Dacheux, "Design and Properties of FSW Tools: a Literature Review", *6th International Symposium on Friction Stir Welding*, (2006), 1-16.

4. M.Z.H. Khandkar and J.A. Khan, "Thermal Modeling of Overlap Friction Stir Welding for Al-Alloys", *Journal of Materials Processing & Manufacturing Science,* 10 (2001), 91-105.

5. M.Z.H. Khandkar, J.A. Khan and A.P. Reynolds, "Prediction of Temperature Distribution and Thermal History During Friction Stir Welding: Input Torque Based Model", *Science and Technology of Welding and Joining,* 8 (3) (2003), 165-174.

6. Song, M. and R. Kovacevic: 'Thermal Modeling of Friction Stir Welding in a Moving Coordinate System and its Validation'. *International Journal of Machine Tools & Manufacture,* 2003 (43) 605-615.

7. H.N.B Schmidt, and J. Hattel, "Heat Source Models in Simulation of Heat Flow in Friction Stir Welding", *International Journal of Offshore and Polar Engineering*, 14 (4) (2004), 296-304.

8. P.A. Colegrove and H.R. Shercliff, "3-Dimensional CFD Modelling of Flow Round a Threaded Friction Stir Welding Tool Profile", *Journal of Materials Processing Technology*, 169 (2005), 320-327.

9. P. Dong, F. Lu, J.K. Hong and Z. Cao, "Coupled Thermomechanical Analysis of Friction Stir Welding Process Using Simplified Models", *Science and Technology of Welding and Joining,* 6 (5) (2001), 281-287.

10. P. Heurtier, C. Desrayaud and F. Montheillet, "A Thermomechanical Analysis of the Friction Stir Welding Process", *Materials Science Forum*, 346-402 (2002), 1537–1542.

11. C.M. Chen and R. Kovacevic, "Finite Element Modeling of Friction Stir Welding. Thermal and Thermomechanical Analysis", *International Journal of Machine Tools & Manufacture*, 43 (2003), 1319-1326.

12. P. Heurtier, M.J. Jones, C. Desrayaud, J.H. Driver, F. Montheillet and D. Allehaux, "Mechanical and Thermal Modelling of Friction Stir Welding", *Journal of Materials Processing Technology*, 171 (2006), 348-357.

13. P. Ulysse, "Three-Dimensional Modeling of the Friction Stir-Welding Process", *International Journal of Machine Tools & Manufacture*, 42 (2002), 1549-1557.

14. H. Schmidt, J. Hattel and J. Wert, "An Analytical Model for the Heat Generation in Friction Stir Welding", *Modelling and Simulation in Materials Science and Engineering*, 12 (2004), 143-157.

15. H. Schmidt and J. Hattel, "Modelling Heat Flow Around Tool Probe in Friction Stir Welding", *Science and Technology of Welding and Joining*, 10 (2) (2005), 176-186.

16. Z. Chen, T. Pasang, Y. Qi and R. Perris, "Tool-Workpiece Interface and Shear Layer Formed During FSW", *6th International Symposium on Friction Stir Welding*, (2006), 1-9.

17. R. Zettler, J.F. dos Santos, T. Donath, F. Beckmann and D. Lohwasser, "Material Flow in Friction Stir Butt Welded Aluminium Alloys", *6th International Symposium on Friction Stir Welding*, (2006), 1-6.

18. H. Schmidt and J. Hattel, "Analysis of the Velocity Field in the Shear Layer in FSW – Experimental and Numerical Modelling", *6th International Symposium on Friction Stir Welding*, (2006), 1-11.

19. Z. Chen, T. Pasang, Y. Qi and R. Perris, "Tool-Workpiece Interface and Shear Layer Formed During FSW", *6th International Symposium on Friction Stir Welding*, (2006), 1-9.

20. J.C. McClure, W. Tang, L.E. Murr, X. Guo, Z. Feng and J.E. Gould, "A Thermal Model of Friction Stir Welding", *ASM Proceeding of the International Conference: Trends in Welding Research*, (1998), 590-595.

21. J. Xu, S.P. Vaze, R.J. Ritter, K.J. Colligan, J.R. Pickens, "Experimental and Numerical Study of Thermal Process in Friction Stir Welding", *Proceedings from Joining of Advanced and Specialty Materials*, (2003), 10-19.

22. O. Shi, T. Dickerson and H.R. Shercliff, "Thermo-Mechanical FE Modeling of Friction Stir Welding of Al-2024 Including Tool Loads", *4th International Symposium on Friction Stir Welding*, (2003), 1-12.

23. C. Chen and R. Kovacevic, "Finite Element Modeling of Thermomechanical Performance of Friction Stir Welding", *4th International Symposium on Friction Stir Welding*, (2003), 1-16.

24. L. Fratini, S. Beccari and G. Buffa, "Friction Stir Welding FEM Model Improvement Through Inverse Thermal Characterization", *Transactions of NAMRI/SME*, 33 (2005), 259-266.

25. V. Soundararajan, S. Zekovic, R. Kovacevic, "Thermo-Mechanical Model with Adaptive Boundary Conditions for Friction Stir Welding of 6061", *International Journal of Machine Tools & Manufacture*, 45 (2005), 1577-1587.

26. A.P. Reynolds, Z. Khandkar, T. Long, W. Tang and J. Khan, "Utility of Relatively Simple Models for Understanding Process Parameter Effects on FSW", *Material Science Forum*, 426-432 (2003), 2959-2694.

27. L. Dubourg, F.O. Gagnon, F. Nadeau, L. St-Georges and M. Jahazi, " Process Window Optimization for FSW of Thin and Thick Sheet Al Alloys Using Statistical Methods", *6th International Symposium on Friction Stir Welding*, (2006), 1-12.

PHASE SPACE ANALYSIS OF FRICTION STIR WELD QUALITY

Enkhsaikhan Boldsaikhan[1], Edward Corwin[1], Antonette Logar[1], Jeff McGough[1]
and William Arbegast[1]

[1]South Dakota School of Mines and Technology; 510 East Saint Joseph Street; Rapid City, SD, 57702; USA

Keywords: Phase space, friction stir welding, Poincaré map

Abstract

A plot of the y feedback parameter values versus its derivative defines a trajectory for a dynamical system that appears to be an indicator of weld quality. The ideal trajectory for the dynamical system is defined by the spindle frequency and is characterized by repeated orbits in an elliptical shape. The number of points in each orbit is also defined by the spindle frequency, and each orbit, or sequence of orbits, can be analyzed to determine the stability of the system during a weld. The orbits in phase space of welds without metallurgical defects tend to remain within a neighborhood surrounding the ideal trajectory while welds containing defects produced more variable orbits. Calculation of Lyapunov exponents for the dynamical system and analysis of the Poincaré map generated by successive orbits were evaluated for their ability to accurately predict weld quality from the phase space trajectories. The Lyapunov exponent shows promise as an indicator of weld quality in certain situations, but the Poincaré map approach appears to provide a more flexible, and more computationally efficient, method for predicting the presence of metallurgical defects.

Introduction

Friction stir welding machines are often capable of measuring many variables in the process and providing these values (feedback) to a device capable of processing this information. Examples include the vertical force on the pin tool (z), the force in the direction of the weld (x), and the force perpendicular to the direction of the weld (y). The focus of this research is to design a feedback controller capable of identifying changes in state during the welding process that signal changes in weld quality. The selection of appropriate control parameters has been explored in other research and is only briefly mentioned in the following section describing the data used. The time series of y feedback values generated during a weld provided the basic information from which weld quality predictions were made. Various techniques have been applied to extract quantitative information about weld quality from the system outputs, including correlation analyses of feedback parameter values to quality and a neural network pattern recognition system [1], with good results on limited problem domains. However, as the type of weld or the thickness of the material changes, these methods require updating to maintain their effectiveness. The goal of this research was to identify a technique that is not as sensitive to changes in welding situations; one that is a mathematical abstraction of the process and not tied to the physical details of the weld being made.

The technique selected for this work is based upon a stability analysis of the dynamical system. Ideally, equations describing the system dynamics can be used for stability analysis, however, in many situations, those equations are not available or easily derived. In those cases, including this research, a sampling of the system outputs is used to approximate the system dynamics. Working with the original y feedback time series data generated by the welding process supplies a flood of numerical data, but the details tend to obscure possible patterns. Plotting the y

101

feedback values on the horizontal axis against the change in the y values (y') on the vertical axis provides a mechanism for examining the dynamics of the welding process with respect to the y output value. As with position and velocity, the relationship between y and y' gives information about the behavior of the system. Also note that this behavior is abstracted from time and provides a compact framework within which the system dynamics can be examined and patterns easily recognized. These plots are commonly referred to as phase space diagrams or phase plots.

The curves formed by plotting a variable against its derivative through time are known as trajectories or orbits. In most physical systems, the trajectory will be bounded in some finite region of phase space and often tend to either a rest point or a limit cycle. A rest point is one where all system dynamics cease – the system is at rest. Limit cycles are periodic solutions which attract or repel the orbits. After the transient behavior has died out, the flow is essentially on the limit cycle. In such cases, it is sensible to talk about the orbit as a closed curve since the trajectory will return to the same point and repeat the same sequence of values. For both fixed points and limit cycles we refer to them as stable or unstable if they attract or repel the trajectories. As a simple example, plotting the function $f(t) = \sin(t)$ against its derivative $f'(t) = \cos(t)$ will produce a circle. No matter how large t becomes, the trajectory defined by plotting $f(t)$ vs. $f'(t)$ will always remain on the circle. This is an example of stable dynamics.

An examination of hundreds of phase space plots indicated that the trajectories of the y vs. y' plots tended to have stable orbits for good welds and tended to diverge from the orbits for bad welds. Examples of phase space plots of a good weld and a bad weld can be found in Figure 1. Figure 2 shows the location and size of the wormhole resulting from the bad weld in Figure 1. Note that the phase space plot of the spindle frequency (rpm) produces a stable orbit which can be considered to be the ideal trajectory for a weld made using the given system parameters. The spindle frequency orbit is plotted on the graphs to provide a reference for divergent or convergent behavior. Thus, a visual examination of the system dynamics plots showed a clear pattern for good and bad welds, but, to be useful, this graphical representation needed to be quantified.

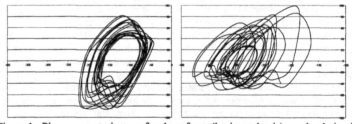

Figure 1. Phase space trajectory for the y force (horizontal axis) vs. the derivative of the y force (vertical axis) measured in foot-pounds (lbf). The phase space plot on the left is for a good weld, the plot on the right is for a bad weld. The trajectory of the spindle frequency is also plotted for reference.

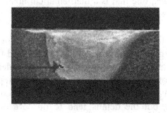

Figure 2. The location of the resulting wormhole at 40x magnification for the bad weld in Figure 1.

The plots are representations of the dynamics in two dimensions, but this does not imply that the overall dynamical system is two-dimensional. It is possible that the orbits exist in a higher dimensional space and that useful information may not be visible in the two-dimensional rendering of the trajectory.

In subsequent sections, methods for quantifying the trajectory patterns that emerge in the phase space analysis of the y feedback value are described. The goal was to assign a numeric value to the trajectory that is easily and cheaply calculated, and thus suitable for real-time processing, and has a high degree of accuracy for predicting weld quality.

The Data

The data used for these experiments was generated by an MTS ISTIR-10 Friction Stir Welding machine. To test the flexibility of the weld quality analysis technique proposed here, several weld parameters were varied. Although all of the welds were made using the same machine with Aluminum 7075 as the metal to be joined, both bead on plate welds and butt joint welds were examined. In addition, weld thicknesses were varied from 0.125 inches to .25 inches, welding speed measured in inches per minute (ipm) varied from 2 to 25, and the rotation speed of the welding pin tool measured in rotations per minute (rpm) varied from 200 to 700. The ISTIR machine provides a wealth of feedback information during the welding process. At least forty-five feedback values, or feedback parameters, can be analyzed in real-time. Observation by experienced welders identified a subset of four values that appeared to be most closely tied to weld quality [2, 3] : the x force, y force, z force, and torque. Welds were tested for strength and metallurgical defects (wormholes) and a correlation analysis was performed to determine the degree to which changes in parameter values were tied to changes in weld quality. No one parameter provided adequate information to be a sufficient indicator of quality, but the y force values were most strongly correlated with the presence of metallurgical defects. Combinations of outputs have not been investigated, in part because the dynamics analysis explained in the introduction is well suited to changes in a single variable. Thus, subsequent investigations have concentrated on exploring ways in which information can be extracted from the y force feedback parameter.

One of the obstacles to evaluating weld quality from the y feedback, or any feedback parameter, is the presence of noise in the output signal. The welding process does not occur in an ideal environment, but rather is influenced by external factors such as temperature, metallurgical variations, edge inconsistencies etc., as well as by the inevitable machine-specific variability. Investigations are on-going to identify and characterize all of the types of noise that contribute to the composition of the feedback signals, but, at present, only mechanical vibration noise is removed by a low-pass filter. The filter was implemented by computing the discrete Fourier transform (DFT) of the time series data, excising the high frequency components from the DFT, and computing the inverse DFT. High frequency components are defined to be those above the spindle frequency, where the spindle frequency index is found by :

index = spindle RPM / (60 * sampling rate (Hz) / number of points in DFT)   (1)

Figure 3(a) shows a sample of the original time series data (FSW06048-5) generated for a butt weld of two pieces of 0.25 inch thick aluminum 7075 welded at 8 ipm, with a pin tool speed of 200 rpm, using forge force control and sampled at a rate of 68.2667 Hertz. Figure 3(b) presents the 256 point discrete Fourier transform of the original time series signal and indicates the

103

spindle frequency for this sample, calculated from equation [1] above, to be at index 12.5 (rounded up 13) in the DFT array. Figure 3(c) shows the filtered time series data. The resulting signal is smoother, but more importantly, the trends in the data are more apparent once the high frequency noise is removed. The appropriateness of using a low pass filter for removing mechanical vibration noise during friction stir welding was noted in [4]. The resulting filtered time series data is used for the phase space analysis described briefly above and in more detail in the next section.

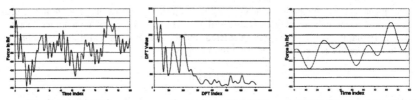

Figure 3. (a) Left: the raw y force data. (b) Middle: the DFT for the data with the spindle frequency indicated (c) Right: the data after the lowpass filter.

Lyapunov Exponents

The friction stir welding process defines a continuous-time dynamical system which is approximated by the discrete-time system found through the sampling process. As explained in the introduction, the dynamics of the system, in particular its stability, can be monitored by tracing the trajectory of appropriate system feedback values in phase space. To be useful as a control signal, however, these trajectories, and in particular the deviations of these trajectories from some desired path, must be quantified. In addition, this quantification must be accomplished in real-time.

To understand the dynamics after the trajectory has settled down, a measure of the "turbulence" of the trajectory is needed. Initial investigations of the plots lead to the supposition that good welds have simple dynamics at the steady state while bad welds exhibit complicated or possibly chaotic behavior. One well-known technique for measuring the complexity of the trajectory is the Lyapunov exponent [5]. Assume two points are initially located near each other in phase space. Their individual trajectories may diverge or converge with successive orbits. The exponential rate of divergence between these orbits is measured by the Lyapunov Exponent. Formally it measures the infinitesimal displacement growth rate. The mathematical definition of the largest Lyapunov exponent [5] is as follows : assume that $\phi(t,x)$ is the flow at time t starting at point x and that the flow satisfies the equation $d\phi/dt = F(\phi)$. The distance between the orbits of two points with nearby initial positions is approximated by v(t) which satisfies the equation $dv/dt = [DF(\phi)]v$. Note that $[DF(\phi)]$ represents the matrix of partial derivatives of the function $F(\phi)$ with respect to ϕ_1, ϕ_2, etc.. The log of the magnitude of the vector, $\ln \|v(t)\|$, extracts the exponential growth rate. The desired exponent is computed by averaging these values over time . Thus, $L = \lim (\ln \|v(t)\|)/t$ as t goes to infinity.

If the largest exponent is positive, the trajectory diverges which indicates chaos. If all of the exponents are negative, the system is evolving toward a fixed point. At a fixed point, the system ceases to change or evolve. If one exponent is zero and the rest are negative, the trajectory is converging to or is on a limit cycle (and when a limit cycle is reached, the system has reached a stable, repetitive, and predictable dynamic). Fixed point behavior will be disregarded in this analysis since it has no physical analog in the active welding process.

A number of techniques have been published for computing the Lyapunov exponent [6], but many were discarded because they require knowledge of the continuous dynamics equations rather than a time series or discrete-time approximation to the continuous dynamics [7]. Other were unsuitable for real-time application due to the complexity of the computation [8]. The technique implemented here is described in [8]. In brief, the spindle frequency phase plot, an ellipse, serves as the reference curve from which the y vs. y' trajectory deviations are measured and also determines the number of points per orbit. For a given starting point, P_0, on the trajectory, a divergence measure is generated based on the distance of that point and the corresponding point in a subsequent orbit from a fixed point on the ellipse. These distances are labeled d_0 and d_1 respectively.

$$\text{Exponent for } P_0 = \ln\left(\frac{d_1}{d_0}\right) \qquad (2)$$

If the orbits were identical, the distances would be equal, the quotient would be one, and the log would give a zero exponent. This process is repeated and generates a sequence of exponent values for each point. The exponent for a single orbit may not give an accurate picture of the system dynamics making it advisable to average the exponents over a number of orbits. The number of orbits to average is a design decision for each problem. The Lyapunov exponent reported below corresponds to the largest average exponent generated for all the points in the orbit. It should be noted that even if insufficient orbits are used to give the Lyapunov exponent accurately, the goal here is to determine a measure of weld quality and not to make a statement about chaos.

Experiments were conducted under both forge force control and position control. Using forge force control, the downward pressure exerted during the welding process remains constant, while position control maintains a constant z position during the weld. Thus, if material thickness varies and forge force control is used, the position of the pin tool will be modified to maintain the desired pressure in the z direction. Using position control, the pin tool position will not change along the z axis. These welding regimes produce markedly different feedback values for identical system parameters and welding conditions. The Lyapunov exponent was calculated for 14 position control experiments and 16 forge force control experiments. For the position control experiments, all of the calculations produced positive exponents (Table I) but a correlation appears to exist between the size of the exponent and the size of the defect. The sample size is small and only two defective welds were made, but the large gap between the exponent size of the good welds and those of the two bad welds suggests further investigation might be warranted if only position control welds were to be made. Similar experiments were conducted for forge force control. The results, also summarized in Table I, show that no correlation exists between defect size (in mm^2) and the Lyapunov exponent.

105

Table I. Position Control and Forge Force Control Lyapunov exponents.

Position Control					Forge Force Control			
RPM	IPM	LE	Defect Size		RPM	IPM	LE	Defect Size
350	6	0.0800	0		400	12	-0.0010	0.0085
425	10	0.0890	0		400	12	0.0260	0.0045
350	2	0.0900	0		400	12	0.0270	0.0472
425	6	0.1000	0		300	3	0.0300	0
350	10	0.1000	0		300	25	0.0450	0.0500
500	10	0.1300	0		450	8	0.0780	0.0007
500	9	0.1300	0		450	8	0.0800	0
500	2	0.1400	0		300	25	0.0800	0.0330
500	6	0.1600	0		400	4	0.0900	0
250	2	0.2000	0		300	25	0.0900	0.0005
250	10	0.2300	0		300	12	0.0900	0.0003
250	6	0.3600	0		300	25	0.0920	0.0160
250	14	0.7000	0.0500		300	25	0.1000	0.0079
250	16	0.8400	0.1000		450	8	0.1000	0.0002
					400	12	0.1100	0.0010
					300	20	0.1100	0.0017

Poincaré Maps

A different solution to the problem of quantifying the stability of the system dynamics in phase space can be found by employing a Poincaré Map. As noted above, the continuous-time dynamics of the friction stir welding process are approximated, through sampling, by a discrete-time model. The drawback with sampling is that the sampling intervals selected may not accurately characterize the dynamics of the underlying continuous system. In addition, uniform sampling intervals, which, in the absence of evidence to the contrary are assumed to be appropriate, may not correspond to consistent locations in phase space. An alternate way to "sample" the system is to locate a hyperplane orthogonal to the direction of the trajectory. If a trajectory is periodic and stable in its orbit, the trajectory will repeatedly cross the hyperplane at the same point. If the trajectory is nearly-periodic or has an orbit that returns to nearby points, the crossings will be tightly clustered. For example, in Figure 4 the trajectory trace begins at point P0 and will return to point P1 on the hyperplane during a subsequent orbit. This defines a mapping h: P0 → P1, from the hyperplane back to itself. This "first return map" is known as the Poincaré map. It is another way to convert the continuous dynamical system to a discrete dynamical system and is a useful tool for demonstrating properties about the trajectory.

Figure 4. A sample "first return map", or Poincaré map, showing the trajectory piercing the hyperplane in a location near to, but distinct from, the first piercing.

As discussed in the introduction, the stability of the welding process is strongly correlated to the formation of metallurgical defects, and the degree of instability is a predictor for the size of the defect formed. The Poincaré Map provides a way to quantify the degree of stability the system exhibits at selected points in time.

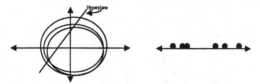

Figure 5. The picture on the left is an example of three orbits in phase space. The picture on the right demonstrates the piercings of the corresponding Poincaré map. The hyperplane for a 2D dynamical system is a line, and the piercings are the points where the trajectory intersects the line. Note that the actual hyperplane used in this research coincides with the horizontal axis. The hyperplane is positioned elsewhere in this figure for visual clarity.

The points of intersection of the trajectory with the Poincaré map, often referred to as "piercings", will be tightly grouped for trajectories with low divergence and more widely dispersed for trajectories in which successive orbits follow more divergent paths (Figure 5). The special case of a perfectly periodic function would then produce a single piercing of the Poincaré map as it passes through in one direction and a single piercing where it passes through the map in the returning direction. The closer the trajectory orbits are to each other, the more tightly grouped the piercings will be.

To use a Poincaré map, the position of the hyperplane that will record the piercings must be selected. As mentioned in the Lyapunov exponent discussion, the system dynamics are approximated by a two dimensional trajectory in phase space. Ideally, the hyperplane should be position to be orthogonal to the trajectory. Using a two dimensional representation of the dynamics implies that the Poincaré map will be a simple line and the piercings will be generated by the intersection of the trajectory with the line.

Two observations led to the selection of the horizontal axis as the location for the hyperplane. Phase space plots of each of the welds were generated and clear patterns emerged. Good welds demonstrated more compact trajectories and bad welds exhibited a greater degree of trajectory dispersion. Figure 1 depicts the phase space plot for a good weld as well as an exemplar of a weld containing a metallurgical defect. In these examples, as in all of the phase space plots, the direction of greatest deviation occurs roughly along the horizontal axis. Thus, the horizontal axis provides a logical position for the hyperplane.

A second reason for selecting the horizontal axis for the location of the hyperplane can be found by comparing a purely periodic function to that of the time series data. In an ideal environment, the time series data would closely match the graph of the spindle frequency. The y force represents the amount of displacement orthogonal to the direction of the weld and, if the weld was perfect, only the rotation of the spindle would affect the y feedback. Thus, the graph of the spindle frequency provides a reasonable baseline for comparison. Examination of the graph in Figure 6 shows that the variations between the perfectly periodic function and the actual time series data evaluated at the critical points provides a measure of the divergence of the two curves. In effect, this sampling of the magnitude of the separation between the curves provides an estimate of the actual difference which could be found by integrating the areas under each curve and computing the numeric differences. The importance of this observation is that the critical points are those points in the time series where the derivative is zero. Thus, if the critical points provide an adequate measure of the time series data's deviation from pure periodicity, it is reasonable to position the hyperplane along the y axis. This assumption also significantly simplifies the calculations needed for the real-time control system.

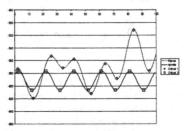

Figure 6. Time series data and the spindle frequency curve. A measure of the degree of separation can be found by calculating the difference between the two curves at the critical points. The *x*-axis is time and the *y*-axis is the force in lbf.

Once the map is generated, an algorithm for measuring the dispersion of the points is needed. The results of applying one such measure is described in the next section.

Results

The Poincaré map technique discussed in the previous section was applied to the *y* force time-series data for 31 weld samples. A phase space diagram and Poincaré map were generated for each sample and the resulting map piercings were analyzed. The number of orbits generated before the results are analyzed is a function of the rotation speed. The calculation used in these experiments was :

$$\text{Number of orbits} = 5 * \left\lfloor \frac{\text{spindle frequency (RPM)}}{200} \right\rfloor \tag{3}$$

The spindle rate of the MTS ISTIR-10 Friction Stir Welding machine ranges from 200 rpm to 2000 rpm, but these experiments only used spindle rates in the range of 200 rpm to 700 rpm. Thus, the number of orbits generated before the map was analyzed varied from 5 to 15. As discussed above, the number of points in the orbit of the *y* vs. *y'* trajectory is assumed to be the same as that of the spindle frequency ellipse which can be calculated by :

$$N_{spindle} = (\text{Sampling Rate (Hz) / Spindle Rate (RPM)}) * 60 \text{ sec/min} \tag{4}$$

where $N_{spindle}$ is the number of points in the spindle frequency ellipse and the sampling rate was between 51.2 at 102.4 Hertz. For example, a weld made at 200 rpm that is sampled at 51.2 Hz. will have 15 points per orbit while a weld made at 700 rpm will have 4.4 (rounded to 5) points per orbit. In either case, 75 *y* values are acquired before the quality check is performed. This variation maintains a constant *y* sampling interval of about 1.5 seconds before the control variable value is re-calculated. It should also be noted that the pin tool geometry has an effect on the *y* force behavior and is held constant across all experiments.

The Poincaré map is generated by the intersection of the *y* vs. *y'* trajectory with the horizontal axis and creates two distinct point clusters (Figure 5). One technique for characterizing the distribution pattern of those points is by calculating the standard deviations of the points in each cluster, σ_1 and σ_2, and combining them using the formula:

108

$$\sigma = \sqrt{\sigma_1^2 + \sigma_2^2} \tag{5}$$

Table II contains a list of the welds used, the system parameters, the size of the resulting wormhole defects where applicable, and the standard deviations computed from the Poincaré map of a one-second sample. These samples correspond to the positions at which the weld was cut and checked for metallurgical defects. A graphical representation is presented in Figure 7, separated by position control vs. forge force control (load control).

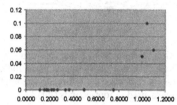 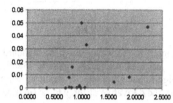

Figure 7. Plot of standard deviation versus defect size for position control (left) and forge force control (right). The x-axis is standard deviation and the y-axis is defect size in mm^2

Table II. Standard Deviation vs. defect size (mm^2)

RPM	IPM	Defect Size	SD	RPM	IPM	Defect Size	SD
250	2	0	0.1200	300	3	0	0.9900
425	6	0	0.1600	450	8	0	0.7088
250	6	0	0.1800	400	4	0	0.3600
425	10	0	0.1900	450	8	0.0002	0.8213
350	2	0	0.1900	300	12	0.0003	1.0675
350	6	0	0.2200	300	25	0.0005	0.9156
500	10	0	0.2400	450	8	0.0007	0.7819
500	2	0	0.2700	400	12	0.0010	0.9667
500	6	0	0.3400	300	20	0.0017	0.9690
350	10	0	0.3700	400	12	0.0045	1.6133
500	9	0	0.5000	300	25	0.0079	0.7776
250	10	0	0.7530	400	12	0.0085	1.8933
250	16	0.1000	1.0000	300	25	0.0160	0.8280
250	14	0.0500	1.0400	300	25	0.0330	1.0884
500	9	0.0600	1.1000	400	12	0.0472	2.2233
				300	25	0.0500	1.0020

Figure 7 shows that this technique is somewhat effective in determining weld quality. For position control, a standard deviation below 0.8 indicates no defect while a standard deviation of 1.0 or above indicates a defect. However, more data needs to be gathered to see if these relationships hold in general. For forge force control, the results are not as clear. While the biggest defects are in welds with standard deviations of 1.0 or above, there are some small defects in welds with standard deviation below 1.0 and some good welds above 1.0. However, visual studies of the phase space plots indicate that there are differences in the character of the trajectories that are not captured by the standard deviation computation. Figure 8 displays the phase plots for a bad weld and a good weld both of which have a standard deviation of approximately 1.0. However, the phase space plots clearly indicate a difference in stability. Further studies are in progress to attempt to improve these results by the addition of information such as forge force, weld pitch, or pseudo heat index. Also, the effectiveness of other clustering measures and curve fitting techniques is currently under investigation.

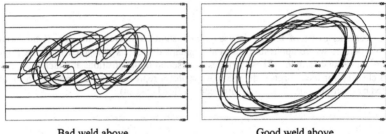

| Bad weld above | Good weld above |

Figure 8. Calculation of the standard deviation of the Poincaré map piercings of both trajectories produced virtually identical values, but the phase space plots indicate a significant difference in stability. [Phase space trajectory for the y force (horizontal axis) vs. the derivative of the y force (vertical axis) measured in foot-pounds (lbf).]

A limitation on this research is dictated by the real constraints placed on the slicing and testing of welding samples. Although the technique is capable of generating results in real-time, the accuracy of the predictions can only be tested at the discrete points selected for metallurgical testing. Thus, the results reported above reflect only the system's ability to match the observed outcome at those testing points.

Conclusion

This research identifies a promising tool for monitoring defect formation during friction stir welding. Traditional approaches to analyzing the time series data generated during the welding process have been somewhat successful in detecting changes in weld quality, but phase space analysis produced more accurate results with a low computational burden. The behavior of the system dynamics in phase space must be quantified if it is to provide useful control information. A variety of techniques have been investigated, but analysis of the dispersion of the trajectory piercings indicated on a Poincaré map produced the best results. The relationship between the dispersal pattern and weld quality provides a simple numeric value that can be used by a feedback controller during the welding process. The procedure is fast, allowing for real-time implementation, and has thus far proven to be accurate for predicting not only weld quality as a class, but for estimating the size of the defect being formed. The next phase of this work will concentrate on identifying an algorithm for adjusting the system parameters, specifically ipm and rpm, to maintain consistent quality throughout a weld.

References

[1] E. Boldsaikhan, E. Corwin, A. Logar, and W. Arbegast, "Neural Network Evaluation of Weld Quality using FSW Feedback Data", 6th International FSW Symposium, Montreal, 2006.

[2] W. J. Arbegast, "Modeling Friction Stir Joining as a Metalworking Process", *Hot Deformation of Aluminum Alloys III*, ed. Z. Jin, TMS (The Minerals, Metals, and Materials Society), 2003.

[3] W. J. Arbegast, "Using Process Forces As A Statistical Process Control Tool For Friction Stir Welds", *Friction Stir Welding and Processing III*, ed. K. V. Jata, et al, TMS (The Minerals, Metals & Materials Society), 2005.

[4] Computational Systems Inc., "Selection of Proper Sensors for Low Frequency Vibration Measurements", *Noise & Vibration Control Worldwide*, Oct 1988, 256.

[5] R. Clark Robinson, *An Introduction to Dynamical Systems: Continuous and Discrete*, Pearson Prentice Hall, Upper Saddle River, New Jersey, USA, 2004.

[6] A.Wolf, J.B. Swift, H.L. Swinney, and J.A. Vastano, "Determining Lyapunov exponents from a time series", PHYSICA D, vol. 16, pp. 285-317, 1985

[7] M. W. Hirsch, S. Smale, and R. L. Devaney, *Differential Equations, Dynamical Systems & Introduction to Chaos*, 2nd edition, Elsevier (USA), 2004, 194-203.

[8] M. T. Rosenstein, J. J. Collins, and C. J. Deluca, "A practical method for calculating largest Lyapunov exponents from small data sets", *Physica D*, 65:117-134, 1993.

SPECIFIC ENERGY AND TEMPERATURE MECHANISTIC MODELS FOR FRICTION STIR PROCESSING OF AL - F357

P. Kalya[1], K. Krishnamurthy[1], R. S. Mishra[2], J. A. Baumann[3]

[1]Mechanical and Aerospace Engineering, University of Missouri-Rolla, Rolla, MO 65409, USA
[2]Center for Friction Stir Processing, Materials Science and Engineering, University of Missouri-Rolla, Rolla, MO 65409, USA
[3]The Boeing Company, St. Louis, MO 63166, USA

Keywords: Friction Stir Processing, Temperature, Mechanistic Model

Abstract

Mechanistic models for process specific energy and surface temperature profile of the work material lateral to the tool motion were investigated in this study. Experiments were conducted with a robotic friction stir welding machine on Al-F357 investment cast plates to measure the axial force, spindle torque and surface temperature to determine the mechanistic model parameters. The results obtained show that the specific energy correlates linearly with the surface temperature of the work material. Also, the temperature mechanistic model based on Rosenthal's moving line source equations can estimate the surface temperature very well.

Introduction

Material flow and evolution of the thermal cycle have been argued to be the two most important factors affecting the properties of the friction stir processed (FSP) material. Modeling the strength and microstructural properties requires appropriate mathematical expressions to predict the heat generation and subsequent dissipation. This would enable one to determine the thermal cycle experienced by the material and hence predict the microstructural properties.

Various methods have been utilized to model the generation and transfer of heat during friction stir welding (FSW). These include analytical techniques based on finite element analysis and computational fluid dynamics, and experimental techniques. Frigaard et al [1] have developed a simple model for energy generated based on interfacial pressure and radius of the tool without considering the pin effect. Based on finite difference technique, they have developed a heat transfer model to determine the temperature distribution on the plate. Khandkar et al [2] have discussed a moving heat input model based on average power input and have developed a torque-based model for determining the heat generated at different contact positions, namely, shoulder and pin surfaces. Schmidt et al [3] have further developed this model to account for conical shouldered tools, by classifying the surface based on vertical and horizontal orientations. Heurtier et al [4] have tried to predict the heat generated based on three sources namely, plastic strain, shoulder friction and pin friction. Schmidt and Hattel [5] have developed several models

113

to account for various possible contact conditions between the tool and the base material. McClure et al [6] have presented thermal field generation model based on Rosenthal's moving point source equation [7] and compared them to experimental results.

The aforementioned modeling studies, however, do not deal with the effect of process parameters on generation of heat. It has been commonly known that rotation speed and traverse rate are the key parameters to determine the "hotness" or "coldness" of a weld/process. To describe heat generated as a function of process parameters, Arbegast has defined a heat index expressed as [8]

$$HI = \frac{\omega_h^2}{v_h * 10^4} \qquad (1)$$

where ω_h is the rotational speed of the tool in revolutions per minute and v_h is the traverse rate in inches per minute. Utilizing this definition, a simple thermal model was developed by Arbegast to define a power relationship ($K * HI^\alpha$) between the homologous temperature (temperature normalized with work piece melting point) and process parameters.

In this study, mechanistic models are developed to study the heat generated during FSW/P. First, a model is presented to calculate the specific energy as a function of the process parameters without any *a priori* assumption of the relationship between the variables. Second, a model to study the surface temperature of the work material lateral to the motion of the tool is presented. Experiments were conducted on Al-F357 investment cast plates to determine the model parameters and study the efficacy of the models.

Mechanistic model development

Specific energy model

Spindle torque is a representative measure of the power input into the process. The constitutive equation for power input can be written as [1]

$$q = \omega M \qquad (2)$$

where ω is the rotational speed and M is the spindle torque. It can also be argued that the spindle torque is a function of process parameters, namely, rotational speed, traverse rate, plunge depth and travel/work angle, high temperature flow stress of the material, surface area of contact between the material and the tool and the interfacial friction conditions. In this study, it is assumed that the material and contact surface area do not change. Therefore, the spindle torque can be modeled as

$$M = A\omega^\alpha v^\beta \qquad (3)$$

where v is the transverse speed and A, α and β are constants, which will be determined in this study using a least squares approach. The structure of eqn (3) was chosen based on the observation that the measured torque reduces with increase in rotational speed, which is due to increased heating and hence softening of the material. This requires α being negative, which would have an effect of reducing the value of M as ω increases. The spindle torque is also observed to increase with an increase in the traverse rate at a given rotational speed, due to faster material feed and reduction in temperature with increasing traverse rate. This behavior requires β

114

to be a positive number, increasing M as v increases. Substituting the expression obtained for spindle torque into eqn. (2) yields the expression for power input as

$$q = A\omega^{\alpha+1}v^{\beta} \tag{4}$$

The process specific energy defined as the ratio of power input and traverse rate is then given by

$$E_{sp} = \frac{q}{v} = A\omega^{\alpha+1}v^{\beta-1} \tag{5}$$

Note that the aforementioned specific energy model precludes the need to *a priori* choosing of the indices as presented by Arbegast [8].

Temperature model

For a plate with pseudo-steady state temperature distribution and a moving line heat source, the temperature profile lateral to the direction of the tool can be calculated using the Rosenthal's thin plate equation [7, 10]

$$T - T_0 = \frac{q_t}{2\pi\lambda}\frac{1}{d}K_0 \tag{6}$$

where T is the temperature distribution, T_0 is initial temperature, q_t is the power input contributing to the temperature increase, λ is the thermal conductivity of the material, d is thickness of the plate and K_0 is the modified Bessel's function of second kind and order zero approximated by [10].

$$K_0 = \sqrt{\frac{\pi a}{vx}}\, e^{\left(\frac{vx}{2a}\right)} \tag{7}$$

Here x is the distance from the centerline of the process and v is the traverse rate and a is the thermal diffusivity of the material. Because the line source is an idealization of the FSP under consideration, the temperature profile in this study is modeled as

$$T - T_0 = B\left[A\omega^{\alpha}v^{\beta}\right]\left(v^{\psi}x^{\gamma}\right)e^{(-\delta xv)} \tag{8}$$

where B, γ, δ, *and* ψ are constants, which will be determined again using a least squares approach.

Note that to calculate the temperature using eqn. (8), it will be necessary to know the value of the initial temperature. In this study, the initial temperature is assumed to be the temperature at the end of the dwell cycle. As the traverse rate during plunge is typically constant, the initial temperature will be estimated as

$$T_0 - T_{amb} = A_{plunge}\omega^{\alpha_{plunge}}e^{\left(-B_{plunge}r^2\right)} \tag{9}$$

where A_{plunge}, α_{plunge} and B_{plunge} are constants that will be determined again in the sense of least squares, T_{amb} is the room temperature, and r is the radial distance from the plunge position to the temperature measurement points.

Experimental details

The experimental setup consists of an IRB-940 Tricept six-axis robot from ABB Inc. (Fig. 1). It has three non-parallel telescopic translational joints and three rotational joints. The robot uses a S4cPlus robot controller with RAPID as the programming language. A teach pendant allows the user to manually control and program the robot. The robot is retrofitted with a FSW head from

115

Friction Stir Link, Inc. which provides the rotational motion to the tool. The friction stir spindle assembly consists of a rotational axis driven by an external 10 HP Exlar SLM115-368 servo motor with speeds up to ± 3000 rpm. The spindle is capable of processing up to 6.125 mm (0.25 in) thick material. The load rating of the spindle is 9 kN (2,023 lb) along the tool axis and 4.5 kN (1,012 lb) along the radial direction.

A JR3 multi-axis force-torque sensor system can determine the loading on the tool by measuring the forces along three orthogonal axes and moments about each of the three axes. The outputs are analog signals in the range of ±10 volts. The rated loads for the sensor are 6 kN (1,348 lb) in x and y directions and 12 kN (2696 lb) in the z-direction (axial direction). The rated moments are 1,150 Nm (848 ft-lb) about all the three axes.

Figure 1. Photographs of IRB 940 (a) Tricept manipulator, and (b) FSW Spindle assembly.

Experiments were conducted on 3 x 80 x 130 mm Al F-357 investment cast plates, the composition of which was Al 92.5 %, Mg 0.5%, Si 7%. Processing was done over a distance of 100 mm. The centerline of the process was placed at 40 mm from the edge of the plate. The tool had a shoulder diameter 12 mm and a conical pin with spiral step features, the length of which was 1.75 mm. Experiments were performed with a travel angle of 2.5°.

Surface thermal measurements were made with the aid of a FLIR P-40 thermal camera. Figure 2 shows the schematic of the experimental setup. The thermal camera detects the infrared radiation emitted by a surface, which is related to the temperature given by

$$P = \varepsilon A_s \sigma_s T^4 \tag{10}$$

where P is the energy radiated per unit time, ε is the surface emissivity, A_s is the surface area of radiating body, $\sigma_s = 5.6703 \times 10^{-8}$ W/(m^2 K^4) is the Stefan Boltzmann's constant, and T is the surface temperature. Emissivity depends on the material, surface texture and color of the radiating surface. For a perfect black body, emissivity is one. On the other hand, it is zero for a completely reflecting surface. Aluminum, unfortunately, is an excellent reflector of radiation in the infrared spectrum and has a very low value of emissivity. Depending on the surface finish, aluminum emissivity varies from 0.05 to 0.35. Another issue complicating the matter is the fact that the camera cannot be placed normal to the surface of the plate. The difficulty is to map the actual distances between tool and points on the surface, to the image captured by the thermal camera. To address both these issues, the procedure followed was to paint the material surface with markings at known distances across the tool path with black paint. In doing so, the emissivity of the surface was standardized and the markings appeared to have a higher temperature compared to the neighboring surface because of the black color (Fig. 3). This helped in conveniently positioning the virtual thermocouples where temperature measurements were required.

To determine accurately the emissivity of the black colored markings, a test plate was heated to a known temperature and the emissivity of the camera was adjusted to read the same temperature as that obtained using a thermocouple. It was found that an emissivity value of 0.95 was satisfactory. This value was further verified for a range of temperatures up to 220°C by heating the plate and measuring the temperature both by the thermal camera and thermocouple. Figure 4 shows a good agreement between the two types of measurements. Therefore, it was concluded that the measurements made by the thermal camera are as good as the temperature measurements obtained using the thermocouple.

The third issue during thermal measurements using an IR sensing device is reflected temperature. The sum of emissivity and reflectivity of an opaque surface is one; hence, any surface with emissivity less than one reflects back a portion of the incident radiation. The FSW head and the tool are good reflectors of thermal radiation. Consequently, it was necessary to reduce their reflectivity by painting them with black paint as well. The markings were made at distances of 10, 15, 20 and 25 mm on either side of the tool path, and the thermal camera was placed at 300 mm from the line of measurement at an angle of 18° to the horizontal as shown in Fig. 2.

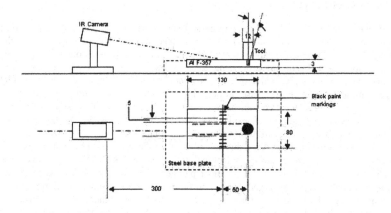

Figure 2. Schematic of the experimental setup (numbers are dimensions in mm).

117

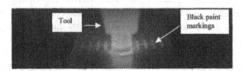

Figure 3. Thermal image of the run showing the irradiated markings.

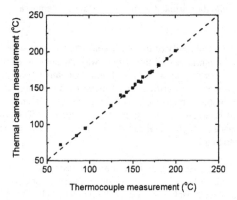

Figure 4. Plot showing temperature measurements using the IR camera and thermocouple.

Results and discussion

Axial force and torque in FSP

Figure 5 shows the measured tool axial force and spindle torque during FSP with a constant traverse rate of 4.233×10^{-3} m/s (10 ipm), rotational speed of 33.33 rps (2000 rpm) and a plunge depth of 1.75 mm. The experimental run can be divided into two distinct zones. The plunge and dwell zone from A to E and the processing zone beyond E. During the plunge sequence, the force was seen to rise to a peak value of 1.71 kN when the pin had plunged to a distance of 0.947 mm. This initial increase is the phase during which the material is experiencing force due to indentation and the interfacial contact between the tool and material is in slip condition as a result of insufficient heat generated, explaining the low torque values [9]. At this juncture, the tool-material interface changes from slip to stick-slip condition, which plasticizes the material and results in the drop of axial force due to softening of the material. The drop in axial force continues until the shoulder comes in contact with the work piece at C. The torque was observed to rapidly rise to around 7 Nm as the shoulder plunged into the material from C to D, due to increased material plasticization and contact surface. The region from D to E represents the dwell time during which the axial force reduces and the torque remains almost constant. The actual processing period is from E to F, where the axial force is observed to increase before stabilizing at around 4 kN. The plunging induced force and moment values in the other directions do not change on the same scale unless the robot arm moves due to vibration.

118

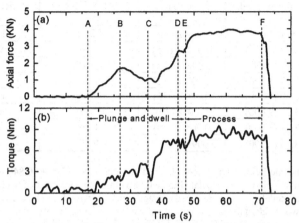

Figure 5. Time histories of (a) axial force and (b) spindle torque for a typical run with rotational speed of 33.33 rps, traverse rate of 4.233×10^{-3} m/s, plunge depth of 1.75×10^{-3} m and travel angle of $2.5°$.

<u>Specific energy</u>

From Fig. 5 it can be seen that the torque remains almost constant during the processing period, suggesting that the average value could be used to calculate the constants in eqn. (3). Because the least squares technique requires the parameters and regression variables to be linearly related, eqn. (3) can be rewritten as

$$\ln M = \ln A + \alpha \ln \omega + \beta \ln \nu \qquad (11)$$

A set of $k = 8$ experiments were performed at various traverse rates and rotational speeds. The parameter estimates for the constants were then determined in the sense of least squares as

$$\boldsymbol{\eta} = [\boldsymbol{\Phi}^T \boldsymbol{\Phi}]^{-1} \boldsymbol{\Phi}^T \mathbf{y} \qquad (12)$$

where

$$\boldsymbol{\eta} = \begin{bmatrix} \ln A & \alpha & \beta \end{bmatrix}^T \qquad (13)$$

$$\boldsymbol{\Phi} = \begin{bmatrix} 1 & \ln(\omega(1)) & \ln(\nu(1)) \\ \cdot & \cdot & \cdot \\ \cdot & \cdot & \cdot \\ \cdot & \cdot & \cdot \\ 1 & \ln(\omega(k)) & \ln(\nu(k)) \end{bmatrix} \qquad (14)$$

$$\mathbf{y} = \begin{bmatrix} ln(M(1)) .... \ln(M(k)) \end{bmatrix}^T \qquad (15)$$

and it was assumed that the matrix $[\boldsymbol{\Phi}^T \boldsymbol{\Phi}]$ was nonsingular. Table I shows the process parameters used to perform the set of eight experiments and the corresponding average values of the measured and modeled spindle torque.

Table I Process parameters along with measured and modeled torque.

Run No.	Rotational speed, rps (rpm)	Traverse rate, m/s (ipm)	Measured torque, Nm	Modeled torque, Nm
1	16.07 (964)	0.00169 (4)	12.75	13.25
2	26.67 (1,600)	0.00233 (5.5)	8.94	9.09
3	16.07 (964)	0.00233 (5.5)	14.18	13.97
4	26.67 (1,600)	0.00106 (2.5)	8.32	7.98
5	26.67 (1,600)	0.00635 (15)	11.29	10.73
6	33.33 (2,000)	0.00381 (9)	7.73	8.16
7	33.33 (2,000)	0.00423 (10)	8.36	8.30
8	33.33 (2,000)	0.00992 (23.44)	9.50	9.56

The values of the coefficient and indices obtained were A = 399.87, α = -0.848 and β = 0.165. The correlation coefficient was calculated to be 0.98 with a standard error of 0.43 Nm. Figure 6 shows the plot of modeled versus measured average spindle torque. It can be clearly seen that the data lie within two standard deviations indicating that the model has a 95% confidence limit.

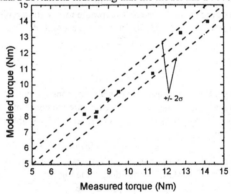

Figure 6. Plot showing the results of the mechanistic model for spindle torque.

Figure 7 shows the contours of torque, power, and specific energy based on the mechanistic model given by eqns. (3-5). Figure 7(a) shows the expected trend of the torque increasing with an increase in traverse rate and torque decreasing with increase in rotational speed. Similarly, Fig. 7(b) shows that as the rotational speed and traverse rate increase, the input power also increases, which is due to increased heating effect of the rotational rate and increased torque due to traverse rate. However, Fig. 7(c) shows that increasing the traverse rate drastically reduces the specific energy compared to that of an increase in the rotational speed. This is because for a unit length of the process, an increase in traverse rate at a given rotational speed has two effects: i) reduces the energy input per unit length due to a faster moving power source; and ii) reduces the total number of revolutions per unit length contributing to a reduction in the magnitude of the power source. However, an increase in the rotational speed has only the effect of increasing the power input.

120

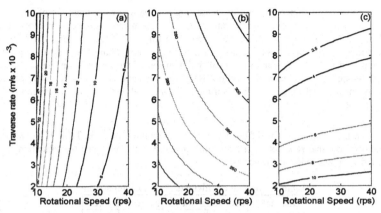

Figure 7. Contours of (a) modeled torque, (b) modeled power, and (c) modeled specific energy ×
10^{-4} as functions of rotational speed and traverse rate.

Temperature profile

To determine if the specific energy has any physical significance, surface temperatures were recorded using a thermal camera for each of the run performed. Figure 8 shows temperature profiles across the process center line for a typical run. The region A is due to the tool interfering with the field of view of the camera and hence changing the temperature recorded. The peak temperature at each point on the plate is a function of the material conductivity, the amount of heat generated at the tool interface and the rate at which the source is traveling.

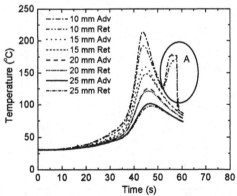

Figure 8. Surface temperature profile on the line perpendicular to the process for run #6 (refer Table I).

121

It was noted that the temperatures on the advancing side were greater than that of temperatures measured at symmetric locations on the retreating side. The maximum temperature recorded was 220 °C at a distance of 10 mm from the process line for rotational speed of 26.667 rps (1600 rpm) and traverse rate of 1.058×10^{-3} m/s (2.5 ipm). The value obtained was lower compared to results reported in [1,2,4,5], due to the smaller dimension of the tool used in the current study. The peak temperatures for each of these runs were measured and plotted as a function of specific energy. A linear correlation between specific energy and peak temperature was found from fig. 9(a) and (b) as

$$T_{peak} = aE_{sp} + b \tag{16}$$

where a and b are tabulated for different positions across the process in Table II. However, fig. 9(c) and (d) show that satisfactory relationship can not be obtained when the same temperatures are plotted with heat index as the abscissa.

Table II Constant coefficients for peak temperature.

Distance (mm)	Advancing side		Retreating side	
	a	b	a	b
25	78.316	2.125	78.781	2.125
20	101.772	2.125	99.843	2.125
15	131.315	2.125	128.84	2.125
10	185.661	2.125	174.392	2.125

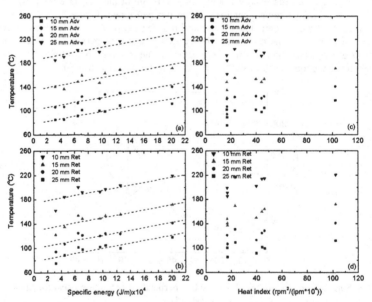

Figure 9. Peak temperatures as a function of specific energy and heat index at various locations on the surface (a) and (c) advancing side (b) and (d) retreating side.

122

<u>Surface temperature estimation</u>

Modeling of the initial temperature, T_0, was done using the data obtained for surface temperatures. The temperature data was selected at a time corresponding to the end of dwell period as seen in the axial force plot. Table III shows the values calculated for the coefficients in eqn. (9) using three different rotational speeds given in Table I.

Table. III Parameter estimates of initial temperature model

A_{plunge}	α_{plunge}	B_{plunge}
31.481	0.7914	-1.1859

The peak temperature at each of the measured points on the plate along with the modeled initial temperature can be used to model the temperature distribution on the surface as well as the temperature at the shoulder-material interface. Table IV shows the calculated values of the constants in eqn. (8) using the least squares technique. Figure (10) shows the temperature was estimated well using the proposed model. The correlation coefficient obtained was 0.98 with a standard error of 6.37 °C.

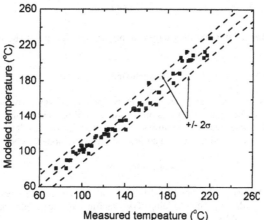

Figure 10. Modeled versus measured temperature.

The peak temperature distribution from the model and the experimental data are plotted in fig. 11. It is evident from the figure that the relationship obtained satisfactorily explains the thermal evolution within the range of experiments performed. Similar results were obtained for all the runs performed in this study. Microstructure evolution is greatly influenced by the thermal cycle the material undergoes. In precipitation strengthening alloys, the microstructural changes depend on the maximum temperature reached along with cooling sequence. The methodology in this paper is a step in the direction of controlled microstructural gradient across the welded/processed region.

Table. IV Parameter estimates of temperature model.

B	ψ	γ	δ
364.40	0.261	-0.883	2256

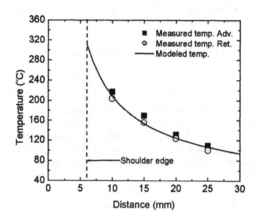

Figure 11. Temperature profiles across the process line for run #1 (Refer Table I).

Conclusions

Mechanistic models for torque and correspondingly specific energy were developed with the assumption that energy per unit length was only a function of process parameters for a given set of tool and material. The peak temperatures at reference points correlated linearly with the mechanistic specific energy model. Also, a surface temperature model has been developed.

Acknowledgement

This work was performed under the NSF-I/UCRC for Friction Stir Processing at UMR. The financial support provided by NSF, Boeing, GM Pacific Northwest National Laboratory and Friction Stir Link is gratefully acknowledged.

References

1. O. Frigaard, Ø. Grong, and O. T. Midling, "Modelling of heat flow phenomena in friction stir welding of aluminium alloys," (Paper presented at the Inalco '98: Seventh International Conference on Joints in Aluminium, Cambridge, UK, 15-17 Apr. 1998), 208-218.

2. M. Z. H. Khandkar, J. A. Khan, and A. P. Reynolds, "Prediction of temperature distribution and thermal history during friction stir welding: input torque based model," Science and Technology of Welding and Joining, 8 (3) (2003), 165-174.

3. H. Schmidt, J. Hattel and J. Wert, "An analytical model for the heat generation in friction stir welding," Modelling and Simulation in Materials Science and Engineering, 12 (2004), 143-157.

4. P. Heurtier et al., "Mechanical and thermal modelling of Friction stir welding," Journal of Materials Processing Technology, 171 (3) (2006), 348-357.

5. H. Schmidt and J. Hattel, "Modelling heat flow around tool probe in friction stir welding," Science and Technology of Welding and Joining, 10 (2) (2005), 176-186.

6. J. C. McClure et al., "A thermal model of friction stir welding," (Paper presented at the 5th International Conference: Trends in Welding Research; Pine Mountain, GA, USA, 1-5 June 1998), 590-595

7. D. Rosenthal, "The theory of moving sources of heat and its application to metal treatments," Trans. ASME, 68 (1946), 849-866.

8. W. J. Arbegast, "Modeling friction stir joining as a metalworking process," (Paper presented at the Third Symposium on Hot Deformation of Aluminum Alloys III as held at the 2003 TMS Annual Meeting; San Diego, CA, USA, 2-6 Mar 2003.) 313-327.

9. S. F. Miller et al., "Experimental and numerical analysis of the friction drilling process," ASME Journal of Manufacturing Science and Engineering, 128 (3) (2006), 802-810.

10. $\varnothing$. Grong, Metallurgical modeling of welding (The institute of materials, 1994), 2-55.

A NUMERICAL STUDY OF THE PLUNGE STAGE IN FRICTION STIR WELDING USING ABAQUS

Saptarshi Mandal, Justin Rice, Abdelmageed Elmustafa

Old Dominion University
Dept. of Mechanical Engineering
238 Kaufman Hall, Norfolk, VA 23529

Keywords: FSW, Plunge, ABAQUS, Johnson-Cook Law

Abstract

Although, there are several finite element (FEA) based models developed for friction stir welding (FSW), most of these models primarily concentrate on the phase of the weld where the tool is moving forward, eliminating the plunge stage. A better understanding of the plunge phase is extremely important with the growing role of friction stir spot welding and also for understanding tool wear in case of FSW of harder materials. This research develops a three dimensional FEA based model of the plunge stage using the commercial code ABAQUS to study the thermomechanical processes involved during the plunge. The strain rate and temperature dependent Johnson-Cook material law is used as the constitutive law. The heat source incorporated in the model involves friction between the material, the probe and the shoulder and also heat generated due to plastic deformation. A good correlation was found between the results obtained from ABAQUS and experimental data.

Introduction

Friction stir welding can essentially be divided into a three step process - the plunge stage, where a hard non-consumable rotating tool penetrates the plates to be welded, a dwell stage where the tool having penetrated the metal rotates without moving forward and the welding stage where the tool moves forward to form a weld bead. The plunge stage in the friction stir welding (FSW) process is extremely critical, as this is where most of the initial thermomechanical conditions are generated and the material undergoes significant transformation due to the high temperatures and stresses involved in the process. The highly dynamic nature of this phase makes it an important albeit challenging research area. A thorough understanding of the plunge stage is also important in the development of tools and processes for successfully stir welding high strength alloys like steel and titanium based alloys, as most of the tool wear occurs during this phase [1-3]. This warrants the need for more experimental and numerical studies to study this part of the process. Although there are a few experimental studies that address this area [4], there are hardly any numerical models which concentrate on the thermomechanical conditions developed during the plunge.

In this paper we simulate this critical phase using the commercial finite element (FE) code ABAQUS/Explicit. There are several models which deal with FE-based numerical modeling of FSW [5-8], however nearly all of them simulate either the welding stage or the welding stage and the dwell stage. One of the primary difficulties in simulating the plunge stage is excessive mesh distortion of the finite element model leading to premature termination of the program. This problem has been successfully dealt with in this research.

127

Schmidt et al [5-7] developed thorough numerical models in their research using ABAQUS/Explicit and the arbitrary Lagrangian Eulerian (ALE) formulation. These models used a 'simplified plunge' wherein the simulation was started at the point where the shoulder already made contact with the work-piece. Goetz et al [9] developed a two dimensional model using DEFORM to simulate the metal flow around the tool and the initial tool plunge. Tool plunge forces and temperatures were predicted using this two-dimensional model. Three dimensional models involving the plunge phase have been developed for Friction Stir Spot Welding (FSSW) [10,11]. The difference between these models and the present work is the plunge depth. Gerlich et al [10] used a computational fluid dynamics (CFD) approach to model FSSW with a plunge depth of 300μm. On the other hand, Kakarla et al [11] used a solid mechanics approach to develop an isothermal model with a plunge depth of 0.0125 in. (0.3175 mm). In comparison the model presented here has a plunge depth of 12 mm, as it is primarily designed to study the plunge region in a regular FSW. However, this model could fairly easily be modified to accommodate smaller plunge depths.

Numerical Model

The numerical modeling of FSW poses a challenge due to the high strain rates and temperatures involved in the process making it a complicated non-linear problem. ABAQUS is used for the purposes of numerical modeling due to its strong capabilities of handling non-linear problems and the built-in Johnson-Cook material law which is used in the present problem. The use of remeshing is possible in the ABAQUS explicit solver using the ALE approach. This is crucial to avoid unacceptable element distortion which could lead to premature termination of the problem in cases involving large deformations.

The model consists of a deformable workpiece and a rigid stir welding tool. The threads on the pin of the tool are avoided to simplify the computation process. The workpiece is meshed using 8 node brick elements and the enhanced hour glass control formulation is adopted. The mesh consists of 1600 elements and 2206 nodes and is graded in a way such that there is a higher mesh density around the tool plunge area. This improves the accuracy of the solution around the tool without tremendously increasing the computational cost. The graded meshes are obtained by partitioning the workpiece into smaller cells. The mesh is shown in Fig. 1. The workpiece is 20mm thick. The tool dimensions used for the model are shown in Fig. 2.

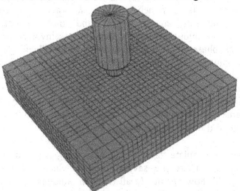

Fig.1. Model assembly and finite element mesh

The workpiece is constrained at the bottom surface to prevent the bending of the surface and the sides are constrained such that there is no deformation along the boundary other than compression along the tool plunge direction. The tool is modeled as a rigid surface with no thermal degrees of freedom. For the contact conditions between the tool and the workpiece, the tool is modeled as a master surface and the workpiece as a slave. A constant friction coefficient of 0.3 is assumed between the tool and the workpiece [7] and the penalty contact method is used to model the contact interaction between the two surfaces. The tool rotational speed is set at 300 RPM and the tool plunge velocity is set to a uniform 4mm/s.

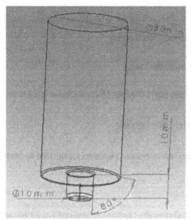

Fig 2. Tool dimensions used in simulation

Material Law

The selection of an appropriate constitutive law to reflect the interaction of flow stress with temperature, plastic strain and strain rate is important for modeling the FSW process. For this reason the temperature and strain rate dependent elastic-plastic Johnson-Cook law is selected for this model. The constitutive law in this case, calculates the flow stress as a function of temperature and strain rate upto the melting point or solidus temperature at which point the stress is reduced to zero as can be seen from the equations below [12]. For Al 2024, the solidus temperature is set to 502°C.

$$\bar{\sigma} = \left[A + B\left(\bar{\varepsilon}^{pl}\right)^n \right]\left[1 + C \ln\left(\frac{\dot{\bar{\varepsilon}}^{pl}}{\dot{\varepsilon}_0}\right) \right]\left(1 - \hat{\theta}^m\right) \tag{1}$$

$\hat{\theta}$ is the nondimensional temperature defined as

$$\hat{\theta} \equiv \begin{cases} 0 & for \quad \theta < \theta_{transition} \\ \left(\theta - \theta_{transition}\right)/\left(\theta_{melt} - \theta_{transition}\right) & for \quad \theta_{transition} \le \theta \le \theta_{melt} \\ 1 & for \quad \theta > \theta_{melt} \end{cases} \tag{2}$$

Johnson-Cook strain rate dependence assumes that

$$\bar{\sigma} = \sigma^0\left(\bar{\varepsilon}^{pl}, \theta\right) R\left(\dot{\bar{\varepsilon}}^{pl}\right) \tag{3}$$

and

$$\dot{\bar{\varepsilon}}^{pl} = \dot{\varepsilon}_0 \exp\left[\frac{1}{C}(R-1)\right] \qquad for \qquad \bar{\sigma} \geq \sigma^0, \tag{4}$$

where
$\bar{\sigma}$ is the yield stress at nonzero strain rate;

$\dot{\bar{\varepsilon}}^{pl}$ is the equivalent plastic strain rate;

$\dot{\varepsilon}_0$ and C are material parameters measured at or below the transition temperature, $\theta_{transition}$;

$\sigma^0(\bar{\varepsilon}^{pl}, \theta)$ is the static yield stress; and

$R(\dot{\bar{\varepsilon}}^{pl})$ is the ratio of the yield stress at nonzero strain rate to the static yield stress (so that

$R(\dot{\varepsilon}_0) = 1.0$).

A, B, C, n, m are material parameters that are measured at or below the transition temperature.

In case of Al 2024-T3 these values are [7]

A = 369 MPa, B = 684 MPa, C = 0.0083, n = 0.73 and m = 1.7

Results and Discussion

The workpiece was modeled as per the dimensions specified earlier. The tool was plunged in for a period of 3 sec upto a distance of 12 mm. The primary challenge in this simulation was the premature termination of the solution due to excessive element distortion. The Arbitrary Lagrangian Eulerian (ALE) feature was adopted for these cases and the material was remeshed constantly. Increasing the frequency of remeshing did not solve the issue of excessive element distortion. The next approach was to remove the elements which were excessively distorted from the calculation, thus preventing a premature termination. This was done using the 'shear failure' criterion built into ABAQUS/Explicit. This method has been previously used for ABAQUS based finite element modeling of machining problems [13]. The basic principle behind it is the elimination of the elements which have reached a preset damage parameter. The damage parameter in the element is measured by ABAQUS based on the Johnson-Cook shear failure criterion. The use of this approach prevented a premature termination of the solution and the material deformation was closer to reality. However, far too many elements were being removed from the model resulting in the presence of large voids. This was probably due to the fact that the parameters for the Johnson-Cook shear failure criterion were designed for ballistic purposes [14] which involve a very high strain rate unlike FSW.

130

Fig.3 shows the temperature distribution in the Al 2024 workpiece at the end of the plunge period. As expected the temperatures are comparatively higher at the trailing edge. The temperature history shown in Fig.4 was recorded at the tool tip for the entire duration of the plunge and was compared with experimental data obtained from Gerlich et al [15], who measured the temperature at the tool tip during a plunge into Al 6111. The ABAQUS data correlated well with the experimental data apart from around the 2.5 sec mark where the ABAQUS temperatures were lower than the experimental data. A possible reason for this discrepancy could be due to the different materials used for the simulation and the experiment. A difference in hardening properties in the two materials around that temperature range could make a difference in the measured values.

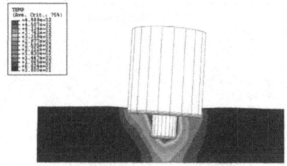

Fig.3 Longitudinal cross-section showing temperature distribution at the end of the plunge stage

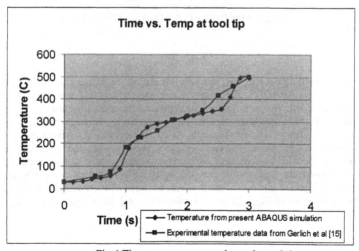

Fig.4 Time vs. temperature plot at the tool tip

Fig.5 shows the axial force on the tool as a function of time. As is evident from the graph, the axial forces calculated from the ABAQUS simulation do not closely match the experimental

131

data obtained from literature. A possible reason for this could be the relatively low mesh density along the element thickness. However, there is a similar trend that can be observed in the two cases. During the initial part of the plunge in case of the simulation, the force rises upto 5.5 kN at approximately 1 sec and then the load drops down to below 2 kN and then rises back up consistently to the peak load at the end of the plunge of 3 sec. Similarly in the experimental data obtained by Gerlich et al [15] the axial force rises a little over 2 kN around the 1 sec mark and then drops down before rising back to its peak value. A possible cause of this peak could be the high axial force experienced while plunging into a relatively cold metal with significantly higher flow stresses and hardness. As the workpiece heats up slightly the load drops and then again rises with the rise in plunge depth.

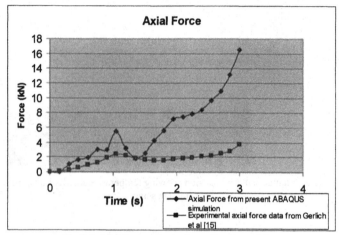

Fig.5 Axial force on the tool as a function of time

Conclusion

A finite element based model was developed for simulating the plunge process in FSW using commercially available ABAQUS. The Johnson-Cook material law was used as the constitutive model to simulate the temperature and strain rate dependent material properties. It was found that with the use of the ALE approach for simulating the plunge, the solution was prematurely terminated due to excessive element distortion. Also, the deformation obtained in the material was unrealistic. These problems however were not present when the same problem was simulated without the ALE approach and solution convergence was obtained. Temperatures at the material immediately beneath the pin are around 500°C, which correlates well with experimental data from the literature. Future effects are directed towards developing the model for different materials especially high strength alloys like steel, to understand the effect on tool wear.

References

1. 1. Lienert T.J., Stellwag W.L. Jr., Grimmett, B.B., Warke, R.W., "Friction stir welding studies on mild steel", *Supplement to the Welding Journal*, 2003, pp.1s-9s.

2. Thomas, W., 1999 "Friction stir welding of ferrous materials; A feasibility study", *Proceedings of First International Symposium on Friction Stir Welding,* Thousand Oaks, California.

3. Mandal, S., Williamson, K. "A thermomechanical hot channel approach to friction stir welding", *Journal of Materials Processing Technology,* 2006, pp. 190-194.

4. Santella, M., "Plunge testing to evaluate tool materials for friction stir welding of 6061+20wt% Al_2O_3 composite", *4th International Friction Stir Symposium, Park City, Utah, 2003.*

5. Schmidt, H., Hattel, J., Wert, J., "An analytical model for heat generation in friction stir welding", *Modeling and Simulation in Materials Science and Engineering,* 2004, pp. 143-157

6. Schmidt, H., Hattel, J., "Modelling thermomechanical conditions at the tool/matrix interface in Friction Stir Welding", *Proceedings of the 5th International Friction Stir Welding Symposium* Metz, France, 2004

7. Schmidt, H., Hattel, J., Wert, J., "A local model for the thermomechanical conditions in friction stir welding", *Modeling and Simulation in Materials Science and Engineering,* 2005, pp. 77-93

8. Ulysse, P., "Three-dimensional modeling of the friction stir-welding process", *International Journal of Machine Tools and Manufacture,* 2002, pp. 1549-1557.

9. Goetz, R.L., Jata, K.V., "Modeling friction stir welding of titanium and aluminum alloys", *Friction Stir Welding and Processing,* TMS, 2001.

10. Gerlich, A., Su, P., Bendzsak, G.J., North, T.H., "Numerical modeling of FSW spot welding: preliminary results", *Friction Stir Welding and Processing III,* TMS, 2005.

11. Kakarla, S.T., Muci-Kuchler, K.H., Arbegast, W.J., Allen, C.D., "Three dimensional finite element model of the friction stir spot welding process', *Friction Stir Welding and Processing III,* TMS, 2005.

12. ABAQUS v6.5 Documentation.

13. Wen, Q., Guo,Y.B., Todd, B.A., "An adaptive FEA method to predict surface quality in hard machining", *Journal of Materials Processing Technology,* 2006, pp. 21-28.

14. Lesuer, D.R., "Experimental investigations of material models for Ti-6Al-4V Titanium and 2024-T3 Aluminum", DOT/FAA/AR-00/25, 2000.

15. Gerlich, A., Su, P., Bendzsak, G.J., North, T.H., "Tool penetration during friction stir spot welding of Al and Mg alloys", *Journal of Materials Science,* 2005, pp. 6473–6481.

MINIMIZING LACK OF CONSOLIDATION DEFECTS IN FRICTION STIR WELDS

S.K.Chimbli[1], D.J.Medlin[2], W.J.Arbegast[1]

[1]Advanced Materials Processing and Joining Center
[2]Department of Materials and Metallurgical Engineering
South Dakota School of Mines and Technology,
501 East St. Joseph Street, Rapid City, South Dakota, 57701

Keywords: Wormhole Defects, Lack of Consolidation Defects, Forge Force

Abstract

Minimizing lack of consolidation defects in friction stir welds is important to develop maximum material properties within the weld, as well as maintaining consistent material properties along the length of a continuous weld. Process parameters such as travel speed, forge force, rotational speed, heel plunge depth, and tilt angle are important to control the thermal energy in the weld zone and minimize weld zone defects. In this evaluation, several weld process parameters were varied on butt welded 0.250 and 0.125 inch thick 7075-T7351 aluminum alloy panels and the resulting weld zone defects were destructively characterized by traditional metallographic techniques. It was found that welds made with higher heat indexes had fewer and smaller lack of consolidation defects. The most important variable to minimize lack of consolidation defects in this particular investigation was the forge force. The information learned from this evaluation will be used for future non-destructive evaluations of friction stir butt welds.

Introduction

Friction stir welding (FSW) was invented by The Welding Institute (TWI) and involves stirring of the weld zone that maintains a sold state joining of the weld zone instead of a liquid state joining process [1]. In the last decade many scientists and engineers have contributed to the research and development of FSW and have significantly advanced the application of this technology to other related processes such as friction stir processing (FSP) and friction stir joining (FSJ) [2]. Aerospace industries adopted this technology very quickly due to the difficulties associated with welding aluminum alloys with conventional welding techniques (liquid state joining) and gradually the automobile industries started applying FSW. Friction stir welding is a complicated process that involves several solid state processing methods such as deformation, extrusion, and forging [3] and evaluating and understanding the material flow mechanisms that occur during this joining process are critical to improving the weld quality and consistency. High joint strengths, fewer weld zone defects, lower processing costs, no filler materials, improved mechanical properties, low energy consumption, and reduced weld gases or hazardous fumes are some of the advantages of FSW [2].

Wormhole defects, or lack of consolidation defects, are volumetric defects that occur in welds due to the cold processing parameters when compared to liquid state joining processes [4]. The mechanisms that causes a wormhole defect to form is complicated, but the main reason is insufficient metal flow behind the pin tool and the factors for insufficient metal flow include improper pin tool design, high travel speeds, high spindle (pin tool) rotation speed, and insufficient forge force [4]. In an effort to evaluate the formation of lack of consolidation defects, welds were produced by FSW on 0.250" and 0.125" thick Al 7075-T7351 panels and varying the traveling speed, spindle rotational speed, and forge force, and destructively evaluating the quality and strength of the welds.

Experimentation

Friction stir butt welds 5 inches length were produced on 0.250 and 0.125 inch thick Al 7075-T7351 panels. Pin tool "A", Figure-1, was designed to FSW 0.250 inch thick panels and pin tool "B", Figure-2, was designed to FSW 0.125 inch thick panels. These pin tools were selected based on prior research that resulted in weld zone defect formation [5]. These two pin tool designs have a larger pin diameter and smaller shoulder diameter compared to pin tool designs that are typically used.

AMP-250-750

Figure 1. Pin tool "A"
for 0.250" thick plates

AMP-04003

Figure 2. Pin tool "B"
for 0.125" thick plates

Position Control FSW Using Pin Tool "A":

A total of 14 welds were processed on 0.250 inch thick panels using position control (controlling plunge depth of the pin) and changing the travel speed and spindle rotational speed of the pin tool. The first 7 welds were processed with a continuous change in the pin tool tilt angle from 3^0 to 1^0 along the length of the 5 inch weld. The next 7 welds were processed by maintaining a constant tilt angle at 3^0 along the length of the welds. Metallography was performed at four different locations along the length of the welds to evaluate whether the change in tilt angle affected the size of the defects.

Position Control FSW Using Pin Tool "B":

A total of 14 welds were processed on 0.125 inch thick panels using position control and maintaining the tilt angle constant at 3^0 along the length of the weld. Processing welds with a continuous tilt angle change was not performed with this pin tool.

Load Control FSW Using Pin Tool "A":

A total of 18 welds were processed on 0.250 inch thick panels using load control (maintaining a constant load on the pin tool during the weld) and a constant 3^0 tilt angle along the length of the welds. Five welds were processed by maintaining a constant travel speed and a constant spindle rotation speed and changing the forge force from 11,400 to 12,000 lbs. One weld was cut longitudinally along the advancing side of the weld to verify the spacing between the defects in a weld. The remaining 13 welds were processed by changing the travel speed, spindle speed, and forge force based on the data obtained from the position controlled welds previously produced.

<u>Load Control FSW Using Pin Tool "B":</u>
A total of 25 welds were produced on 0.125 inch thick panels using load control and a constant 3^0 tilt angle along the length of the welds. The forge force, travel speed, and spindle rotation speed were selected based on the position control weld data obtained from the previous welds. Based on the data obtained from the first 10 welds, the last 15 welds used higher forge forces in an effort to minimize the formation of wormhole defects.

Results and Discussions

Continuously changing the tilt angle from 3^0 to 1^0 along the length of the weld using pin tool "A" increased the wormhole defect size. Also, as the tilt angle decreased along the weld, the applied forge force decreased and this increases the size of the wormhole defects. Figure-3 shows a typical example of how a decrease in tilt angle increases the wormhole defect size along the length of a weld.

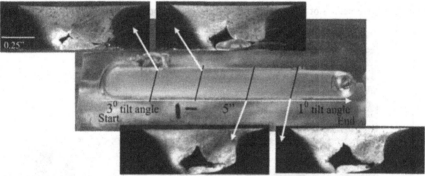

Figure-3. Macrostructures of the weld (35 inches per minute (IPM), 200 revolutions per minute (RPM), position control) showing increase in the size of the wormhole defect along the length of the weld with a change in tilt angle from 3^0 to 1^0.

The welds processed using pin tool "A" also showed that as the travel speed increased, the resulting defect size increased. Small wormhole defects tend to be discontinuous and periodic along the length of the welds, and as the defect size increases, the periodicity decreases to the point to where the wormhole defects are continuous along the length of the weld. If the wormhole defects are large, as shown in Figure-3, then they become a continuous, interconnected pipe or tube along the weld. When the rotational speeds are 200 RPM and the travel speeds greater than 20 IPM, the defects are more likely to form a continuous pipe along the length of the weld, as shown in Table-I. When the spindle speed is 300 RPM and the traveling speed is 25 IPM, the wormhole defects are discontinuous (periodic) along the welds. The smaller discontinuous (periodic) wormhole defects along the weld are termed a lack of consolidation (LOC) defects. The probability of forming a LOC defect is decreased with increased spindle speeds and decreased travel speeds due to increased heat and material flow in the weld zone.

The position control welds processed using pin tool "B" resulted in welds with very, large, continuous, wormhole defects. Consequently, we were not able to develop any LOC defects with these process parameters and pin tool design. The macrostructures of these welds are shown in Table-II.

The load controlled welds processed using pin tool "A" resulted in decreased LOC defect sizes when the forge force was increased and the travel and rotational speeds where held constant. Figure-4 shows how the defect size decreases when the forge force is increased while maintaining constant travel and rotation speeds. For example, when the travel speed is maintained at 25 IPM, the rotation speed is held at 300 RPM, and the forge force is varied from 11,400 lbs to 12,000 lbs, the resulting defect size decreases from 35×10^{-5} in^2 to 0.7×10^{-5} in^2. One of these welds was cut longitudinally along the entire advancing side of the weld to verify the spacing between the LOC defects. This weld was processed at 25 IPM, 300 RPM, and a forge force of 11,400 lbs. The average distance between each LOC defect along the weld was measured to be 0.0833 inches and the weld pitch (curls on the top surface of the weld) was also measured to be 0.0833 inches. The distance between the curls on top surface of the weld can be calculated by taking the ratio of travel speed (IPM) and rotation speed (RPM). Therefore, the average distance between each LOC defect along the length of a weld can be estimated by calculating the ratio of travel speed and rotation speed (see Figure-5).

Table I. Macrostructures of position control welds using pin tool "A" on 0.250 inch thick panels.

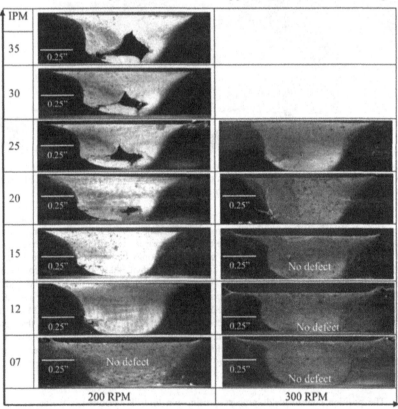

Table II. Macrostructures of the position control welds using pin tool "B" on 0.125 inch thick panels.

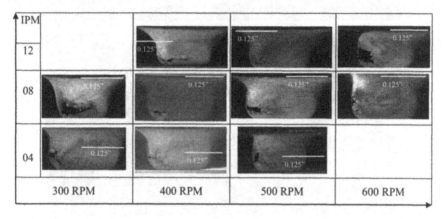

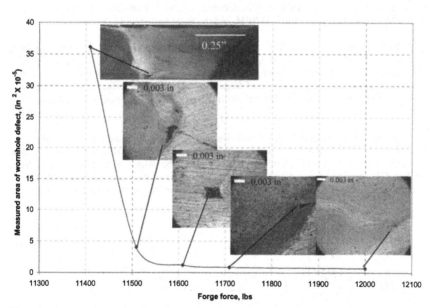

Figure-4. The measured area of wormhole/LOC defects versus forge force when the travel speed and rotation speed are held constant (25 IPM and 300 RPM).

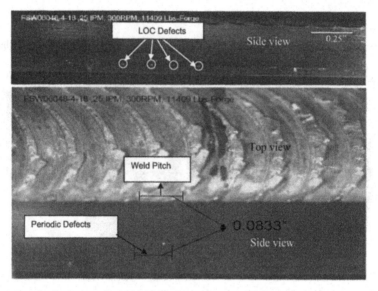

Figure-5. Longitudinal cut weld showing the distance between LOC defects. The travel speed was 25 IPM, the rotation speed was 300 RPM, and the forge force was 11,400 lbs.

The load controlled welds processed using both pin tool designs determined that the forge force was the most important factor to minimize the size of LOC defects. In addition, increasing the spindle rotation speed and decreasing the travel speed resulted in decreased wormhole defect size. Figure-6 illustrates the importance of forge force on wormhole defect size. Increasing the forge force will reduce a large continuous wormhole defect into a small discontinuous LOC defect.

All the weld data using both pin tool designs were used to create a process map with a pseudo heat index (PHI) on abscissa and defect diameter on ordinate. PHI is the measure of heat input in the material during the welding process and can be calculated as a function of forge force, Eqn (1), or plunge depth, Eqn (2) [6,7]. The diameters of the wormhole and LOC defects were calculated from the surface area measurements by assuming that the defects were spherical.

$$PHI_{IIA} = \frac{(RPM)^2}{IPM} \times (Forge) \qquad \text{Eqn (1)}$$

$$PHI_{IC} = \frac{(RPM)^2}{IPM} \times \left(EPL - \left(PD \times cosine(\alpha) \right) \right) \qquad \text{Eqn (2)}$$

Forge = Forge Force (lbs);   EPL = Effective Pin Length (in);   PD = Plunge Depth (in);
α = Tilt Angle (degrees).

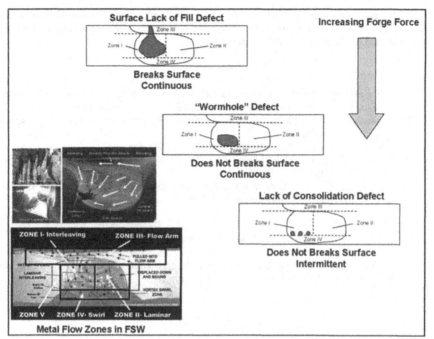

Figure-6. An illustration showing how increased forge force decreases the wormhole defect size.

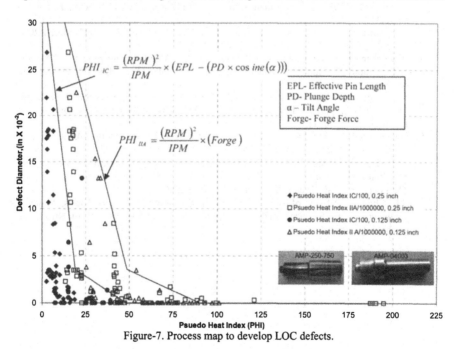

Figure-7. Process map to develop LOC defects.

Figure-7 shows the process map that was developed using all the weld data in this evaluation. The process map shows that when the heat input in the weld zone is high, the wormhole and LOC defect sizes will be minimized.

Conclusions

1) As the tilt angle is continuously decreased from 3° to 1° along a weld, while maintaining a constant travel speed and spindle rotation speed, the defect size increased.
2) As the traveling speed is increased, while maintaining a constant spindle rotation speed, the defect size increased.
3) Increasing in the forge force minimized the LOC defect size.
4) Welds produced with high pseudo heat index values (PHI) will result in discontinuous wormhole defects (LOC defects).
5) LOC defects are periodic in nature and can be estimated by taking the ratio of travel speed (IPM) and spindle rotation speed (RPM).

Acknowledgements

The authors express their gratitude to National Science Foundation (NSF) and Center for Friction Stir Processing (CFSP) for supporting this project.

References

1. W.M. Thomas. et al., "Friction Stir Butt Welding", International Patent App. No. PCT/GB92/02203 and GB Patent App. No. 9125978.8, Dec. 1991, U.S. Patent No. 5,460,317.
2. W.J. Arbegast, "Friction Stir Welding After a Decade of Development", March 2006, "The Welding Journal", vol 85, No. 3, pp 28-35.
3. W.J.Arbegast "Modeling Friction Stir Joining as a Metalworking process", 2003 Hot Deformation of Aluminum Alloys, Z. Jin, ed., TMS (The Minerals, Metals, and Materials Society), 2003.
4. W.J. Arbegast, E.R. Coletta and Z. Li, "Characterization of Friction Stir Weld Defect Types", (Paper presented at TMS 2001, Annual Spring Meeting, New Orleans, LA February 11-15, 2001).
5. Unpublished Research, Advanced Materials Processing Center, South Dakota School of Mines and Technology, Rapid City, South Dakota.
6. W.J. Arbegast, "Using Process Forces as a Statistical Process Control Tool for Friction Stir Welds", *Friction Stir Welding and Processing-III*, Edited by K. V. Jata, et. al. TMS (The Minerals, Metals & Materials Society), 2005.
7. S.K Chimbli, "Effects of Defects in Friction Stir Welds", Masters Thesis, Materials Engineering and Science Graduate Program, South Dakota School of Mines and Technology, 2006.

MICROSTRUCTURES AND MECHANICAL PROPERTIES OF FRICTION-STIR-WELDED AA5754-O AND AA5182-O ALUMINUM TAILOR-WELDED BLANKS

Yen-Lung Chen

General Motors Research and Development Center
MC: 480-106-212, 30500 Mound Road, Warren, MI 48090, USA

Keywords: friction stir welding, tailor-welded blanks, aluminum, microstructure, mechanical properties

Abstract

The microstructures and mechanical properties of friction-stir-welded AA5754-O and AA5182-O aluminum tailor-welded blanks have been characterized. All samples failed in the base metal region except in those cases where there was a severe undercut at the weld toe. Microhardness measurements showed that there was a maximum hardness increase of 16-30% in the thermomechanically affected zone (TMAZ) over the base metal, mainly due to work hardening effect of the friction stir welding process and grain refining effect in the weld nugget. The weld nuggets consisted of small, equiaxed grains, about 8 μm in diameter. In this study the higher hardness of the friction-stir welds than the base metals should help prevent failure in the weld during forming of the tailor-welded blanks. Furthermore, the fine grains in the weld nugget should be beneficial to the ductility of the weld.

Introduction

With increasing demands for improved automobile fuel economy and reduced vehicle emissions, the automobile manufacturers are continually striving to fabricate a lighter vehicle, reduce vehicle air drag, produce a more efficient powertrain, or achieve a combination of the above. The reduction in vehicle weight is done without sacrificing the performance and safety of the vehicle. Aluminum alloys have a specific weight of about one third that of steels, good strength, acceptable formability, and excellent corrosion resistance. Therefore, aluminum is one of the leading candidate materials for the construction of a lightweight vehicle.

To make the most of the lightweight characteristic of aluminum alloys, instead of part-for-part substitution, aluminum tailor-welded blanks (TWB) can be used to further reduce vehicle weight. In addition, TWBs will allow design engineers to use appropriate materials and thicknesses in places where their properties are best utilized in a component. This may also bring about design flexibility and part consolidation. Therefore, the application of TWBs may result in a cost reduction both from material utilization and vehicle assembly considerations.

Although steel TWBs have been in volume production for automotive stamping applications, it appears that aluminum TWB technology still remains in the research and development stage. Fusion welding techniques such as gas tungsten arc welding (GTAW) [1], laser beam welding (LBW) [2-4], non-vacuum electron beam welding (NVEBW) [5] and variable polarity plasma arc welding (VPPAW) [6] have been used to make aluminum TWBs. Common concerns with fusion welds are the presence of porosity and solidification cracks. These defects would reduce weld ductility and thus formability of TWBs.

Friction stir welding (FSW) is a solid-state welding process, invented and patented in 1991 by TWI in the United Kingdom [7-8]. A rotating, profiled tool of harder material than the work pieces being welded makes contact and is plunged into the joint region. The initial plunging friction heats and plasticizes the immediate region of the work pieces. Material flows around the tool and coalesces behind the tool as the rotating tool moves forward.

FSW is an autogenous welding process; it does not require filler metal, shielding gas (for low melting metals such as aluminum and magnesium), and edge preparation. It produces a solid-phase weld with little or low distortion, high mechanical properties and good weld appearance at a relatively low cost [9]. Since it does not involve fusion of the base metals, it does not have the welding defects usually associated with fusion welding: porosity, solidification cracking, loss of magnesium (laser welding for 5xxx alloys), etc. Therefore, FSW welds have better and more consistent mechanical properties than fusion welds. Furthermore, FSW can be used to join a large range of dissimilar materials, especially useful for those materials difficult to join by fusion welding processes [10].

A friction stir weld generally consists of four different regions [10]. At the center of the weld is the "weld nugget", characterized by fine, equiaxed recrystallized grains. This weld nugget is a part of the thermomechanically affected zone, in which the material has been plastically deformed by the FSW tool and also affected by the heat generated from the welding process. Outside the TMAZ is the heat-affected zone (HAZ), in which the material has experienced a thermal cycle. Although no apparent plastic deformation can be detected by light optical microscopy, the microstructure and mechanical properties of HAZ have been modified by the thermal cycle of the FSW process. The rest of the parent material is not affected by the welding process.

Since its invention, FSW research and development has primarily been focused in the areas of thick ($\geq$ 3 mm) aluminum sheet or plate alloys for aerospace, marine and other non-automotive applications [10-11]. Recently, several applications of FSW to the automotive industry have been made. FSW was used to produce suspension links from aluminum extrusions [12-13]. Promising results were also obtained on the friction stir welding and forming of AA6181A-T4 door inner panels (1.5 mm sheet welded to 3.0 mm sheet) [14]. The formability of AA5754-O TWBs and AA6111-T4 TWBs (1 mm sheet joined to 2 mm sheet in both cases), made by using LBW, NVEBW, FSW and GTAW processes, was investigated for USAMP (United States Automotive Materials Partnership) [15]. The objective of this research was to produce differential-gauge aluminum TWBs of AA5754 and AA5182 alloys by the FSW process to evaluate their mechanical properties and microstructures for lightweight automotive vehicle applications.

Experimental Procedure

AA5754-O and AA5182-O aluminum sheet materials, 1.0 to 3.1 mm thick, were cut into 152 mm x 305 mm (6" x 12") weld coupons. The chemical composition of these alloys were determined by chemical analysis in the GM R&D Chemical and Environmental Sciences Lab, as shown in Table 1. These coupons were friction stir welded in the as-received condition along the long edge in a butt joint configuration. TWBs, 305 mm x 305 mm (12" x 12"), of four different material and thickness combinations were made: (a) 1.2 mm 5182-O with 1.5 mm 5754-O, (b) 1.5 mm 5754-O with 2.6 mm 5182-O, (c) 1.0 mm 5754-O with 2.1 mm 5754-O and (d) 2.1 mm 5754-O with 3.1 mm 5754-O.

Table I. Chemical Composition of the Alloys Used (wt. %)

Alloy	Mg	Mn	Fe	Si	Cr	Cu	Zn	Ti
1.0 mm 5754	3.1	0.24	0.19	<0.1	<0.01	<0.01	<0.01	0.01
1.5 mm 5754	3.0	0.19	0.15	<0.1	0.02	0.01	0.01	0.01
2.1 mm 5754	3.1	0.25	0.18	<0.1	0.01	0.01	0.01	0.01
3.1 mm 5754	3.2	0.23	0.17	<0.1	0.01	<0.01	0.01	0.01
1.2 mm 5182	4.6	0.32	0.21	<0.1	<0.01	<0.01	<0.01	0.01
2.6 mm 5182	4.0	0.23	0.23	<0.1	<0.01	0.01	<0.01	0.03

Tensile test specimen blanks, 19 mm x 203 mm (3/4" x 8"), were sheared from these TWBs and machined into standard flat tensile test specimens [16]. At least three specimens for each sample configuration were tested. The weld was placed at the center of each tensile test specimen with the welding direction perpendicular to the tensile axis. Shims of appropriate gauges were used in the grips during testing to minimize bending stresses. All tests were conducted using a crosshead speed of 5 mm/min. Tensile tests of the base materials were performed similarly.

Cross-sectional samples of the welds were prepared for macro- and micro-structural examinations using standard metallographic practices. Average grain size was determined by using the line intercept method. Microhardness values were measured across the weld by using a microhardness tester. For all the samples examined, microhardness measurements were made progressing from the thin sheet toward the thick sheet. Electron probe microanalysis was used to examine solute distribution in the samples.

Results and Discussion

Tensile Properties

Tensile properties and average grain size of the base metals are given in Table 2, where σ_y is the 0.2% offset yield strength, σ_u is the ultimate tensile strength, ε, the elongation, and d, the average grain size. Table 3 lists the ultimate tensile strength and elongation of the welded blanks, along with the fracture locations in the tensile test specimens. All tensile test specimens failed in the base metal region except in those cases where there was a severe undercut at the weld toe, at which the sample failed (e.g., Specimen C4 in Figure 1). When the severe undercut was removed by grinding, the tensile specimens again failed in the base metal.

The ultimate tensile strengths of the TWB samples agreed with those of the corresponding base metals; however, the elongation values of the TWB specimens were less than those of the respective base metals. This is attributed to the fact that most of the deformation in a tensile test specimen occurred in the thinner metal during testing, due to the higher stresses existing in the thinner metal for the same load. The fact that Sample A failed in the 1.2 mm 5182 base metal region, despite the same load-carrying capacity on both sides of the weld, was probably due to the same reason mentioned above.

Weld Structure

The cross-sectional macrostructures of the welds are shown in Figure 2. The thickness transition from the thicker sheet down to the thinner sheet was gradual and fairly smooth as opposed to the abrupt change in thickness in TWBs welded from the backside. This smooth transition would minimize the stress concentration effect on the weldment due to the change in cross-sectional area.

The microstructures of all the samples studied were similar. Figure 3 shows the typical microstructural features of the welds. The average grain size of the base metals varied from 20 to 30 μm (Table 2); however, in the weld nugget, the recrystallized equiaxed grains were about 8 μm in size (Figure 3d). The large difference in grain size between the weld nugget and the base metal is quite evident. As mentioned before, the fine, equiaxed grains in the weld nugget resulted from the recrystallization process that occurred during friction stir welding. Surrounding the weld nugget is part of the thermomechanically affected zone (TMAZ), characterized by the curved flow lines of grains (Figures 3a-3c). Note that the weld nugget is part of TMAZ (see Figure 2).

Table II. Tensile Properties and Average Grain Size of the Base Metals

Alloy	σ_y (MPa)	σ_u (MPa)	ε (%)	d (μm)
1.0 mm 5754-O	100	227	20.6	30
1.5 mm 5754-O	104	226	21.5	20
2.1 mm 5754-O	100	224	24.8	26
3.1 mm 5754-O	104	233	24.6	27
1.2 mm 5182-O	120	282	23.9	24
2.6 mm 5182-O	110	260	22.4	23

Table III. Tensile Properties of the Tailor-Welded Blanks

Sample	σ_u (MPa)	ε (%)	Fracture Location
A (1.2 mm 5182-O/1.5 mm 5754-O)	274	8.1	1.2 mm 5182-O
B (1.5 mm 5754-O/2.6 mm 5182-O)	227	11.8	1.5 mm 5754-O
C (1.0 mm 5754-O/2.1 mm 5754-O)	224	7.6	1.0 mm 5754-O
D (2.1 mm 5754-O/3.1 mm 5754-O)	226	14.2	2.1 mm 5754-O

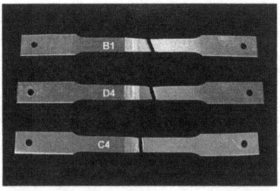

Figure 1.   Photograph showing three tensile specimens after having been tested. Specimens B1 and D4 failed in the base metal whereas specimen C4 failed at the undercut of the weld.

146

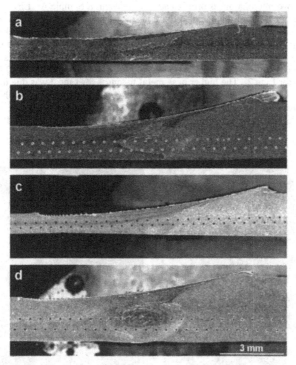

Figure 2.   Photographs showing the cross-sectional macrostructures of welds from Blank A (a), Blank B (b), Blank C (c) and Blank D (d).   The arrays of small dots are microhardness indentations.

On the top surface of the welds, there was a thin layer of material, 50-150 μm thick, consisting of very small 2-5 μm grains. The microstructure in this layer is determined by the rubbing action of the rear face of the tool shoulder [4]. This layer of material is part of the TMAZ.

Figure 4 shows the magnesium x-ray maps of Samples A-D. It can be seen from these pictures that all the aluminum materials had a lower magnesium content in the mid-thickness region. This difference in magnesium concentration would be expected to cause hardness value to vary across the thickness of sheet, as shown in Table 4. In Figure 4a, the thin sheet on the left is 1.2 mm 5182-O aluminum, and the thick sheet on the right is 1.5 mm 5754-O material. From the distribution of magnesium across the weld, it is quite evident that there was large scale mass transport occurring during friction stir welding of these samples. Magnesium differential in this case served as a very effective internal marker to show material flow during the FSW process.

<u>Hardness Profiles</u>

Figures 5-8 show the microhardness profiles of the weld samples. At least two hardness profiles across the weld were determined for each specimen, one at about the centerline of the thin piece and another about one quarter to one third of the sheet thickness from the bottom surface of the thin sheet (Figure 2).

147

Figure 5 shows two microhardness profiles across the weld in Sample A. Line 1 was measured approximately along the mid-thickness of the thin sheet while Line 2 was measured along a line located at about one quarter sheet thickness from the bottom surface of the thin sheet (Figure 2a). From Figure 5, it is seen that the FSW operation caused a hardness increase in both the thin (1.2 mm 5182-O) and thick (1.5 mm 5754-O) sheets across the weld. On the left side of the weld nugget, a maximum hardness increase from that of the base metal was about 10 HV for Line 1, and about 6 HV for Line 2. The higher base metal hardness for Line 2 was due to the fact that it was closer to the bottom surface of the sheet, where the magnesium content was higher than Line 1. In fact, the maximum magnesium concentration occurred between about one quarter and one third of sheet thickness from the surface, as can be clearly seen in Figs. 4a and 4b in the 1.5 mm 5754 sheet.

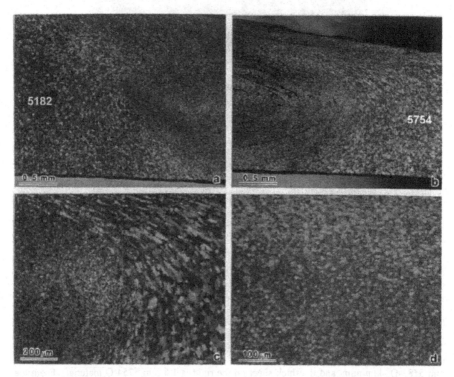

Figure 3. Microstructures of Sample B: (a) TMAZ region on the 5182 side, area including part of the weld nugget, (b) TMAZ on the 5754 side, area including part of the weld nugget, (c) a higher magnification micrograph of (b) showing more clearly the weld nugget and deformed grains in TMAZ, (d) a higher magnification micrograph showing the fine, equiaxed recrystallized grains (~ 8 μm) in the center of the weld nugget.

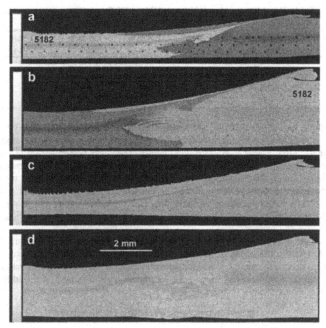

Figure 4. Magnesium x-ray maps of Sample A (a), Sample B (b), Sample C (c) and Sample D
(d). Note that for Samples C and D, both thin and thick sheets were AA5754-O
materials (see Table 3).

Microhardness Profiles of Sample A

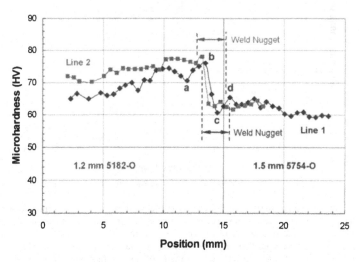

Figure 5. Microhardness profiles across a weld in Sample A.

149

Table IV. Microhardness Values (HV) at Different Thickness Locations

Location	1.2 mm 5182-O	2.6 mm 5182-O	1.0 mm 5754-O	1.5 mm 5754-O	2.1 mm 5754-O	3.1 mm 5754-O
~ Top	74	65	61	64	64	63
~ Center	68	63	57	55	56	57
~ Bottom	74	68	61	64	65	63

The big drop in hardness across the weld nugget from the 1.2 mm 5182 to the 1.5 mm 5754 was due to the large difference in magnesium content between the two aluminum materials (see Table 1 and Table 4). On the right side of the weld nugget, there was no significant change in hardness along Line 2; however, there was a maximum hardness increase of 5 HV along Line 1 in this region. Along Line 1, the hardness drop in the vicinity of Point a and Point c was mainly due to a lower magnesium content (4.0% at Point a versus 4.6% at Point b; 2.8% at Point c versus 3.2% at Point d). The larger hardness increase for Line 1 on both sides of the weld nugget was attributed to more severe plastic deformation, thus more strain hardening (see Figure 2).

The width of the weld nugget in Sample A was estimated from its microstructure to be about 2.5 mm. For estimating the width of TMAZ, the range of observable microstructural change (i.e., grain flow or grain deformation) was considerably smaller than that can be estimated from the corresponding microhardness profiles. As estimated from Figure 5, the width of TMAZ in Sample A was about 14 mm.

Microhardness Profiles of Sample B

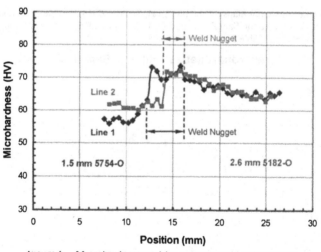

Figure 6. Microhardness profiles across a weld in Sample B.

Figure 6 shows two microhardness profiles across the weld in Sample B. Line 1 was measured approximately along the mid-thickness of the thin sheet, while Line 2 was measured along a line positioned at about one quarter sheet thickness above the bottom surface of the same sheet. From Figure 6, it can be seen that the peak hardness of the weld nugget (72-74 HV) was 16-18 HV (about 30%) higher than that in the base metal of the 1.5 mm 5754-O sheet and 9-11 HV (about 16%) harder than that of the 2.6 mm 5182-O material. For Line 2, the hardness had little change in the TMAZ next to the left side of weld nugget; however, there was a hardness increase of about 8 HV in the TMAZ on the right side of the weld nugget. On the left side of the weld nugget, the hardness of the base metal was higher along Line 2 (61-63 HV) than Line 1 (56-57 HV) due to the higher magnesium concentration along Line 2 than Line 1 (see Figure 4b).

From Figure 4b, the mixing pattern of the AA5754 and AA5182 materials can be clearly seen in the weld nugget. The plastic deformation in the TMAZ region surrounding the weld nugget can also be distinguished in this magnesium map. It is also obvious that the magnesium distribution across the thickness of the 5182 sheet was much more uniform than the 5754 sheet. Therefore, there was no significant difference between Lines 1 and 2 in hardness in the 5182 base metal region on the right side of the weld nugget (Figure 6). Within the weld nugget, the variation in hardness values was mainly due to the differences in local magnesium content, especially for Line 1.

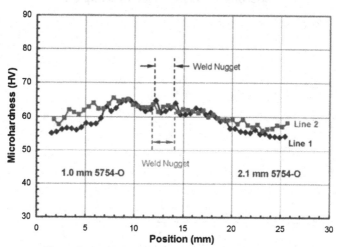

Figure 7. Microhardness profiles across a weld in Sample C.

Sample C was made by friction stir welding one 1.0 mm 5754-O sheet to one 2.1 mm 5754-O sheet of almost identical overall chemical composition (see Table 1). As can be seen in Figure 4c, the magnesium content in the center portion of the two base metals was also lower than the regions closer to the top and bottom surfaces. Figure 7 shows two microhardness profiles across the weld in Sample C. Line 1 was measured approximately along the mid-thickness of the 1.0 mm sheet, whereas Line 2 was located at about one third sheet thickness from the bottom surface of the same sheet. The peak hardness of the weld nugget was 65-66 HV, about 16% higher than the base metal. The hardness fluctuations in the TMAZ, including the weld nugget, are also attributed mainly to variations in the local magnesium content. As discussed before, the hardness increase in the TMAZ was due to the work hardening effect (grain refining effect in the weld nugget) of the friction stir welding operation.

Similar to Sample C, Sample D was also manufactured by friction stir welding two 5754-O aluminum sheets of almost identical overall chemical composition, one 2.1 mm sheet welded to one 3.1 mm sheet. Figure 8 shows two microhardness profiles across the weld in Sample D. Line 1 was measured approximately along the mid-thickness of the 2.1 mm sheet. Line 2 was located about one quarter sheet thickness from the bottom surface of the 2.1 mm sheet and about

151

one sixth sheet thickness from the bottom surface of the 3.1 mm sheet. The peak hardness in the TMAZ was considerably higher than that of the base metals, 27% higher for Line 1 and 16% higher for Line 2. The decrease in hardness within the weld nugget again is attributed to the lower local magnesium content.

The formability of a tailor-welded blank is determined by the mechanical properties of base metals, base metal gauge differential, and the stresses being imposed onto the weld. The formability of the weld in a TWB is related to the weld hardness and weld geometry [18]. The strength of an aluminum weld made by a fusion welding process generally is no better than the base material. In this study the higher hardness of the friction-stir welds than the base metals should help prevent failure in the weld during forming of the tailor-welded blanks. Furthermore, the fine grains in the weld nugget should be beneficial to the ductility of the weld. Formability tests (e.g., limiting dome height test) need to be conducted to substantiate the above points.

Microhardness Profiles of Sample D

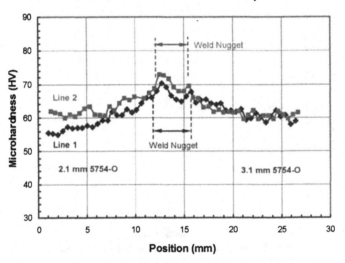

Figure 8. Microhardness profiles across a weld in Sample D.

Summary and Conclusions

1. The mechanical properties and microstructures of tailor-welded blanks of the following combinations of aluminum sheet alloys have been studied: (a) 1.2 mm 5182-O with 1.5 mm 5754-O, (b) 1.5 mm 5754-O with 2.6 mm 5182-O, (c) 1.0 mm 5754-O with 2.1 mm 5754-O, and (d) 2.1 mm 5754-O with 3.1 mm 5754-O.
2. All samples failed in the base metal region except in those cases where there was a severe undercut at the weld toe.
3. The weld nuggets consisted of small, equiaxed, recrystallized grains about 8 μm in diameter, whereas the grain size in the base metals was about 20-30 μm. Within the top surface layer, 50-150 μm thick, the grain size was very small, 2-5 μm in diameter.
4. Microhardness measurements showed that there was a maximum hardness increase of 16-30% in the thermomechanically affected zone (TMAZ) over the base metal, mainly due to the work hardening effect of the friction stir welding process and the grain refining effect in the weld nugget.

5. Hardness in the regions near the top and bottom surfaces was higher than that in the central region of the aluminum sheet alloys studied. This was attributed to the higher magnesium content in these regions than the mid-sections.
6. The hardness fluctuations in the weld nugget and TMAZ were attributed to the local variations in magnesium content resulted from the material transport action of the friction stir welding process.
7. In this study the higher hardness, therefore higher strength, of the friction-stir welds than the base metals should help prevent failure in the weld during forming of the tailor-welded blanks. Furthermore, the fine grains in the weld nugget should be beneficial to the ductility of the weld.

Acknowledgments

I wish to express my appreciation to the following GM R&D colleagues: Noel M. Potter for his chemical analysis of the aluminum alloys, Richard A. Waldo for his electron probe microanalysis of the welds, Spurgeon M. Willett, Jr. for his assistance in metallographic sample preparation, Roy J. Sexton for his assistance in the tensile tests of base metals and James G. Cross of Ricardo-MEDA for his assistance in microhardness measurements. The aluminum tailor-welded blanks were welded by Friction Stir Link, Inc., Waukesha, Wisconsin.

References

1. E. R. Pickering et al.: SAE Technical Paper No. 950722, SAE International, Warrendale, PA, 1995.
2. R. Mueller, H. Gu and N. Ferguson: SAE Technical Paper No. 1999-01-3148, SAE International, Warrendale, PA, 1999.
3. S. Venkat et al.: Welding Journal, 1997, vol. 76, pp. 275-s – 282-s.
4. R. P. Martukanitz et al.: SAE Technical Paper No. 960168, SAE International, Warrendale, PA, 1996.
5. P. Martin et al.: Proc. Int. Symp. Light Metals, M. Sahoo and C. Fradet, eds., The Metallurgical Society of CIM,1998, pp. 409-423.
6. E. Mayne: WARD's Auto World, February 2001, p. 63.
7. W. M. Thomas et al.: GB Patent Application No. 9125978.8, December, 1991, International Patent Application No. PCT/GB92/02203, June 10, 1993.
8. W. M. Thomas et al.: U. S. Patent No. 5460317, 1993.
9. W, M. Thomas and E. D. Nicholas: *Materials and Design*, 1997, vol. 18, pp. 269-273.
10. P. L. Threadgill and A. J. Leonard: TWI Members Research Report No. 693, 1999.
11. C. G. Rhodes et al.: *Scripta. Mater.*, 1997, vol. 36, pp. 69-75.
12. C. B. Smith: *Proc. 2nd Int. Symp. Friction Stir Welding*, Gothenburg, Sweden, 2000, Session 1, Paper 2.
13. M. Enomoto: *Proc. 3rd Int. Symp. Friction Stir Welding*, Kobe, Japan, 2001, Session 8, Paper 2.
14. J. Gould et al.: EWI/TWI Document 40700CPQ/FR/99, 1999.
15. M. J. Worswick et al.: "Formability of Multi-Gauge Aluminum Alloy Sheet Tailor Welded Blanks," a report to the United States Automotive Materials Partnership, ALCAN Aluminum Corporation, Aluminum Company of America and Reynolds Metal Company, August 29, 1999.
16. The American Society for Testing and Materials, "B 557-94, Standard Test Methods of Tension Testing Wrought and Cast Aluminum- and Magnesium-Alloy Products," 1994.
17. B. London et al.: *Friction Stir Welding and Processing II*, K. V. Jata, M. W. Mahoney, R. S. Mishra, S. L. Semiatin and T. Lienert, eds., TMS, Warrendale, PA, 2003, pp. 3-12.
18. W. Waddell, S. Jackson and E. R. Wallach, SAE Technical Paper No. 982396, SAE International, Warrendale, PA, 1998.

FORMABILITY AND SPRINGBACK EVALUATION OF FRICTION STIR WELDED AUTOMOTIVE SHEETS

Wonoh Lee[1], Daeyong Kim[2], Junehyung Kim[1], Chongmin Kim[3],
Michael L. Wenner[4], Kazutaka Okamoto[5], R.H. Wagoner[6] and Kwansoo Chung[1]

[1]Department of Materials Science and Engineering; Intelligent Textile System Research Center;
Seoul National University; 56-1 Shinlim-dong; Kwanak-gu; Seoul 151-742; South Korea
[2]Corporate Research & Development Division; Hyundai-Kia Motors;
772-1 Jangduk-dong; Hwaseong-si; Gyeonggi-do 445-706; South Korea
[3]Materials & Processes Lab.; GM R&D and Planning, General Motors Corporation;
Warren; MI 48090-9055; USA
[4]Manufacturing Systems Research Lab.; GM R&D and Planning, General Motors Corporation;
Warren; MI 48090-9055; USA
[5]Research & Development Division; Hitachi America Ltd.;
34500 Grand River Ave.; Farmington Hills; MI 48335; USA
[6]Department of Materials Science & Engineering; Ohio State University;
2041 College Road; Columbus; OH 43210; USA

Keywords: Friction stir welding, Combined isotropic-kinematic hardening,
Anisotropic yield functions, FLD, Formability, Springback

Abstract

Formability and springback of automotive friction stir welded TWB (tailor welded blank) sheets were experimentally and numerically investigated for aluminum alloy 6111-T4, 5083-H18 and dual-phase steel, DP 590 sheets, which were friction-stir welded with the same and different thicknesses. To represent mechanical properties, anisotropic yield function Yld2004-18p and the combined isotropic-kinematic hardening law were utilized. In order to examine formability, deformation until failure in three tests (the simple tension test with various weld line directions, the hemisphere dome stretching test and the cylindrical cup drawing test) were measured and simulated. For springback, two types of weld line combinations (longitudinal and transverse TWBs), were prepared for three springback tests (the unconstrained cylindrical bending, 2-D draw bending and OSU draw-bend tests). The numerical code performed reasonably well in analyzing all verification tests, while the magnitude of springback was largely dependent on the ratio of the yield strength with respect to Young's modulus and thickness.

Introduction

A newly emerging and highly promising welding technology is friction stir welding (FSW), which was developed primarily for aluminum alloys in 1991 by the Welding Institute (TWI), in Cambridge, U.K. [1]. As a solid-state welding, FSW has various advantages over conventional fusion welding techniques such as its low capital investment, extremely low energy use and its capability to weld very thick plates with little or no porosity. Joining in FSW is achieved by heat and material flow generated by the FSW tool, which rotates as it moves along the joint interface [2]. The solid state nature of joining often induces minimal residual stress and distortion [3], retaining the dimensional stability of joined elements. Initially developed for the aerospace industry, the friction stir welding has been further applied for aluminum structures including

155

decks for ferry boats and fuel/oxidizer tanks for space launch vehicles [3] and, most recently, for the automotive industry.

Here, a systematic study of the macro-performance of friction stir welded automotive sheets is the main scope of the work. In particular, to develop a numerical technique to simulate forming of friction stir welded TWB automotive sheets in conjunction with formability and springback is the objective of the work. For the purpose of accurate prediction, the combined isotropic-kinematic hardening law based on the modified Chaboche model [4,5] with the anisotropic yield function, Yld2004-18p [6] was utilized. The constitutive laws have been implemented into both ABAQUS standard/explicit codes using the user-subroutines.

In this work, three automotive sheets, AA6111-T4 and AA5083-H18 aluminum alloys and DP 590 sheets, each having two different thicknesses, are considered. The base sheets with the same and different thicknesses were friction-stir welded for tailor-welded blank samples. In order to characterize mechanical properties, the uni-axial tension, hydraulic bulge and disk compression tests have been performed for anisotropic yield properties. The hardening behavior has been measured using the uni-axial tests, while, for the Bauschinger and transient behavior during unloading, uni-axial tension/compression and uni-axial compression/tension tests have been performed by providing guides along the both sides of the sheet specimen on which clamping forces are applied to prevent buckling. The hardening properties of the weld zones have been calculated using the rule of mixture and also directly measured using subsized specimens machined from the weld zone. Forming limit diagrams have been measured using hemispherical dome stretching tests for the base materials and those of the weld zones have been calculated based on Hill's bifurcation theory and the M-K theory.

To numerically predict formability, simulations have been performed for the FSW sheets and results have been compared with experiments for three cases with gradual complexity: the simple tension test with various weld line directions, the hemisphere dome stretching test and the cylindrical cup deep drawing test over a cylindrical punch. In order to validate the springback prediction capability, a comparison of simulation and experimental results has been performed for three springback tests: unconstrained cylindrical bending, 2-D draw bending and OSU draw-bend tests. The unconstrained cylindrical bending does not involve blank holders, therefore, the experiment mainly shows the material difference effect but not the process parameter effect. The 2-D draw bending test involves two dimensional blank holders so that it would show the effects of the material difference as well as process parameters. Furthermore, the sensitivity of springback for the bend/unbend operation can be examined in the 2-D draw bending test. Recently, the OSU draw-bend test has been developed to study springback under the direct control of process parameters such as the friction coefficient, sheet tension and tool radius [7,8].

Theory

3-D Yield Stress Function: Yld2004-18p

In order to describe the initial anisotropic yield stress surface, a yield stress function recently proposed by Barlat et al. [6] was considered: Yld2004-18p. Sixteen mechanical measurements to represent orthogonal anisotropy are typically σ_0, σ_{15}, σ_{30}, σ_{45}, σ_{60}, σ_{75}, σ_{90}, R_0, R_{15}, R_{30}, R_{45}, R_{60}, R_{75}, R_{90}, σ_b and R_b, which are simple tension yield stresses and R-values at every $15°$ off the rolling direction, and yield stress and in-plane principal strain ratio under the balanced biaxial tension condition, respectively.

The yield function, Yld2004-18p, is defined in the 3-D space as the following specific form of Eq. (1); i.e., for the effective stress $\bar{\sigma}$,

$$f^{1/M} = \{\Phi/4\}^{1/M} = \bar{\sigma} \tag{1}$$

where

$$\Phi = \left|\tilde{S}'_I - \tilde{S}''_I\right|^M + \left|\tilde{S}'_I - \tilde{S}''_{II}\right|^M + \left|\tilde{S}'_I - \tilde{S}''_{III}\right|^M + \left|\tilde{S}'_{II} - \tilde{S}''_I\right|^M + \left|\tilde{S}'_{II} - \tilde{S}''_{II}\right|^M + \left|\tilde{S}'_{II} - \tilde{S}''_{III}\right|^M$$
$$+ \left|\tilde{S}'_{III} - \tilde{S}''_I\right|^M + \left|\tilde{S}'_{III} - \tilde{S}''_{II}\right|^M + \left|\tilde{S}'_{III} - \tilde{S}''_{III}\right|^M . \tag{2}$$

In Eq. (2), $\tilde{S}'_k$ and $\tilde{S}''_k$ ($k = I \sim III$) are the principal vales of tensor $\tilde{s}'$ and $\tilde{s}''$, which are deviatoric stresses modified by linear transformations, respectively. The associated linear transformations on the stress deviator provide eighteen anisotropic coefficients to represent orthotropic anisotropy. Based on crystal plasticity, the exponent M is recommended to be 6 and 8 for BCC and FCC materials, respectively. For calculations under the plane stress condition, the reduced form for the plane stress condition of Yld2004-18p was utilized.

Combined Isotropic-kinematic Hardening Law

The combined type isotropic-kinematic hardening constitutive law based on the modified Chaboche model is given by

$$f(\boldsymbol{\sigma} - \boldsymbol{\alpha}) - \bar{\sigma}_{iso}^M = 0 \tag{3}$$

where $\boldsymbol{\alpha}$ is the back stress for the kinematic hardening and the effective stress, $\bar{\sigma}_{iso}$, is the size of the yield surface as a function of the accumulative effective strain. In the Chaboche model, the back-stress increment is composed of two terms, $d\boldsymbol{\alpha} = d\boldsymbol{\alpha}_1 - d\boldsymbol{\alpha}_2$ to differentiate transient hardening behaviors during loading and reverse loading. To complete the constitutive law, hardening behavior describing back-stress movements and the change in yield surface size should be provided for $d\bar{\alpha}_1$, $d\alpha_2$ and $d\bar{\sigma}_{iso}$, respectively.

In order to describe the permanent softening in reverse loading, hardening parameter, $h_1 (= d\bar{\alpha}_1/d\varepsilon)$, was modified by introducing the softening parameter, ξ; i.e.,

$$h_1^s = h_1 \cdot \{\xi(\bar{\varepsilon}^*)\}^n \quad \text{when } n < m \tag{4}$$

where $\bar{\varepsilon}^*$ is the accumulative effective strain measured during the n-th reverse loading and $0.0 < \xi(\bar{\varepsilon}^*) \le 1.0$, while $\xi(\bar{\varepsilon}^* = 0.0) = 1.0$. Here, the constant m is introduced for the case that permanent softening does not occur after any specific number of reverse loading m. The softening parameter was considered as

$$\xi = a_s + b_s \exp(-c_s \bar{\varepsilon}^*) \tag{5}$$

where a_s, b_s, c_s are values dependent on the total accumulative effective strain during the previous reverse loading, $\bar{\varepsilon}_{pre}^*$. The values of a_s, b_s, c_s were parameterized as

$$a_s = a_s^1 + a_s^2 \exp(-a_s^3 \bar{\varepsilon}_{pre}^*), \ b_s = b_s^1\left(1 - \exp(-b_s^2 \bar{\varepsilon}_{pre}^*)\right), \ c_s = c_s^1\left(1 - \exp(-c_s^2 \bar{\varepsilon}_{pre}^*)\right). \tag{6}$$

Details on the combined isotropic-kinematic hardening law with the consideration of softening behavior are referred to previous works [9,10].

Numerical Formulation for Stress Update

For the numerical formulation for large deformation, the incremental deformation theory [11] was applied to the elasto-plastic formulation. Under this scheme, the strain increment in the flow formulation becomes the discrete true strain increment while a material rotates by incremental rotation obtained from the polar decomposition at each discrete step.

As for the stress update scheme, the updated stress is initially assumed to be elastic for a given discrete strain increment $\Delta\varepsilon$. Therefore,

$$\sigma_{n+1}^T = \sigma_n + \mathbf{C} \cdot \Delta\varepsilon, \ \bar{\varepsilon}_{n+1}^T = \bar{\varepsilon}_n, \ \alpha_{n+1}^T = \alpha_n. \tag{7}$$

where the superscript 'T' stands for a trial state and the subscript denotes the process time step. If the following yield condition is satisfied with trial values for a prescribed elastic tolerance Tol^e;

$$f\ (\sigma_{n+1}^T - \alpha_{n+1}^T) - \bar{\sigma}_{iso}(\bar{\varepsilon}_{n+1}^T) < Tol^e \tag{8}$$

then the process time step $n+1$ is considered elastic. However, if the above yield condition is violated, the step is considered elasto-plastic and the trial elastic stress state is taken as an initial value for the solution of the plastic corrector problem. The nonlinear equation to solve for $\Delta\bar{\varepsilon}$, which enables resulting stresses to stay on the hardening curves at the new step $n+1$, is

$$f\ (\sigma_n - \alpha_n + \Delta\sigma - \Delta\alpha_1 + \Delta\alpha_2) = \bar{\sigma}_{iso}(\bar{\varepsilon}_n + \Delta\bar{\varepsilon}). \tag{9}$$

The predictor-corrector scheme based on the Newton-Raphson method was used to solve Eq. (9). After obtaining the converged solution of Eq. (9), the stresses, back stresses and equivalent plastic strain are updated for the next step.

Forming Limit Diagram

Forming limit diagrams were measured using hemispherical dome stretching tests. However, when their measurements were not available, they were calculated utilizing Hill's bifurcation theory [12] and the Marciniak-Kuczynski (M-K) theory [13] based on rigid-plasticity with isotropic hardening for simplicity. The forming limit criterion based on Hill's bifurcation theory was applied to the strain field whose minor strain is non-positive and the M-K theory was applied to the strain field whose minor strain is positive.

Material Characterization

Three automotive sheets were considered in this work: AA6111-T4 sheets (with 1.5t and 2.6t), AA5083-H18 sheets (with 1.2t and 1.6t) and DP 590 sheets (with 1.5t and 2.0t). These sheets were friction-stir welded along the rolling direction. In this work, same materials with same thickness (similar gauges, SG) and different thickness (dissimilar gauges, DG) were welded together. Whole results are discussed elsewhere [9] and results are selectively discussed here.

For base materials, anisotropic plastic behavior was characterized using uni-axial and balanced biaxial tension tests as well as the disk compression test. These results were subsequently used to calculate three-dimensional yield surface shapes based on the yield function, Yld2004-18p. Yield characteristics were selectively shown in Figure 1 for AA6111-T4 (1.5t) material. In this work, the weld zone was assumed to have the isotropic property. Therefore, all anisotropic coefficients of Yld2004-18p were chosen to be 1.0. Yield function exponents M were chosen to be the same as those of their base materials.

In order to measure isotropic-kinematic hardening behavior, uni-axial tension-compression and uni-axial compression-tension tests were performed. DP 590 base sheets and AA5083-H18 (SG) showed softening behavior and the hardening behavior calculated with the softening parameters for DP 590 (1.5t) base sheets are selectively plotted in Figure 2, which confirms that the modified Chaboche model with softening parameters well represents permanent softening as well as Bauschinger and transient behaviors.

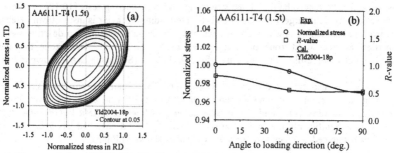

Figure 1. Characteristics of Yld2004-18p for AA6111-T4 (1.5t): (a) yield surface contour and (b) anisotropies of normalized stress and R-value.

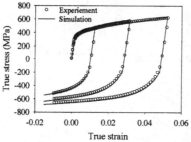

Figure 2. Calculated and measured hardening behavior in tension-compression tests with softening for the DP 590 (1.5t) base material.

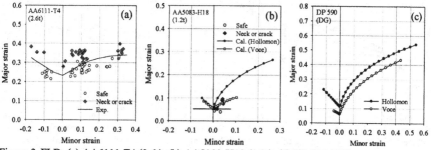

Figure 3. FLD: (a) AA6111-T4 (2.6t), (b) AA5083-H18 (1.2t), (c) DP 590 (DG).

The hemispherical dome stretching test was carried out to obtain FLD of base materials. FLD of AA6111-T4 and DP 590 base sheets were obtained as selectively shown in Figure 3(a). AA5083-H18 samples were so brittle that FLD was successfully measured only near the plane stain condition. Since there is not enough data for this material, forming limit strains were also calculated using the M-K theory as shown in Figure 3(b). Also note that because of difficulty in performing measurement, all forming limit diagrams of weld zones were calculated based on Hill's bifurcation and M-K theory. Figure 3(c) selectively shows the calculated FLD of DP 590 (DG) weld zone.

Formability Evaluation

In order to verify the performance of the developed numerical code, the formability of the similar and dissimilar gauge welded sheets was predicted for the uni-axial tension test and also selectively for the hemispherical dome stretching and cup drawing tests [10,12]. In these tests, the stretching mode is largely dominant and reverse loading behavior is insignificant. Hence, the pure isotropic hardening law was utilized without kinematic hardening. To describe weld zone properties, average weld zone properties were used with varying weld zone thickness as measured. Selective results of failure onsets for AA6111-T4 (SG) samples are shown in Figure 4.

Formability was also evaluated between the base and TWB materials based on simple tension hardening behavior. In weld zone properties, the AA6111-T4 weld zone had lower flow stress with reduced ductility compared to the base material, while the AA5083-H18 zone improved ductility with significantly lower flow stress. The DP 590 weld zone had larger flow stress and ductility was reduced compared to the base material as shown in Figure 5.

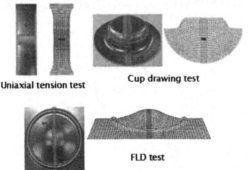

Uniaxial tension test

Cup drawing test

FLD test

Figure 4. Experimental and simulated failure of AA6111-T4 (SG) specimens for formability tests.

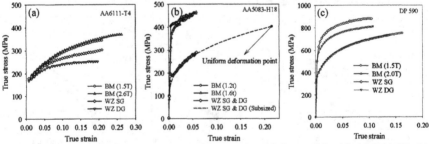

Figure 5. Comparison of hardening curves of base and welded materials: (a) AA6111-T4, (b) AA5083-H18, (c) DP 590

160

Table I. Measured and simulated limit dome height (mm) in hemispherical dome stretching tests

Materials		Experiment		Simulation					
		200×200	200×120	200×200			200×120		
				w FLD		w/o	w FLD		w/o
				Voce	Hol.	FLD	Voce	Hol.	FLD
AA6111	SG	17.9 (±1.6)	25.1 (±1.1)	17.6	17.6	17.5	22.9	24.5	25.6
-T4	DG	12.7 (±2.3)	16.9 (±1.6)	10.5	10.8	11.3	15.5	15.9	16.1
AA5083	SG	-	17.0 (±1.6)	-		-	15.1	15.1	17.9
-H18	DG	-	16.9 (±1.4)	-		-	14.4	14.4	17.7

In the hemispherical dome test, formability of the welded TWB samples improved slightly for AA5083-H18 but reduced for the other two. However, formability in forming processes might be dependent on weld zone line arrangement as well as weld zone ductility. Formability in forming processes is dependent on weld zone line arrangement as well as weld zone ductility: weld zone ductility was more important if the major principal loading direction is aligned with the weld zone line, while the thickness and strength of the weld zone were more important if the major principal loading direction is vertical to the weld zone line, as confirmed in the three verification tests for formability.

Load and displacement for all cases have very good agreement with experimental results, which implies that the FEM simulation can accurately predict strain distribution. As for the onset of failure, numerical prediction based on FLD usually under-predicted failure points, while predictions based on inflection points without FLD usually worked better. Especially for uni-axial tension and cup drawing tests, however, predictions without FLD failed severely for AA5083-H18, which is so brittle. Table I shows experimental and simulated results of limit dome height in hemispherical dome stretching tests. Note that the Hollomon type FLD showed better agreement than the Voce type FLD and this discrepancy was similar to the cup drawing test. Formability tests of AA6111-T4 (SG) also showed that the retreating side within the weld zone might be weaker than the advancing side.

Springback Evaluation

For verification purposes, cylindrical unconstrained bending, 2-D draw bending and OSU draw bending tests have been performed for DP 590 and aluminum alloy 5083-H18 and 6111-T4 sheets, each for base materials with two thicknesses and TWB samples welded longitudinally and transversely with similar and dissimilar gauges, respectively [10,13]. Using obtained material properties, the springback of FSW sheets was numerically investigated for three example problems and results were then compared with experimental results. Simulations were carried out using the average weld zone properties with varying weld zone thickness, which gave most accurate results as verified in formability evaluation. Results are selectively shown in Figures 6-8 for three springback tests.

For the unconstrained bending test, experiment results agreed well with the order of ratios of the yield stress with respect to Young's modulus and thickness ($\Delta \kappa \propto Y/Et$).The springback amount order of base materials for the 2-D draw bending test was very similar to that observed in the unconstrained cylindrical bending test. In order to investigate the role of tension in springback, OSU draw-bend tests were carried out with three or four different back force conditions and only the longitudinal (RD‖RD) type of TWB specimen was considered. As expected, the sheet tension has a clear effect on reducing total springback. Similarly to results from the unconstrained cylindrical bending and the 2-D draw bending tests, springback amount results followed the order in their magnitude determined by Y/Et for welded materials.

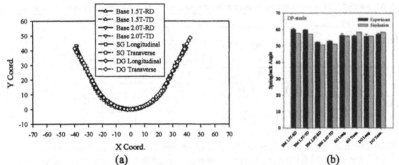

(a)   (b)

Figure 6. (a) Experimental springback profiles and (b) measured/simulated springback angles of DP 590 materials for the unconstrained cylindrical bending test.

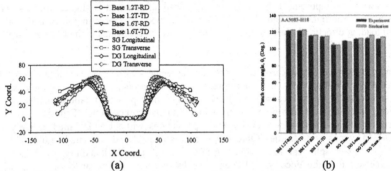

(a)   (b)

Figure 7. (a) Experimental springback profiles and (b) measured/simulated springback angles of AA5083-H18 materials for the 2-D draw bending test.

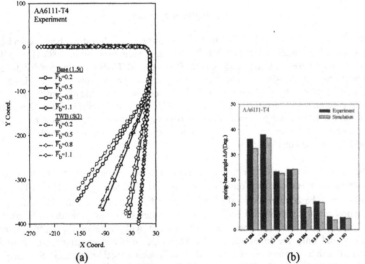

(a)   (b)

Figure 8. (a) Experimental springback profiles and (b) measured/simulated springback angles of AA6111-T4 materials for the OSU draw-bend test.

Conclusions

The numerical method to predict the formability and springback of automotive friction stir welded TWB sheets in sheet forming processes was developed. To represent mechanical properties, anisotropic yield function Yld2004-18p and the combined isotropic-kinematic hardening law were utilized, while the forming limit diagram was considered as a failure criterion. Numerical formulations based on the incremental deformation theory were implemented into ABAQUS commercial codes using user-subroutines. Three automotive sheets, AA6111-T4 and AA5083-H18 aluminum alloy and DP 590 sheets, each having two different thicknesses, were considered. To characterize mechanical properties, the uni-axial tension, hydraulic bulge and disk compression tests were performed for anisotropic yield properties. Hardening behavior was measured using the uni-axial tests, while uni-axial tension/compression and compression/tension tests were performed for the Bauschinger and transient behavior as well as for softening during reverse loading. Hardening properties of weld zones were calculated by using the rule of mixture or by direct measurement using sub-sized specimens machined out of the weld zone. The combined isotropic-kinematic hardening law based on the modified Chaboche model along with the non-quadratic anisotropic yield function Yld2004-18p successfully described the anisotropy and hardening behavior of all material samples including the Bauschinger and transient behavior with/without softening during reverse loading. Softening was observed for DP 590 base materials and the AA5083-H18 similar gauge weld zone. Forming limit diagrams were measured using hemispherical dome stretching tests or were calculated based on Hill's bifurcation theory and the M-K theory.

In order to verify the performance of the numerical codes, failure in three applications including the simple tension test with various weld line directions, the hemisphere dome stretching test and the cylindrical cup drawing test were measured and simulated. Also, to validate the prediction capabilities of the developed numerical code for springback, three springback analysis have been performed: unconstrained cylindrical bending, 2-D draw bending and OSU draw-bend tests. The numerical codes performed reasonably well in analyzing all verification tests for formability and springback. In weld zone properties compared to those of base materials, AA6111-T4 had lower flow stress with reduced ductility, while AA5083-H18 zone improved ductility with significantly lower flow stress. DP 590 had larger flow stress with reduced ductility. Formability in forming processes was dependent on weld zone line arrangement as well as weld zone ductility: weld zone ductility was more important if the major principal loading direction is aligned with the weld zone line, while the thickness and strength of the weld zone were more important if the major principal loading direction is vertical to the weld zone line. As for springback, the order of its amount was mostly dependent on the ratio of the yield strength with respect to Young's modulus and thickness.

Acknowledgement

The work has been performed under the joint project between GM and SNU. The authors greatly appreciate the support of GM. The work has been also partially supported through the SRC/ERC Program of MOST/KOSEF (R11-2005-065).

References

1. Thomas, M.W., Nicholas, E.D., Needham, J.C., Murch, M.G.., Templesmith, and P., Dawes, C.J., GB Patent Applications No. 9125978.8, Dec. 1991; US Patent No. 5460317, Oct. 1995.
2. London, B., Mahoney, M., Bingel, W., Calabrese, M., Bossi, R.H., and Waldron, D., 2003, "Material Flow in Friction Stir Welding Monitored with Al-SiC and Al-W Composite

Markers," *Friction Stir Welding and Processing II*, K.V. Jata et al., eds., A Publication of TMS, Warrendale, PA, pp. 3-10.

3. Reynolds, A.P., and Tang, W., 2003, "Alloy, Tool Geometry, and Process Parameter Effects on Friction Stir Weld Energies and Resultant FSW Joint Properties," *Friction Stir Welding and Processing I*, K.V. Jata et al., eds., A Publication of TMS, Warrendale, PA, pp. 15-23.

4. Chaboche, J.L., 1986, "Time Independent Constitutive Theories for Cyclic Plasticity," Int. J. Plasticity, **2**(2), pp. 149-188.

5. Chung, K. Lee, M.-G., Kim, D., Kim, C., Wenner, M.L., and Barlat, F., 2005, "Spring-back Evaluation of Automotive Sheets Based on Isotropic-kinematic Hardening Laws and Non-quadratic Anisotropic Yield Functions, Part I: Theory and Formulation," Int. J. Plasticity, **21**(5), pp. 861-882.

6. Barlat, F., Aretz, H., Yoon, J.W., Karabin, M.E., Brem, J.C., and Dick, R.E., 2005, "Linear Transformation-based Anisotropic Yield Functions," Int. J. Plasticity, **21**(5), pp. 1009-1039.

7. Carden, W.D., Geng, L.M., Matlock, D.K., and Wagoner, R.H., 2002, "Measurement of Springback," Int. J. Mech. Sci., **44**(1), pp. 79-101.

8. Wang, J.F., Wagoner, R.H., Carden, W.D., Matlock, D.K., and Barlat, F., 2004, "Creep and Anelasticity in the Springback of Aluminum," Int. J. Plasticity, **20**(12), pp. 2209-2232.

9. Kim, D., Lee, W., Kim, J., Kim, C., Wenner, M.L., Okamoto, K., Wagoner, R.H., and Chung, K. 2006, "Material Characterization of Friction Stir Welded TWB Automotive Sheets," J. Eng. Mater. Technol.-Trans. ASME, submitted.

10. Lee, W., Kim, J., Kim, D., Kim, C., Wenner, M.L., and Chung, K, 2006, "Numerical Sheet Forming Simulation of Friction Stir Welded TWB Automotive Sheets," Int. J. Plasticity, submitted.

11. Chung, K., and Richmond, O., 1993, "A Deformation Theory of Plasticity Based on Minimum Work Paths," Int. J. Plasticity, **9**(8), pp. 907-920.

12. Hill, R., 1952, "On Discontinuous Plastic States with Special Reference to Localized Necking in Thin Sheets," J. Mech. Phys. Solids, **1**(1), pp.19-30.

13. Marciniak, Z., and Kuczynski, K., 1967, "Limits Strains in the Processes of Stretch-forming Sheet Metal," Int. J. Mech. Sci., **9**(9), pp. 609-620.

14. Lee, W., Kim, J., Kim, D., Kim, C., Wenner, M.L., Okamoto, K., Wagoner, R.H., and Chung, K, 2006, "Formability Evaluation of Friction Stir Welded TWB Automotive Sheets," J. Eng. Mater. Technol.-Trans. ASME, submitted.

15. Kim, J., Lee, W., Ryou, H., Kim, D., Kim, C., Wenner, M.L., Okamoto, K., Wagoner, R.H., and Chung, K, 2006, "Springback Evaluation of Friction Stir Welded TWB Automotive Sheets," J. Eng. Mater. Technol.-Trans. ASME, submitted.

Formability Analysis of Locally Surface-Modified Al 5052 Sheets

Chang Gil Lee, Suk Hoon Kang*, Heung Nam Han*, Sung-Joon Kim, and Kyu Hwan Oh*

Korea Institute of Machinery & Materials, 66 Sangnam,

Changwon, Kyeongnam, 641-010 Korea

*School of Materials Science and Engineering, Seoul National University,

San 56-1 Shillim 9 dong, Kwankak, Seoul, 151-744 Korea

ABSTRACT

In the present study, the 5052-H32 Al sheets were locally surface-modified by the concept of SFJ (Surface Friction Joining) under the various conditions. Severe shear deformation and the heat generated by friction cause the microstructural change at the surface-modified region, so that the mechanical property in that region also changes subsequently. It is noteworthy that the formability of the surface-modified sheets is greatly improved compared with as-received sheets. The formability is improved as the tool diameter is increased or the modifying speed is decreased, and these results are indicated by LDH (Limited Dome Height) test. Yield and tensile strength of the surface-modified region are lower than those of the base metal, but uniform elongation and strain hardening exponent of the surface-modified region are superior to those of the base metal.

Keywords: surface friction joining, local surface-modification, 5052-H32 Al sheet, formability, pattern quality

* Corresponding author

Tel: +82-55-280-3433, Fax: +82-55-280-3599, Email: cglee@kmail.kimm.re.kr

1. Introduction

Friction stir welding (FSW) developed by TWI is a solid-state joining process, which uses frictional heat and severe plastic deformation generated by a rotating and traversing cylindrical tool with a probe-pin penetrated into the workpiece [1,2]. Recently, Surface Friction Joining (SFJ), using a tool without probe-pin, has been developed by KIMM for joining thin sheets [3,4]. In this process, friction is occurred only at the surface of the workpiece.

During SFJ process, severe shear deformation and the heat generated by friction cause the microstructural change at the joined region, similar to the case of FSW [5,6], so that the

165

mechanical property in that region also changes subsequently. In the present study, formability of locally surface-modified 5052 Al sheets using the concept of SFJ under the various conditions were investigated, and technical possibility of local surface-modification for application to industry is discussed.

2. Experimental procedures

2.1. Local surface-modification

Figure 1 describes the concept of local surface-modification by SFJ. When the rotating tool (marked 2) is contacted with a sheet (marked 1) the frictional heat causes the softening of the material in contacted region, and rotating of the tool generates the plastic flow. As the tool moves forward, the locally surface-modified region (marked 3) is formed. In Figure 1, the region marked 4 indicates the plastic zone in the locally surface-modified region.

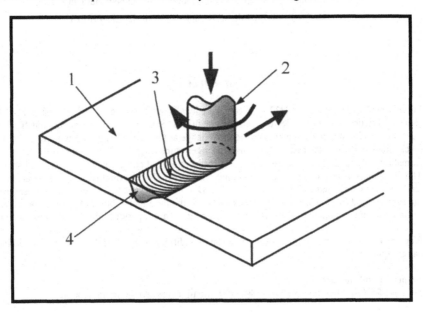

Figure 1. Concept of local surface-modification.

1: sheet, 2: tool, 3: locally surface-modified region

4: plastic zone

In the present study, the 5052-H32 (Al-2.5Mg-0.26Cr-0.13Si-0.33Fe in wt.%) Al sheets (1.5mm thickness) were used. Local surface-modification was carried out under the various conditions: 5~13mm tool diameter, 1,400~2,000RPM tool rotating speed, 100~400mm/min modifying speed.

2.2. Limiting dome height (LDH) test and tensile test

LDH tests were carried out to investigate the formability of the locally surface-modified sheets compared to the as-received sheet. Dimensions of the specimens for LDH test were 200mm length and 100mm width. The longitudinal direction of the specimens was parallel to the rolling direction of the sheets. The direction of local surface-modification was parallel or perpendicular to the longitudinal direction of the specimens. Figure 2 describes the LDH tests, and the test conditions were 2mm/min. punch speed and 60kN blanking pressure without lubrication.

For tensile tests, specimens with 25mm gage length and 6.3mm gage width were prepared from the locally surface-modified. The crosshead speed was 1.5mm/min, and the engineering stress-strain curves were transformed to true stress-strain curves.

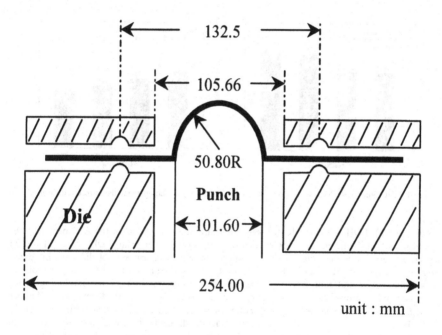

Figure 2. Schematic diagram of LDH test.

2.3. Measurement of plastic zone depth

Plastic zone depth affects directly the mechanical properties of the locally surface-modified sheet. Plastic zone depths of the locally surface-modified sheets under the various conditions were measured by optical microscope and image analysis method.

167

3. Results and Discussions

3.1. Formability of locally surface-modified sheets

Figure 3 is the results of LDH tests for the as-received and locally surface-modified sheets. In Figure 3, 'AR' is the as-received sheet, 'RR' means that the surface-modifying direction is parallel to the rolling direction, and 'RT' means that the surface-modifying direction is perpendicular to the rolling direction. During the LDH tests, direction of the maximum principal stress is equal to longitudinal direction of the specimen.

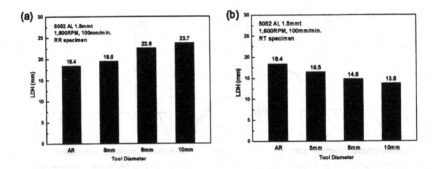

Figure 3. LDH values of the locally surface-modified 5052-H32 Al sheets compared to the as-received sheet.

In Figure 3(a), LDH values of the RR specimens are higher than that of the AR specimen, and increase as tool diameter increases. In Figure 3(b), on the other hand, LDH values of the RT specimens are lower than that of the AR specimen, and decrease as tool diameter increases. Figure 4 is the results of LDH tests for the locally surface-modified sheets as functions of modifying speed and tool rotating speed. Formability is reduced as the modifying speed increases, and improved as the tool rotating speed increases. However, the formability difference is negligible with the change of tool rotating speed. These results show that formability of the 5052-H32 sheets can be improved by local surface-modification by coinciding the direction of the local surface-modification with the direction of maximum principal stress. Also, tool diameter and modifying speed are confirmed as important processing factors for local surface-modification.

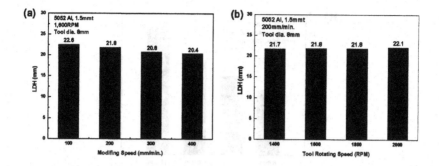

Figure 4. LDH values of the locally surface-modified 5052-H32 Al sheets as (a) modifying speed and (b) tool rotating speed.

Formability of the locally surface-modified sheets is related directly to depth of plastic zone formed during local surface-modification. Figure 5 shows the results of measuring the plastic zone depth under various conditions: tool diameter, modifying speed, and tool rotating speed. When tool diameter and tool rotating speed increase, the depth of plastic zone increases as shown in Figure 5(a) and (c). The depth of plastic zone decreases as modifying speed increases, but the decreasing amount is very small. (Figure 5(b)) The depth of plastic zone can be changed by generation of heat during local surface-modification. Friction between the tool and sheet and plastic deformation occurred by tool rotation are the source of heat generation. So, when tool diameter and tool rotating speed increase, heat can be more generated by increase of frictional area and strain rate, and the depth of plastic zone increases subsequently. On the other hand, when modifying speed increases, passing time through a modified region decreases, and the amount of heat also decreases. These trends of the change of plastic zone depth correspond with those of the change of formability in Figure 3 and 4.

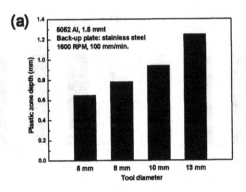

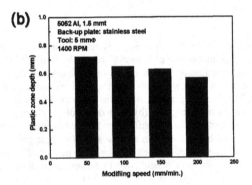

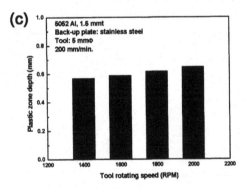

Figure 5. Changes of the plastic zone depth as (a) tool diameter, (b) modifying speed, and (c) tool rotating speed in the 5052-H32 Al sheet.

3.2. Characteristics of locally surface-modified region

Figure 6 shows true stress-true strain curves of the specimens prepared form the locally surface-modified region and base metal ('AR'). Yield and tensile stress of the locally surface-modified region are lower than those of the base metal, but elongation of the locally surface-modified region is increased about 1.5 times or more than that of the base metal. Particularly, it can be seen that strain hardening goes on vigorously as strain increases.

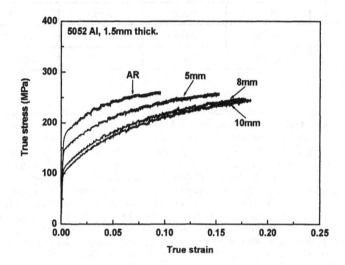

Figure 6. True stress-true strain curves of the locally surface-modified regions and the base metal (AR).

Table 1 is the results of tensile tests for the locally surface-modified region and base metal. When tool diameter increases, yield and tensile stress decrease, and uniform strain as well as strain hardening exponent, n, increases. In the case of 10mm tool diameter, uniform strain and strain hardening exponent are about 2 times when compared to the case of the base metal.

Table1. Tensile properties of the locally surface-modified 5052-H32 Al sheets compared to the as-received (AR) sheet as true stress-true strain.

Tool diameter	Yield Stress (MPa)	Tensile Stress (MPa)	Uniform Strain	Strain Hardening Exponent, n
AR	174.3	261.4	0.096	0.140
5mm	148.1	256.0	0.153	0.194
8mm	112.6	248.8	0.178	0.257
10mm	102.8	244.8	0.184	0.269

The results of Figure 6 and Table 1 mean that the locally surface-modified region can accumulate greater amount of strain than the base metal. EBSD pattern quality (PQ) was measured for the comparison between strain accumulated in the locally surface-modified region and that in the base metal during LDH test. It is known that the mean PQ label gets lower as a material accumulates more strain [8]. Figure 7 is the results of PQ label measured from the base metal and locally surface-modified sheets (10mm tool diameter) at the failed regions before and after LDH test. Rectangles in Figure 7 indicate the mean PQ label of each specimen. In the case of the base metal, the mean PQ labels before and after LDH test are similar. The mean PQ label of the locally surface-modified region is higher than that of the base metal before LDH test. However, after LDH test, the mean PQ labels of the locally surface-modified region and the base metal are almost same. It means that more amount of strain accumulated in the locally surface-modified region than the base metal during LDH test. So, the reason why formability of the locally surface-modified sheets is superior to the base metal is that the locally surface-modified region can accumulate more strain than the base metal.

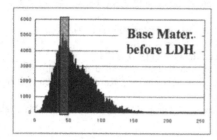
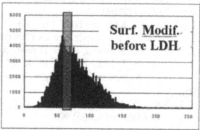
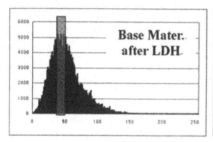
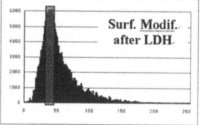

Figure 7. EBSD pattern quality of the as-received and locally surface-modified sheets (tool diameter 10mm) before and after LDH test.

4. Summary

Formability of the locally surface-modified 5052 aluminum alloy sheets is superior to that of the base metal, due to the fact that the locally surface-modified region can accumulate strain more than the base metal. Local surface-modification using the concept of Surface Friction Joining is considered to be the useful tool for the aluminum alloy sheet forming.

5. Acknowledgement

This work has been financially supported by the 21C Frontier R&D Program, which was funded by the Ministry of Commerce, Industry, and Energy of Korea.

Reference

[1] W. M. Thomas, E. D. Nicholas, J. C. Needham, M. G. Murch, P. Temple-Smith, and C. J. Dawes, International Patent WO93/10935 (1993)
[2] J. H. Cho, D. E. Boyce, and P. R. Dawson, Mater. Sci. Eng. A 398 (2005) 146

[3] H. N. Han, C. G. Lee, and S. J. Kim, Korea Patent 2003-68113, USA. Patent 10/717,334, UK Patent 0326324.0 (2003)

[4] C. G. Lee, S. J. Kim, and H. N. Han, J. Kor. Inst. Met & Mater. 43 (2005) 718

[5] D. L. Davidson, Int. Metals Rev. 29 (1984) 75

[6] J. H. Driver, D. J. Jensen, and N. Hansen, Acta Metall. 42 (1994) 3105

[7] C. G. Lee, S. J. Kim, H. N. Han, K. Chung, and S. Park, "Method for Locally Surface-Modifying of Aluminum Alloy Sheet To Improve Formability", Korea Patent 2005-82666 (2005)

[8] S. H. Kang, W. H. Bang, J. H. Cho, H. N. Han, K. H. Oh, C. G. Lee, and S. J. Kim, Materials Science Forum 495-497 (2005) 901

NUMERICAL SIMULATION OF THE STATIC AND DYNAMIC RESPONSE OF CORRUGATED SANDWICH STRUCTURES MADE WITH FRICTION STIR WELDING AND SUPERPLASTIC FORMING

Karim H. Muci-Küchler[1], Sajith K. Annamaneni[1], Darrell R. Herling[2] and William J. Arbegast[3]

[1]Mechanical Engineering Department, Computational Mechanics Laboratory (CML),
South Dakota School of Mines and Technology;
501 East Saint Joseph Street; Rapid City, SD 57701-3995, USA
[2]Pacific Northwest National Laboratory (PNNL);
P.O. Box 999 / K2-03; Richland, WA 99352, USA
[3]Advanced Materials Processing and Joining Laboratory (AMP),
South Dakota School of Mines and Technology;
501 East Saint Joseph Street; Rapid City, SD 57701-3995, USA

Keywords: Friction Stir Welding, Friction Stir Spot Welding, Superplastic Forming, Corrugated Sandwich Structures

Abstract

Corrugated sandwich structures have the potential to find applications in vehicles and vessels subjected to static, dynamic or impulsive loading. A key element of those structures is the geometry of their top and bottom surfaces and the geometry of their core. To justify the use of a corrugated sandwich structure instead of a monolithic plate made of the same material, it is necessary to demonstrate that it will have a superior performance under the expected service conditions while maintaining the same overall ratio of weight per projected flat surface area. Also, an adequate manufacturing process must be established to produce the sandwich structure at a reasonable cost. It has been shown that through the combination of friction stir welding or friction stir spot welding and superplastic forming different types of corrugated sandwich structures can be obtained. In this paper, numerical simulations are used to perform an initial comparison of the static and the basic dynamic characteristics of two of those structures against a monolithic plate of the same material and weight. Rectangular panels simply supported at opposite sides are considered and two different types of analysis are performed: a static analysis in which one of the surfaces of the panels is subjected to a uniform pressure and a normal mode dynamics analysis aimed at obtaining the first natural frequencies and mode shapes. Based on the results of the simulations, preliminary conclusions are drawn regarding the possible use of the sandwich structures under consideration instead of monolithic plates.

Introduction

In general terms, a sandwich structure consists of a core contained between two plates. Different options can be used for the core ranging from tri-dimensional truss arrangements [1], [2], [3] to the use of metallic foams [4]. Among them, cores with different geometric shapes such as "X" shape cores, "Y" shape cores, or those resembling a honeycomb have been proposed [5]. An important aspect associated with this type of structures is how they can be manufactured in a practical and cost effective way.

175

Recently, the SDSM&T Advanced Materials Processing and Joining Laboratory (AMP) and Pacific Northwest National Laboratory (PNNL) used friction stir welding (FSW) [6] or re-fill friction stir spot welding (FSSW) [7], [8] in conjunction with superplastic forming (SPF) (see for example [9], [10], [11]) to produce the two corrugated sandwich structures shown in Fig. 1. In both cases, three flat sheets of aluminum 5083 stacked over each other were used as the starting point to manufacture the panels. The top and bottom sheets had a thickness of 0.079 inches whereas the middle one had a thickness of 0.049 inches. In the case of the sandwich structure shown in Fig. 1(a), the top and middle sheets and the bottom and middle sheets were joined using FSW with a spacing of 2 inches between the weld lines. In addition, the welds performed on the top and bottom surfaces were staggered 1 inch apart from each other. Then, without constraining the top and bottom surfaces, a pressurized gas was injected between the sheets to obtain the final shape via SPF. To make the sandwich structure shown in Fig. 1(b), a similar approach was followed but FSSW was used instead of FSW and the top and bottom sheets were constrained during the SPF operation. The spot welds to join the top and middle sheets and the bottom and middle sheets were made in a 2-inch by 2-inch square pattern. In addition, the welds on the top and bottom surfaces were staggered so that a spot weld in one surface was at the center of the square defined by four spot welds on the other. For convenience, in what follows the sandwich structures corresponding to Fig. 1(a) and Fig. 1(b) will be referred to as "2-D core" and "3-D core", respectively.

(a) FSW sandwich structure: 2-D core *(b) FSSW sandwich structure: 3-D core*
Figure 1. "2-D core" and "3-D core" corrugated sandwich structures

The final geometry corresponding to the 2-D core sandwich structure can be visualized as three two-dimensional profiles (one per sheet) "extruded" in the third dimensio n. However, in the case of the 3-D core sandwich structure, the core geometry follows a three-dimensional pattern that is more difficult to visualize. As can be expected, for both types of structures their final shape depends on several aspects including the initial thickness of each one of the three sheets, the spacing between the welds, and the process parameters associated with the SPF operation.

The corrugated sandwich structures manufactured with FSW or FSSW and SPF may be useful in different types of applications such as space structures and panels for ground vehicles or vessels subjected to impulsive loading. In particular, the use of sandwich structures for cases involving blast loading in air and water has been considered [12], [13]. To determine if they are a good alternative for a specific case, it is necessary to show that their response to the expected service conditions is better than the one of the structures that they intend to replace.

Comparison Methodology

To perform meaningful comparisons, it is necessary to subject structures that are equivalent according to some criteria to the same loads, constraints and, if applicable, initial conditions. For that purpose, the concept of a "basic cell" illustrated in Fig. 2 was introduced based on the geometric characteristics of the 3-D core corrugated sandwich structure. For the 3-D core, a basic cell was defined as the minimum three-dimensional geometry needed to generate a panel by repeating that cell in a rectangular pattern. As can be seen in Fig. 2(a), the length (l) and the width (w) of the basic cell for the 3-D core are determined by the spacing of the spot welds in two perpendicular directions on the top or bottom sheets, and the height (h) is the vertical distance between those sheets. For the 2-D core, the length (l) of the basic cell corresponds to the spacing between the linear welds on the top or bottom sheets and the height (h) is defined as the vertical distance between two imaginary planes tangent to those sheets that do not intersect the geometry of the basic cell. The basic cells for the 2-D core and the 3-D core are such that they have the same length, width and height. In the case of the monolithic plate, the height is not used for the definition of the basic cell which is a rectangle that has the same length and width as the basic cells for the corrugated sandwich structures.

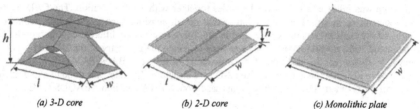

(a) 3-D core   (b) 2-D core   (c) Monolithic plate

Figure 2. Basic cells corresponding to the corrugated sandwich structures and monolithic plate

Besides the geometric aspects mentioned above, one also needs to consider the thickness of the top, middle and bottom sheets of the 2-D and 3-D core panels and the thickness of the monolithic plate in order to define equivalent structures. In this regard, it was decided to use the following approach in the numerical simulations to obtain basic cells with the same total weight and, for the case of the 2-D core and 3-D core, with the same weight distribution among the sheets used to manufacture them. Once the thickness of the top, middle and bottom sheets of the 3-D core are specified, the total volume of the basic cell for the 3-D core is obtained by multiplying the surface area of each sheet by its corresponding thickness and then adding those values. Then the equivalent thickness of the monolithic plate is found by simply dividing that volume by the surface area of its basic cell (i.e., by the product of the length and the width of the basic cell). For the 2-D core, the thickness of each sheet is obtained dividing the volume of the corresponding sheet in the basic cell for the 3-D core by the surface area for the same sheet in the basic cell for the 2-D core. Notice that the proposed procedure gives rise to basic cells that have the same ratio of weight to projected surface area of the top (or bottom) sheet. One characteristic that obviously remains as a difference between the sandwich structures and the monolithic plate is the space requirements since there is no height associated with the later.

As can be expected, for the comparative studies the panels corresponding to the monolithic plate and the 2-D core and 3-D core corrugated sandwich structures need to be made of the same material and have the same total length and width. In this regard, for convenience those dimensions are selected so that they are, respectively, exact integer multiples of the length and width of the basic cells considered.

In this paper the mechanical performance of the panels is compared based on two different aspects: their response to static loading and their natural frequencies and mode shapes. Regarding the later, the effects of damping are not included in the computer models. Two different damping mechanisms can be present during the dynamic loading of the sandwich structures: hysteretic damping due to the behavior of the material and Coulomb damping if there is any dry-friction surface-to-surface contact between the core and the top and bottom sheets.

177

The relevance of each of those energy dissipation mechanisms and parameters to characterize them need to be determined experimentally before a decision is made to try to include them in the simulations.

CAD Representation of the Corrugated Sandwich Structures

The first step in the modeling effort is to generate adequate CAD representations of the 2-D and 3-D core corrugated sandwich structures. For the models presented here, only an approximate geometry was employed. The friction stir welds were represented as a line and the friction spot-welds as a point. Also, the two-dimensional profiles employed to construct the required surfaces were generated using only straight lines and circular arcs but making sure that each profile had a continuous tangent vector. For the 2-D core, three profiles (one per sheet) connected at the welds were used in conjunction with an extrude operation. For the 3-D core, a more elaborate procedure was followed using several profiles together with a loft operation. The CAD models were constructed in a parametric fashion so that key dimensions such as the length, width, and height, of the basic cells could be changed in the future to generate different configurations. The commercial software SolidWorks [14] was used to create the geometry of the 2-D and 3-D core sandwich structures. In addition, the commercial software ABAQUS/CAE [15] was also employed to generate the geometry of the 2-D core structure. The later was done to avoid having to import the geometry corresponding to this case from SolidWorks into ABAQUS/CAE.

The shape and dimensions of the basic cells used to create the CAD models presented here were estimated from sample panels manufactured by PNNL. Using the geometry of the 3-D core sample panel as the reference, the length, width, and height, for the basic cells were selected as 2 inches, 2 inches, and 0.488 inches, respectively. The total length and width of the panels in the CAD models were arbitrarily taken as 12 inches. Figure 3 shows the CAD model for the 2-D core panel, which is based on the picture shown in Fig. 1(a), and Fig. 4 shows the CAD model for the 3-D core panel, which is based on the picture shown in Fig. 1(b). In addition, Fig. 5 presents a side-by-side comparison between the geometry of the actual panel and the CAD model corresponding to the 3-D core.

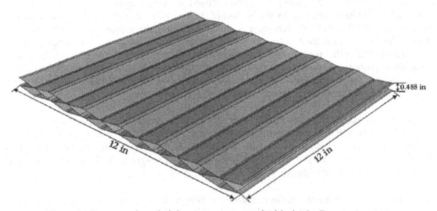

Figure 3. Corrugated sandwich structure generated with the 2 -D core geometry

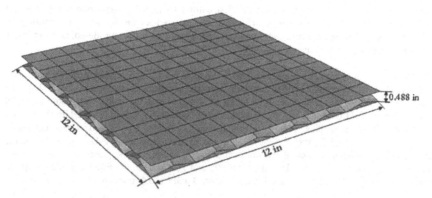

Figure 4. Corrugated sandwich structure generated with the 3-D core geometry

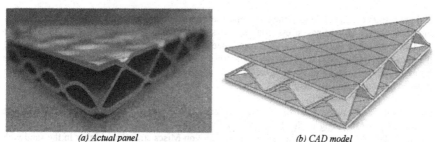

(a) Actual panel   *(b) CAD model*

Figure 5. Comparison of the actual geometry and the CAD model for the 3-D core geometry

Description of the Finite Element Models

Finite element analyses were performed using the commercial software ABAQUS/Standard [15] to compare the static response and the first natural frequencies and mode shapes of the three structures under consideration. For the static analyses, the same uniform pressure was applied over the top surface of all the structures. For the essential boundary conditions, two opposite edges of the panels were simply supported. Regarding the application of the load and constraints, two important aspects need to be mentioned. The total area of the top surface of the 2-D core panel is greater than the one of the other two panels under consideration and, in addition, the direction of the normal vector changes from point to point. Thus, there is a difference in the way the 2-D core structure is being loaded since the pressure is applied to the entire surface in the direction of the normal vector. For the essential boundary conditions, care was taken to apply the constraint only to edges associated with the bottom sheet (surface) of the sandwich structures. Applying it to all the free edges corresponding to one side of the panel would have the net effect of restricting the rotation of that side, making it more similar to a clamped condition than a simply supported one. For the analyses used to determine the natural frequencies and mode shapes, the same essential boundary conditions used for the static analyses were employed.

For the static analyses, only one quarter of the panels is used in the models due to symmetry. The material used for the structures is assumed to be isotropic and is represented using a linear-elastic perfectly-plastic constitutive relation. Non-linear geometric effects due to the deformation of the panels are taken into consideration. In the case of the 3-D core panel, the possible surface-to-surface contact between the sheets of the sandwich structures is taken into account in the

179

simulations. For the analyses to obtain the first natural frequencies and mode shapes the full panels are employed, the generalized Hooke's law is used as the constitutive relation, the possible contact between the core and the top and bottom sheets is not considered, and non-linear geometric effects are not taken into account.

For all the numerical results presented below, 12-inch by 12-inch panels made of aluminum 5083-H111 were considered and the required material properties were obtained from [16]. The length and width of the basic cells was taken as 2 inches and the height was taken as 0.488 inches. For the 3-D core, the top and bottom sheets were 0.079-inches thick while the middle sheet was 0.049-inches thick. Following the approach described earlier, equivalent thicknesses were obtained and used for the monolithic plate and for the sheets of the 2-D core. For the static analyses, four different values of the uniform pressure applied to the plates were used in the numerical simulations: 50 kPa, 100 kPa, 150 kPa and 200 kPa. In the following section, those load cases will be referred to as Case 1, Case 2, Case 3 and Case 4, respectively.

Numerical Results: Static Loading

Sample results corresponding to the static analyses are presented in Table 1 and Figs. 6, 7 and 8. Figure 6 shows von Mises stress contour plots corresponding to Case 4. For the 3-D core sandwich structure, the results for the stresses need to be interpreted with caution since modeling the spot-welds as a point gave rise to high-stress concentration at some weld locations. In the future, models representing the spot-welds as an area instead of a point need to be developed in order to overcome this problem.

From Table 1 it can be seen that, for the constraint considered, the maximum displacement magnitude corresponding to the 2-D core panel was more than the one for the monolithic plate and the 3-D core panel. The value of the maximum von Mises stress was higher in the sandwich structures than in the monolithic plate. For the applied pressures of 100 kPa, 150 kPa and 200 kPa, localized yielding took place in both sandwich structures and, due to the use of a linear-elastic perfectly-plastic constitutive relation, the maximum value of the von Mises stress was the yield stress of the material (250 MPa). As can be observed in Fig. 6, for the 2-D core and 3-D core panels the maximum von Mises stresses occurred at locations corresponding to the welds, which acted as stress raisers.

Table 1. Comparison of results for the static analyses

Panel	Maximum Displacement Magnitude (mm)				Maximum Von Mises Stress (MPa)			
	Case 1	Case 2	Case 3	Case 4	Case 1	Case 2	Case 3	Case 4
Monolithic Plate	3.345	4.581	5.402	6.036	83.7	123.8	153.1	177.2
2-D Core	4.054	8.123	12.34	15.41	208.2	Yield	Yield	Yield
3-D Core	1.755	5.136	6.968	8.316	Yield	Yield	Yield	Yield

Figure 7 presents plots of the total external work, total strain energy, and total plastic dissipation energy vs. the fraction of applied load for the 2-D core and 3-D core panels subjected to a pressure of 200 kPa. Based on the results shown in the graphs, the amount of work converted into plastic dissipation is greater for the 3-D core sandwich structure than for the 2-D one. It must be pointed out that in the case of the monolithic plate plastic dissipation does not take place since the material does not yield. The capability of a structure to absorb energy via plastic deformation without breaking is of great importance in applications involving impulsive loadings conditions.

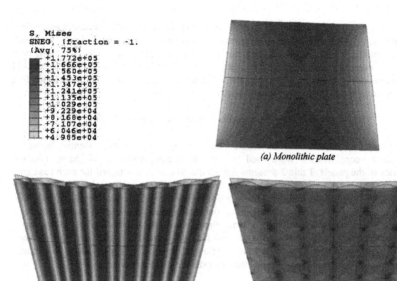

(a) Monolithic plate

(b) 2-D core geometry

(c) 3-D core geometry

Figure 6. Von Mises stress contour plots for the static analyses with a pressure load of 200 kPa

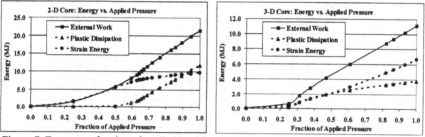

Figure 7. Energy vs. fraction of applied load for the static analyses with a pressure of 200 kPa

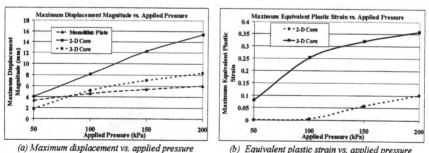

(a) Maximum displacement vs. applied pressure   *(b) Equivalent plastic strain vs. applied pressure*

Figure 8. Plots of maximum displacement magnitude and maximum equivalent plastic strain vs. applied pressure

181

Figure 8 shows plots of the maximum displacement magnitude and the maximum equivalent plastic strain vs. the applied pressure for the monolithic plate, 2-D core and 3-D core sandwich structures. From Fig. 8(a) it can be seen that maximum displacement magnitude for the 3-D core panel and the monolithic plate are relatively close and less than the one for the 2-D core panel. In addition, based on Fig. 8(b), the 3-D core panel has higher values for the maximum equivalent plastic strain than the 2-D core panel due to the stress concentration at the spot welds.

Numerical Results: Natural Frequencies and Mode Shapes

The first ten natural frequencies and mode shapes for the structures and essential boundary conditions under consideration were used to perform a comparison of the basic dynamic characteristics of the panels. Table 2 presents the natural frequencies obtained for each case and, as an example, Figs. 8 to 10 show plots of the first two mode shapes for the monolithic plate, the 2-D core, and the 3-D core sandwich structure, respectively. As can be seen from the results, for similar mode shapes the natural frequencies of the monolithic plate are lower than the ones corresponding to the sandwich structures. Also, although not shown in the figures, the geometry of the sandwich structures gave rise to additional natural frequencies and mode shapes when compared to the monolithic panel. To perform additional comparisons, it is necessary to consider the frequency spectrum associated with a particular external load excitation.

Table 2. Numerical results for the first ten natural frequencies

Natural Frequency	Value (kHz)		
	Monolithic Plate	2-D Core	3-D Core
1	0.13750	0.38278	0.41229
2	0.22618	0.57722	0.49074
3	0.51828	1.0219	0.83599
4	0.55668	1.1228	0.84011
5	0.66164	1.3054	0.91072
6	0.99657	1.8889	1.1463
7	1.0694	1.9003	1.1944
8	1.2565	1.9122	1.2796
9	1.3630	2.1048	1.3184
10	1.5641	2.2539	1.3550

It must be pointed out that for the monolithic plate the results for the first six natural frequencies provided by the finite element model were compared against the analytical solution given in Blevins [17]. The error in the numerical results was less than 3% in all cases indicating that the mesh size used was adequate. For the corrugated sandwich structures, a convergence study was performed but it was not possible to validate the results since experimental results would be needed for that purpose.

(a) First mode shape   *(b) Second mode shape*
Figure 9. First and second mode shapes for the monolithic plate

182

(a) First mode shape   *(b) Second mode shape*

Figure 10. First and second mode shapes for the 2-D core sandwich structure

(a) First mode shape   *(b) Second mode shape*

Figure 11. First and second mode shapes for the 3-D core sandwich structure

Conclusions

The results presented here constitute an initial effort to compare the performance of corrugated sandwich structures made with FSW or FSSW and SPF against the one of a monolithic plate of the same material and weight. Additional simulations including the response of these structures under impulsive loading are required and experiments need to be conducted to validate the results provided by the finite element models. For the case of the 2-D core sandwich structure, the possible contact interaction between the core and the top and bottom sheets needs to be considered. Also, for all the structures, a more suitable constitutive relation that takes into account strain hardening effects needs to be used when the loads imposed are such that a substantial portion of the panel experiences yielding. Furthermore, in the case of impulsive loading, strain rate effects also need to be taken into account.

The numerical results presented here show that corrugated sandwich structures corresponding to the 3-D core geometry have the potential to be used instead of monolithic plates in applications in which the additional space requirement imposed by the height of the sandwich structures is not a concern.

Finally, since the behavior of the corrugated sandwich structures under consideration is controlled by their shape, parametric studies need to be performed in the future to determine optimum values for variables such as the thickness of the sheets, the spacing between the welds, and the separation between the top and bottom sheets.

Acknowledgements

This research was sponsored by the Army Research Laboratory and was accomplished under Cooperative Agreement Number DAAD19-02-2-0011. The views and conclusions contained in this document are those of the authors and should not be interpreted as representing the official policies, either expressed or implied, of the Army Research Laboratory or the U.S. Government. The U.S. Government is authorized to reproduce and distribute reprints for government purposes notwithstanding any copyright notation hereon.

References

1. Wicks, N. and Hutchinson, J.W. "Optimal Truss Plates." *International Journal of Solids and Structures*, vol. **38**, pp. 5165–5183, 2001.
2. Chiras, S.; Mumm, D.R.; Evans, A.G.; Wicks, N.; Hutchinson, J.W.; Dharmasena, K.; Wadley, H.N.G. and Fichter, S. "The Structural Performance of Near-Optimized Truss Core Panels." *International Journal of Solids and Structures*, vol. **39**, pp. 4093–4115, 2002.
3. Wicks, N. and Hutchinson, J.W. "Performance of Sandwich Plates with Truss Cores." *Mechanics of Materials*, vol. **36**, pp. 739–751, 2004.
4. Bart-Smith, H.; Hutchinson, J.W. and Evans, A.G. "Measurement and Analysis of the Structural Performance of Cellular Metal Sandwich Construction." *International Journal of Mechanical Sciences*, vol. **43**, pp. 1945-1963, 2001.
5. Wadley, H.N.G.; Fleck, N.A. and Evans, A.G. "Fabrication and Structural Performance of Periodic Cellular Metal Sandwich Structures." *Composites Science and Technology*, vol. **63**, pp. 2331-2343, 2003.
6. Thomas, W.M.; Nicholas, J.; Murch, M.; Templesmith, P. and Dawes, C. "Friction Stir Welding." G. B. Patent Application No. 9125978.8; US Patent No. 5460317, Oct. 1995.
7. Schilling, C. and dos Santos, J. "Method and Device for Joining at Least Two Adjoining Work Pieces by Friction Welding." US Patent Application 2002/0179 682.
8. Allen, C.D. and Arbegast, W.J. "Evaluation of Friction Spot Welds in Aluminum Alloys." SAE Paper 2005-01-1252.
9. Herling, D.R. and Smith, M.T. "Improvements in Superplastic Performance of Commercial AA5083 Aluminum by Equal Channel Angular Extrusion." *Superplasticity in Advanced Materials*. Switzerland: Trans Tech, pp. 465-470, 2001.
10. Smith, M.T.; Vetrano, J.S.; Nyberg, E.A. and Herling, D.R. "Effects of Mg and Mn Content on the Superplastic Deformation of 5000-Series Alloys." *Superplasticity in Superplastic Forming* 1998 (Eds. Ghosh, A.K. and Bieler, T.R.), pp. 99-108, TMS, 1998.
11. Vetrano, J.S.; Lavender, C.A.; Hamilton, C.H.; Smith, M.T.; and Bruemmer, S.M. "Superplastic Behavior in a Commercial 5083 Al Alloy." *Scripta Metallurgica*, vol. **30**, pp. 565-570, 1994.
12. Xue, Z. and Hutchinson, J.W. "Preliminary Assessment of Sandwich Plates Subjected to Blast Loads." *International Journal of Mechanical Sciences*, Vol. **45**, pp. 687–705, 2003.
13. Xue, Z. and Hutchinson, J.W. "A Comparative Study of Impulsive-Resistant Metal Sandwich Plates." *International Journal of Impact Engineering*, vol. **30**, pp. 1283–1305, 2004.
14. SolidWorks Education Edition, SolidWorks SP4.1, 2006.
15. ABAQUS Version 6.6 User Manuals. ABAQUS Inc., 2006.
16. MIL-HDBK-5H: Military Handbook, Metallic Materials and Elements for Aerospace Vehicle Structures, December 1998.
17. Blevins, R.D. "Formulas for Natural Frequencies and Mode Shapes." Van Nostrand Reinhold Company, 1979.

EFFECT OF PROCESSING PARAMETERS ON MICROSTRUCTURE OF THE FSW NUGGET

J. A. Querin, A.M. Davis, J. A. Schneider

Department of Mechanical Engineering
Mississippi State University, Mississippi State, MS 39762 USA

Keywords: Friction Stir Weld, 2219-T87, EBSD/OIM.

Abstract

Friction stir welding (FSW) is a thermo-mechanical process that utilizes a non-consumable rotating pin tool to stir the edges of a weld seam together. Within the refined zone of the weld nugget, contrasting bands are commonly observed using optical microscopy. Although the mechanisms resulting in the formation of these contrasting bands are unclear, they may result from variations in the thermo-mechanical processing that the material experiences. Orientation image mapping (OIM) is used in this study to investigate the effect of process parameters on the resulting grain size and crystallographic orientations of the refined grains associated with the banded regions.

Introduction

Because the resulting mechanical properties of the friction stir welding (FSW) joint depends on the evolution of the microstructure, many studies have reported on the characterization of the grain morphology and texture in this region. A central nugget of refined grains is reported to display banding of varying thickness in the transverse section much like 'onion rings' [1]. Presence of this optical contrast has been either correlated to grain size variations [2-4] or a non-homogenous distribution of second-phase particles [2-7]. Grain sizes within the FSW nugget have been reported to range from 2-10 μm in various aluminum alloys, independent of the parent material size and morphology.

More recent studies have looked at the texture within the refined weld nugget and report gradients of shear fiber texture across both the width and through the thickness of the refined nugget [2, 8-13]. Grain boundary misorientation angles have been reported in the range of 2 to 60° and display either a bimodal [2, 8, 11, 13] or a flat distribution [10, 12]. Post weld heat treatments to evaluate grain growth and changes in the misorientation angle have provided insight into the stability of the refined weld nugget microstructure [13].

It has been suggested that the FSW process imparts a non-homogenous deformation on the resulting weld nugget [10, 14]. Since the grains in a metal evolve in response to the thermo-mechanical processing history, variations in the grain morphology, misorientation angle and texture may provide insight to variations in metal flow paths as influenced by the process parameters. However since the variations in the FSW flow paths may also be affected by the FSW tool geometry, process parameters, and materials joined, it is difficult to compare the resulting metal flow paths reported in the literature. This study documents the microstructural differences in the weld nugget as the rotational speed of the pin tool is varied.

Procedure

The material used in this study was AA2219-T87. Full penetration welds were made in a 6.35 mm (0.25") thick plate with a pin tool 5.74 mm (0.23") long. Pin diameter was 12.7 mm (0.5"); shoulder diameter, 30.5 mm (1.2"). A double scrolled surface was used on the shoulder of the pin tool and the weld was made with a 0° tilt angle. A left-handed 1.27 mm (0.05") pitch UNF single thread on the pin induced a downward flow (away from the shoulder) close to the pin when rotated by the spindle. Table I summarizes the process parameter variation used in this study. The weld was made along the rolling direction of the 610 mm (24") long plate. A 25.4 mm (1") transition region separated the RPM change, resulting in a 165 mm (6.5") weld length for each sample as illustrated in Figure 1. The FSW weld was started in a 90 mm (3.5") run on tab, which was tack welded to join the weld panels.

Table I. FSW Process Parameters.

Specimen ID	Rotation (RPM)	Plunge Force (kN)	Travel (mm/m)	Tool advance/rotation mm/rev
C29-150	150	31	114	0.76
C29-200	200	31	114	0.57
C29-300	300	31	114	0.38

Transverse sections for microstructural studies were cut at the end of each steady-state weld segment with a diamond blade, prior to being mounted and polished. To reveal the macrostructure optically, the samples were etched with Keller's reagent. A Nikon D1x camera with a 60 mm AF Micro Nikkor lens was used to document the macrostructure. For OIM imaging, a final polish was made with 0.02 µm colloidal silica first on a padded polishing wheel and then on the padded lapping film of a vibratory polisher.

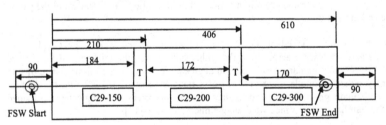

Figure 1. Panel layout for FSW with varying pin tool rotation. All dimensions presented are in mm. 'T' is used to indicate the transition region between changing weld parameters.

186

A JEOL 6500 F field emission, scanning electron microscope (FE-SEM) was used to generate secondary electron images (SEI) of the weld nugget. The OIM maps were obtained using an Oxford electron backscatter detector (EBSD) mounted on the FE-SEM. Analysis was performed in 0.4 μm steps over 215 μm x 161 μm rectangular areas in the banded regions of the transverse sections. All OIM scans were obtained using the same excitation conditions of 20 kV with a working distance of 20 mm.

The EBSD analysis for two samples, C29-150 and C29-300, are discussed in this report. EBSD/OIM was used to determine grain size based on a 5°-misorientation angle. A value of 3°-misorientation angle was used to determine the range of misorientations within a banded region.

Results

Optical images of the transverse sections of the 3 samples investigated are shown in Figure 2 with the advancing side (AS) on the left and the retreating side (RS) on the right. On Specimen 29-150, Figure 2a, the banding 'onion rings' are only optically observed on the AS. This is in contrast to specimens C29-200 and C29-300, which show banding 'onion rings' on both the AS and RS. Furthermore, as the RPM was increased from 150 to 300, the sharpness of the bands on the AS fades while increasing on the RS. Characteristics of the FSW nugget are summarized in Table II using the method of Zettler, et al. [15].

(a) C29-150　　　　　　(b) C29-200　　　　　　(c) C29-300

Figure 2. Optical micrographs of the transverse sections of the specimens investigated. All images show the advancing side (AS) on the left and the retreating side (RS) on the right.

Table II. Comparison of the refined grain region in the weld nugget.

Specimen ID	Area of refined grain region (cm²)	Angle between AS and PM α	Angle between RS and PM β	Ratio α/β
C29-150	1.07	38	31	1.24
C29-200	0.97	33	32	1.03
C29-300	1.29	36	34	1.06

Using the EBSD data with a step size of 0.4 μm, the average grain sizes in the single band regions are summarized in Table III. In addition to the grain sizes given in Table III, the percent of low angle grain boundaries (LAGB) and high angle grain boundaries (HAGB) are also reported. Detailed histograms of the grain boundary misorientations are shown in Figure 3. For C29-150 RS, where no bands were observed optically, the scanned region was chosen to be symmetric to the scan on the AS with respect to the weld centerline.

187

Table III. Comparison of the grain size and misorientation angle within a single band.

Specimen ID	Grain Size (μm)	N sample size	LAGB <15°	HAGB >15°
C29-150 AS	2.5	2159	17	83
C29-150 RS	1.8	7493	14	86
C29-300 AS	4.1	1129	18	82
C29-300 RS	4.2	2019	24	76

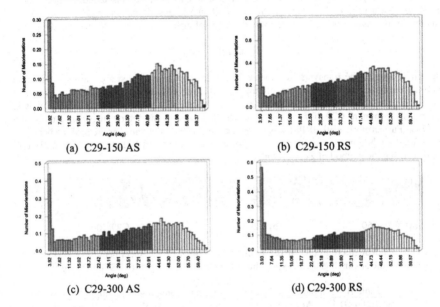

(a) C29-150 AS

(b) C29-150 RS

(c) C29-300 AS

(d) C29-300 RS

Figure 3. Misorientation angles within a single band of the FSW nugget.

Figure 4 relates the EBSD/OIM directions on the transverse metallographic sample to the sample geometry directions. The terminology for the sample geometry directions is independent of the crystallographic directions in the rolled plate parent material. Pole figures provide a comparison of the texture developed within the FSW nugget as shown in Figure 5.

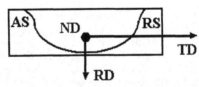

Figure 4. Schematic of the transverse metallographic sample with terminology used for the reference directions (shown in red) in the EBSD analysis.

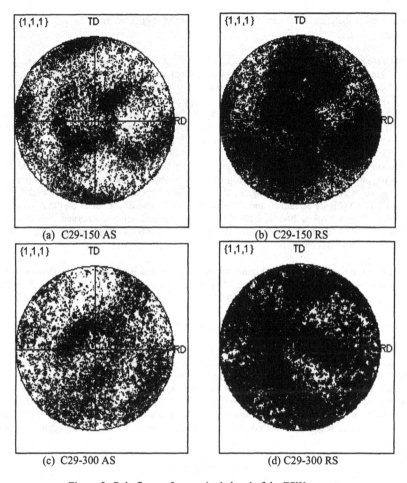

Figure 5. Pole figures from a single band of the FSW nugget.

Discussion

'Hotter welds', as reported in the literature, are typically associated with process parameters of higher rotational speeds, or slower tool advance/rotation rates [4, 16, 17]. Larger grain sizes observed at 300 RPM are in agreement with grain growth occurring during a higher temperature weld. Variations in temperature between the AS and RS have also been reported, with the higher temperatures predicted to be on the AS [18]. Larger grain sizes were observed on the AS for the 150 RPM case, but very little difference in grain size was observed at 300 RPM. This suggests the temperature differential between the AS and RS of the nugget maybe mitigated during higher temperature welds.

Previous studies have reported a population of 70-80% of HAGB within the weld nugget [2, 7]. This is consistent with the welds in this study when considering misorientations above 3°. Approximately 14 and 17% of LAGBs are observed at the slower rotation of 150 RPM on both the AS and RS, respectively. When the rotational speed is increased to 300 RPM, the number of LAGBs on the RS is observed to increase to approximately 24%, showing more of a variation between the AS and RS. At the higher RPM, the banding pattern can also be optically resolved on both the AS and RS of the transverse section of the metallographic sample.

Most texture studies on FSW have indicated a fiber texture associated with the metal flow around the pin tool [8-10]. In this study, a variation is observed between the textures resulting from low and higher RPM processing conditions. Figure 5 summarizes the {111} pole figures for the C29-150 (Figure 5a and b) and C29-300 (Figure 5c and d) for bands on the AS and RS, respectively. On the AS of 29-150, a {101}<111> texture orientation is observed. For the RS of specimen C29-150 RS, in which the banded region could not be resolved optically, an 'A' fiber texture ({111}<hkl>) is observed.

For C29-300, the pole figures in Figure 5c and 5d, AS and RS respectively, show a mirror symmetry of a weak {101}<001> texture. The texture is more diffuse on the AS of C29-300 as compared to the RS. In contrast to C29-150, the texture observed is similar on the AS and RS.

Summary

A comparison of FSWs processed at a high and low RPM showed variations in the grain size and the texture. Although the changes are not large, they do suggest the expected response of a refined microstructure to higher processing temperatures. This is consistent with hotter welds occurring at higher RPM. The grain sizes suggest a temperature differential may exist in the FSW process at lower RPM which is minimized at the higher RPM. Variations in the texture as the RPM changes, may suggest differences in the resulting metal flow path. Additional characterization is required to include band to band variations in the weld nugget transverse section.

Acknowledgements

The authors wish to thank Mr. Sam Clark and Mr. Ronnie Renfroe at the NASA-MSFC for working together to produce the welds. Electron microscopy equipment and detectors were obtained with NSF Grant #DMR.0216703 02070615. Funding was provided by the NASA-Cooperative Agreement #NNM04AA14A and the MSU Center for Advanced Vehicular Studies (CAVS).

References

1.   K.N. Krishnan, "On the formation of onion rings in friction stir welding," *Mater. Sci. Engr.*, A327 (2002), 246-251.

2.   H. Jin et al., "Characterization of microstructure and texture in friction stir welded joints of 5754 and 5182 aluminum alloy sheets," *Mater. Sci. Technol.*, 17 (2001), 1605-1614.

3.   M.W. Mahoney et al., "Properties of friction-stir-welded 7075 T651 aluminum," *Metall. Mater. Trans. A*, 29A (1998), 1955-1964.

4.   B. Yang et al., "Banded microstructure in AA2024-T351 and AA2524-T351 aluminum friction stir welds. Part I. Metallurgical Studies," *Mater. Sci. Engr.*, A364 (2004), 55-65.

5.   M.A. Sutton et al., "Microstructural studies of friction stir welds in 2024-T3 aluminum," *Mater. Sci. Engr.*, A323 (2002), 160-166.

6.   P.L. Threadgill, (TWI report-678/1999).

7.   A.F. Norman, I. Brough, and P.B. Prangnell, "High resolution EBSD analysis of the grain structure in an AA2024 friction stir weld," *Mater. Sci. Forum*, 331 (2000), 1713-1718.

8.   D.P. Field et al., "Heterogeneity of crystallographic texture in friction stir welds of aluminum," *Metall. Mater. Trans. A*, 32A (2001), 2869-2877.

9.   Y.S. Sato et al., "Microtexture in the friction-stir weld of an aluminum alloy," *Metall. Mater. Trans. A*, 32A (2001), 941-948.

10.   J.A. Schneider and A.C. Nunes Jr., "Characterization of plastic flow and resulting microtextures in a friction stir weld," *Metall. Mater. Trans. B.*, 35B (2004), 777-783.

11.   K.V. Jata and S.L. Semiatin, "Continuous dynamic recrystallization during friction stir welding of high strength aluminum alloys," *Scripta Mater.*, 43 (2000), 743-749.

12.   K.V. Jata, "Friction stir welding of high strength aluminum alloys," *Mater. Sci. Forum*, 331 (2000), 1701-1712.

13.   I.C. Hsiao, S.W. Su, and J.C. Huang, "Evolution of texture and grain misorientation in an Al-Mg alloy exhibiting low-temperature superplasticity," *Metall. Mater. Trans. A*, 31A (2000), 2169-2180.

14.   R.W. Fonda, J.F. Bingert, and K.J. Colligan, "Development of grain structure during friction stir welding," *Scripta Mater.*, 51 (2004), 243-248.

15.   R. Zettler et al., "A study on material flow in FSW of AA2024-T351 and AA6056-T4 alloys," *Proc. 5th Int'l Conf. on Trends in Welding Research,* ed. J.M. Vitek, S.A. David, D. Johnson (Materials Park, Ohio: ASM Int'l, 1999).

16.   Y.S. Sato, M. Urata, and H. Kokawa, "Parameters controlling microstructure and hardness during friction stir welding of precipitation-hardenable aluminum alloy 6063," *Metall. Mater. Trans. A*, 33A (2002), 625-635.

17.   L.E. Murr et al., "Intercalation vortices and related microstructural features in the friction stir welding of dissimilar metals," *Mater. Res. Innovations,* 2 (1998), 150-163.

18.   G.E. Gould and Z. Feng, "Heat flow model for friction stir welding of aluminum alloys," *J. Mater. Process. Manuf. Sci.,* 7 (1998), 185-194.

FRICTION STIR LAP JOINING OF Al 7075 WITH SEALANT

Tim Li[1], George Ritter[1], Nick Kapustka[1], Richard Lederich[2]
[1]Edison Welding Institute, 1250 Arthur E. Adams Dr. Columbus, OH 43221, USA
[2] The Boeing Company, PO Box 516, St. Louis, MO 63166, USA

Keywords: Friction stir welding, lap joint, Al 7075, sealant.

Abstract

Friction stir lap welding with sealant was conducted on 0.080-in.-thick Al 7075-T6 sheet. PRC 1750 sealant was applied prior to welding in an attempt to prevent the crevice corrosion for friction stir welding (FSW) lap joints. The sealant was applied in two ways: in the weld path and adjacent to the weld path. FSW lap joining was carried out on the sheet with uncured and with already cured sealant. Two pin tools were used for welding (conventional and modified tools). No crevice corrosion was observed at the faying surfaces of FSW lap joints after exposure to salt fog spray for 500 hrs. However, corrosion attack occurred at the thermal mechanical-affected zone (TMAZ) and the heat-affected (HAZ) zone on the FSW top surface for all corrosive-exposed specimens. This feasibility study has shown promising results for resolving the inherent crevice corrosion issue for bringing FSW lap joints into production.

Introduction

The implementation of friction stir welding (FSW) as a rivet replacement technology has drawn tremendous interest from the aerospace industry in recent years. The traditional riveting process is labor intensive and costly. Cost analysis in the Metals Affordability Initiative (MAI) program has shown that replacing riveting with FSW will reduce costs by 20% and shorten manufacturing lead time by 25%. Moreover, the mechanical properties of FSW lap joints are superior to the riveted ones. Static lap-shear strength of FSW lap joints shows 2.4 times the equivalent strength over riveted joints, and the fatigue strength is equivalent to or better than riveted joints [1]. There are a number of publications investigating FSW lap joints; however, no work has been published to address the crevice corrosion issue that is inherent with FSW lap joints [2-5]. This issue must be resolved before fully implementing the FSW process onto aircraft structures. A desirable approach to this problem is to apply an adhesive or sealant between two metal sheets prior to FSW to act as a sealant and improve overall joint performance. If successful, this in-situ friction stir lap welding process will be the simplest approach to the prevention of crevice corrosion on FSW lap joints.

A typical aircraft aluminum alloy, Al 7075, was selected for this study. FSW lap joints were made with the selected sealant at or between the joints with the intent to produce sound quality lap joints while keeping the sealant intact adjacent to these joints to address the issue of crevice-corrosion prevention. FSW lap joints were characterized in terms of microstructure, static lap-shear strength, and corrosion susceptibility.

Experimental Details

The 0.080-in. thick Al 7075-T6 sheet in dimension of 4 × 12-in. was chemically milled prior to FSW. All welds were made in lap joint configuration. The sealant used in this project was PRC

1750. Welds were made with the sealant initially present at the panel interface in one of two configurations; (1) a single bead (1/16-in. diameter) along the weld centerline, and (2) two beads at 3/4 in. distance from the weld centerline. The sealant was applied to the surface of one of the panels using an air-powered applicator. The second panel was then placed directly on top of the first panel. Panels in which the sealant was in the uncured condition were then immediately welded. Panels that required the sealant to be in the cured condition were stacked, and then clamped to minimize the gap between the panels. These panels were then left in the clamped position for 48 hours prior to welding.

Table I is the welding trial test matrix. A total of 20 welds were made. Welds 1-10 were made using the conventional pin tool, while Welds 11-20 were made using the modified pin tool. The pin tool, referred to as the conventional tool, had a scroll shoulder and tapered pin with a left-hand thread and three flats. The modified tool had a scroll shoulder and a left-hand threaded pin with a larger diameter at the pin tip. The wider pin diameter at the pin tip was intended to generate a wider stir zone at the faying surface. All welds were made with an 815-rpm spindle rotational speed, and either 5- or 15-ipm travel speeds. The weld schedule corresponding to 815 rpm and 5 ipm is referred as high weld pitch (WP), and the weld schedule with 815 rpm and 15 ipm is referred as low WP. WP is defined as a ratio of spindle rotational speed (rpm) over travel speed (ipm). It is generally true that the weld with a higher WP has a higher heat input, and vice versa. However, it may not be true in extreme cases such as excessively high spindle speeds. Spindle torque and travel load data are required to calculate the heat input for an individual weld. For this reason, WP is used instead of heat input in this report. For each of the two tools, five sealant conditions were examined:

1. Weld without sealant
2. Weld through uncured sealant
3. Weld through cured sealant
4. Weld between uncured sealant
5. Weld between cured sealant.

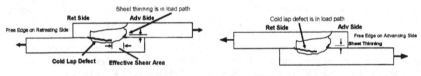

    (a) Advancing specimen;   (b) Retreating specimen

Figure 1. (a) Advancing specimen. Note the top free edge is on retreating side. In this test orientation, the sheet-thinning defect is in the load path. (b) Retreating specimen. Note the top free edge is on advancing side. In this test orientation, the cold-lap defect is in the load path.

Each welded panel was first sectioned into ten 1-in.-wide specimens using a vertical band saw. The static tensile-shear samples were then divided into two groups for further machining. The samples designated as "advancing" and "retreating" are described in Figures 1(a) and 1(b), respectively. These samples were prepared by first machining a 0.060-in.-deep notch 1 in. from the weld centerline, prior to manually fracturing the sheet.

Two macrographs were taken from each weld. One macrograph corresponded to the as-welded condition; the other was taken following the exposure to the neutral salt fog spray. All macrographs were prepared by first sectioning the specimen to an appropriate size, then mounting it in a Bakelite mold. The specimens were polished and etched using Krolls etchant.

An optical microscope was used to obtain measurements of cold lap, sheet thinning, and weld width at the projected interface of the sheet.

Table I. Test Matrix of FSW Lap Joints

Weld ID	FSW Tool Design	Heat Input	Sealant Condition
1	Conventional Tool	High Weld Pitch	No Sealant
2	Conventional Tool	High Weld Pitch	Weld Through Uncured Sealant
3	Conventional Tool	High Weld Pitch	Weld Through Cured Sealant
4	Conventional Tool	High Weld Pitch	Weld Between Uncured Sealant
5	Conventional Tool	High Weld Pitch	Weld Between Cured Sealant
6	Conventional Tool	Low Weld Pitch	No Sealant
7	Conventional Tool	Low Weld Pitch	Weld Through Uncured Sealant
8	Conventional Tool	Low Weld Pitch	Weld Through Cured Sealant
9	Conventional Tool	Low Weld Pitch	Weld Between Uncured Sealant
10	Conventional Tool	Low Weld Pitch	Weld Between Cured Adhesive
11	Modified Lap Tool	High Weld Pitch	No Sealant
12	Modified Lap Tool	High Weld Pitch	Weld Through Uncured Sealant
13	Modified Lap Tool	High Weld Pitch	Weld Through Cured Sealant
14	Modified Lap Tool	High Weld Pitch	Weld Between Uncured Sealant
15	Modified Lap Tool	High Weld Pitch	Weld Between Cured Sealant
16	Modified Lap Tool	Low Weld Pitch	No Sealant
17	Modified Lap Tool	Low Weld Pitch	Weld Through Uncured Sealant
18	Modified Lap Tool	Low Weld Pitch	Weld Through Cured Sealant
19	Modified Lap Tool	Low Weld Pitch	Weld Between Uncured Sealant
20	Modified Lap Tool	Low Weld Pitch	Weld Between Cured Sealant

Low Weld Pitch – 815 rpm/15 ipm
High Weld Pitch – 815 rpm/5 ipm

Five of the ten specimens sectioned from each weld underwent neutral salt fog exposure for 500 hrs. This was conducted in accordance with ASTM B117. Prior to the neutral salt fog spray the specimens were fatigued at a low load for 500 cycles to open any potential defects or discontinuities. The joint cross sections of fatigued specimens were masked with water-proof tape. These steps were administered in an effort to simulate in-service conditions.

Static tensile testing was conducted in an unguided condition. Ultimate lap-shear tensile loads and failure modes of all tested specimens were recorded.

Results and Discussion

Metallographic Examination

One macrograph from each of the 20 welded panels was sectioned to determine the quality of the joints and the sealant condition at the weld. Figures 2 and 3 are typical macrographs of 0.080-in.-thick Al 7075 FSW lap joints made using the conventional and modified pin tools, respectively. The lap joints made with the modified pin tool exhibited wider stir zone compared to the ones made with the conventional pin tool due to a larger pin diameter at the pin tip on the modified tool.

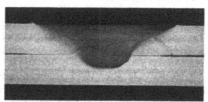

Figure 2. Macrograph of FSW lap joint made using the conventional pin tool.

195

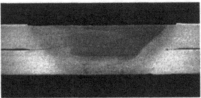

Figure 3. Macrograph of FSW lap joint made using the modified pin tool

Macrographs were prepared after the FSW lap-joined specimens were corroded under a salt fog spray exposure for 500 hrs. Metallographic examinations revealed that there was no evidence of corrosion attack at the faying surfaces near the lap joints for all 20 specimens after 500-hr exposure. A photograph and a macrograph of the FSW lap joint specimens after exposure for 500 hr are shown in Figures 4(a) and 4(b), respectively. However, preferred corrosion attacking occurred at the thermal mechanical-affected zone (TMAZ) and heat-affected zone (HAZ) on the crown side of all 20 lap-joint specimens.

 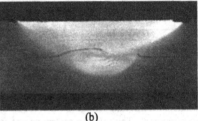

(a)   (b)

Figure 4. Preferred corrosion attack at TMAZ/HAZ on the top surface of FSW lap joints after salt fog spray exposure for 500 hrs. (a) Photograph of FSW lap joint after salt fog spray exposure; (b) Macrograph of FSW lap joint after salt fog spray exposure.

Sheet-Thinning Defect, Cold-Lap Defect, and Effective Shear Area

The sheet-thinning defect, cold-lap defect, and effective shear area are characteristics of FSW lap joints as illustrated in Figures 1(a) and 1(b). The sheet-thinning defect occurs at the advancing side (also referred as hooking). This defect is a result of the material's vertical movement in the stir zone that drags the faying surface of the advancing side up or down. The cold-lap defect appears at the retreating side. It is the remnant of the initial faying surface extending into the stir zone of the retreating side. This defect occurs because the faying surface on this side was not stirred well enough to be completely disrupted. The effective shear area is the actual joined width on the cross section of an FSW lap joint (Figures 1(a) and (1(b)). The lap-shear tensile test results show that the lap joint strength and failure mode are closely related to the magnitude of the sheet-thinning defect, cold-lap defect, and effective shear area.

As shown in Figure 5, for an FSW tool with a left-hand thread spinning clockwise, material near the pin is pushed down then moved up to underneath the tool's shoulder to form a circulating motion inside the stir zone. The initial faying surface of two sheets in the TMAZ adjacent to the stir zone is dragged upward to form a sheet-thinning defect due to the material's circulating motion in the stir zone. A higher weld pitch (WP) generates better circulating motion of materials in the stir zone. As a consequence, it results in a larger sheet-thinning defect but smaller cold lap defect. On the other hand, a lower WP generates less circulating motion of materials in the stir zone. Therefore, the lap joint will have a smaller sheet-thinning defect but larger cold-lap defect. Of course, different pin tool designs will generate different situations of materials movement. However, principally speaking, to obtain a decent FSW lap joint with high joint strength requires a pin tool design combined with weld schedule(s) that produce a weld with a good balance between sheet-thinning and cold lap defects while having decent amount of effective shear area.

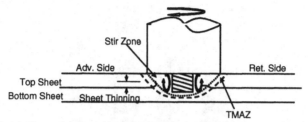

Figure 5. Sketch illustrating formation of sheet-thinning defect due to material's circulating motion during FSW

Figure 6 presents the effects of sealant, pin tool, and WP on sheet-thinning defects. It should be noted that negative sheet thinning reveals the direction of hooking toward the bottom sheet. In this figure and the following figures, the category of "no sealant" indicates welds without sealant. The category of "sealant" includes both welds through sealant and welds between sealant. Due to the difficulty of keeping sealant between the welds, sealant migrated into the weld zone. For this reason, the welds between the sealant were combined with the welds through the sealant as a single category. For the modified tool, the presence of sealant in the lap joint did not affect the amount of sheet-thinning defect at both low and high WP. However, for the conventional tool, the presence of sealant in the lap joints appeared to have some effect on the sheet thinning. The sheet-thinning defect became negative with a smaller absolute value at the low WP but turned to a higher positive value at the high WP. The modified tool appeared to be more sensitive to the WP than the conventional tool. At low WP, both tools produced relatively low sheet thinning. However, at high WP, the modified tool generated much higher sheet thinning than the conventional tool. The sheet-thinning data supports the above-mentioned analysis shown in Figure 5.

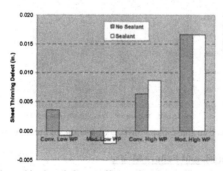

Figure 6. Sheet thinning defect – effects of sealant, pin tool, and weld pitch.

The effects of sealant, pin tool, and WP on the cold-lap defect are shown in Figure 7. For both tools, regardless of the presence of sealant, the cold-lap defect is greater with the high WP than that with the low WP. This is contradictory to the aforementioned analysis in Figure 5. For both tools at low WP, the presence of sealant increased the length of the cold-lap defect, which is expected. At high WP, the presence of sealant did not change the length of the cold-lap defect for the conventional tool, but decreased the cold-lap defect for the modified tool. This is not expected since the presence of sealant should increase the length of a cold-lap defect. Other factors must be affecting the cold-lap defect, but were not identified.

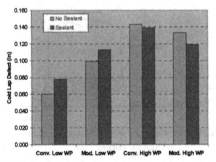

Figure 7. Cold lap defect – effects of sealant, pin tool, and weld pitch.

From macrographs of lap joints, it is seen that the modified tool produced a wider stir zone than the conventional tool. A lap joint with a smaller cold-lap defect does not suggest that it has a greater joint area (effective shear area). Therefore, the effective shear area is a better term to describe the quality of FSW lap joints. Figure 8 presents the effects of sealant, pin tool, and WP on the effective shear area. As expected, the presence of sealant reduced the effective shear area for all cases except the conventional tool with high WP. However, the effect of sealant on the effective shear area is minimal due to the fact that the amount of change is relatively small. It is obvious that the modified tool produced a much larger effective shear area than the conventional tool. For both tools, the high WP generated a lower effective shear area than the low WP. This is contradictory to the aforementioned analysis.

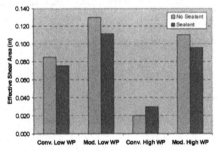

Figure 8. Effective shear area – effects of sealant, pin tool, and weld pitch.

Lap-Shear Tensile Strength and Failure Modes

Lap-shear tensile tests were conducted in two orientations (advancing and retreating) as shown in Figures 1(a) and 1(b), respectively. In the advancing test orientation (Figure 1(a)), the sheet-thinning defect is in the load path. Lower lap-shear strength will be produced if the amount of sheet thinning is high enough to drive the failure along the sheet-thinning defect. In the retreating test orientation (Figure 1(b)), the cold-lap defect is in the load path. Higher lap-shear strength will be obtained if the effective shear area is large enough regardless of the amount of sheet thinning when it is positive. However, lower lap-shear strength will be produced if the negative sheet thinning is high enough to drive the failure along the sheet-thinning defect in the retreating test orientation.

Lap-shear tensile loads are plotted in Figures 9 and 10 as effects of specimen test orientation and WP with and without sealant. Table II summarizes failure modes of lap-shear tensile-tested specimens. Figure 11 shows photographs of the typical failure modes described in Table II. It is evident that failure modes were clearly governed by the characteristics of FSW lap joints (sheet-thinning defect, cold-lap defect, and effective shear area).

Table II. Failure Modes of Lap-Shear Tensile Tested Specimens.

Pin Tool & Weld Pitch	Test Orientation	Failure Mode
Conv. tool, high WP	Adv.	10/10 failed in shear
	Ret.	8/10 failed in shear. 2/10 failed in Nugger@ top sheet
Conv. tool, low WP	Adv.	10/10 failed in shear
	Ret.	9/10 failed in Adv @ bottom sheet. 1/10 failed in Nugget@ top sheet
Mod. Tool, high WP	Adv.	10/10 failed in Adv @top sheet
	Ret.	7/10 failed in Nugget @ top sheet. 2/10 failed in shear. 1/10 failed in mixed mode.
Mod. Tool, low WP	Adv.	10/10 failed in Nugget @ bottom sheet
	Ret.	10/10 failed in Adv @ bottom sheet

Conventional pin tool: In the advancing test orientation at both low and high WP, all advancing test specimens failed in shear as shown in Table II and Figure 11(a). At low WP, the minimal sheet thinning (Figure 6) drove failure away from the advancing side to nugget shear even though there was a reasonable amount of effective shear area (~0.080 in. as seen in Figure 8). At high WP, the small effective shear area (0.020-0.030 in. in Figure 8) has attracted failure to nugget shear from the advancing side although sheet thinning was increased to 0.006-0.008 in. as seen in Figure 6. For advancing test specimens, it is obvious that higher lap-shear tensile loads at low WP is due to a larger effective shear area, compared to high WP.

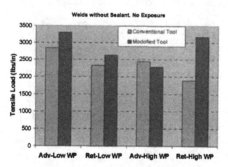

Figure 9. Lap-shear tensile load of the welds with no sealant and no exposure.

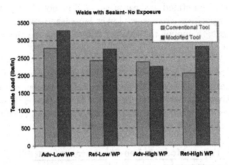

Figure 10. Lap-shear tensile load of the welds with sealant and no Exposure.

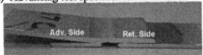

(a) Advancing test specimen failed in shear mode

(b) Advancing test specimen failed in nugget (tensile mode)

(c) Advancing test specimen failed in advancing side (tensile mode)

(d) Retreating test specimen failed in nugget (tensile mode)

(e) Retreating test specimen failed in advancing side (tensile mode)

(f) Retreating test specimen failed in shear mode

(g) Retreating test specimen failed in mixed modes of shear and tensile (in nugget)
Figure 11. Failure modes of lap-shear tensile-tested specimens.

In retreating test orientation at low WP, 9 out of 10 specimens failed in advancing side at the bottom sheet (Figure 11(e)). This is attributed to the negative sheet thinning (Figure 6) that attracted the failure location to the advancing side. Note that negative sheet thinning is in the load path in the retreating test orientation. At high WP, 8 out of 10 retreating test specimens failed in shear (Figure 11(f)). Evidently, positive sheet thinning at high WP drove the failure location away from the advancing side and the small effective shear area attracted failure to nugget shear.

Modified pin tool: In the advancing test orientation at low WP, all advancing test specimens failed in the nugget at the bottom sheet as shown in Figure 11(b). This is attributed to the fact that negative sheet thinning was not in the load path; therefore, the failure location was driven away from the advancing side to the nugget. However, the large effective shear area at low WP has prevented nugget shear failure. The result was a failure in the nugget at the bottom sheet. At high WP, high sheet thinning (~0.017 in.) attracted failure to the advancing side. All advancing test specimens failed in the advancing side at the top sheet as seen in Figure 11(c). Therefore, lap-shear tensile load is higher at low WP than that at high WP in the advancing test orientation.

In the retreating test orientation at low WP, negative sheet thinning was in the load path. All retreating test specimens failed in the advancing side at the bottom sheet as shown in Figure 11(e) due to the negative sheet thinning that attracted the failure location to the advancing side. At high WP, positive and high sheet thinning was not in the load path. That drove the failure location away from the advancing side to the nugget. Seven (7) out of 10 specimens failed in the nugget at the top sheet as shown in Figure 11(d). Two (2) out of 10 failed in shear as shown in Figure 11(f). One (1) out of 10 failed in mixed modes of shear to nugget at the top sheet as shown in Figure 11(g). Lower tensile loads at low WP compared with that at high WP was attributed to negative sheet thinning at low WP.

Salt Fog Spay Exposure

The effect of corrosive exposure on the tensile loads of welds produced with sealant location and tool combinations is presented in Figure 12. For the welds (combining welds made by both tools) without sealant, corrosive exposure produced a higher tensile load. For the welds with sealant made using the conventional tool, corrosive exposure resulted in higher tensile load as well. However, for the welds containing sealant that were produced using the modified tool, corrosive exposure resulted in a slight reduction in the tensile loads. Lap shear tensile tests for the salt fog-sprayed specimens were conducted about 1 month after the non-exposed specimens were tested. Al 7xxx alloys such as 7075 can naturally age for decades in a W temper. The stir zone of as-welded 7075 FSW is in a condition near W temper; therefore, natural aging will occur after FSW. The strength increase for the salt fog-sprayed specimens due to natural aging for about 1 month complicated the effect of corrosive exposure.

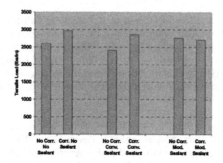

Figure 12. Effect of corrosive exposure on the tensile load of welds produced with different sealant conditions and tool combinations

Conclusions

No crevice corrosion was observed at the faying surfaces of FSW lap joints after exposure to salt fog spray for 500 hrs. However, corrosion attack occurred at the TMAZ and HAZ on the FSW top surface for all corrosive-exposed specimens. This feasibility study has shown promising results for resolving the inherent crevice corrosion issue for bringing FSW lap joints into production.

The modified pin tool having a larger pin diameter at the pin tip resulted in higher lap-shear strength compared to the conventional pin tool.

Lap-shear tensile loads and failure modes were closely related to the characteristics of FSW lap joints (sheet-thinning defect, cold-lap defect, and effective shear area). It is evident that obtaining a high strength FSW lap joint requires a pin tool design combined with weld schedule(s) that produce a lap joint with a good balance of sheet-thinning defect, cold-lap defect, and effective shear area.

References

1. B. Christner, J. McCoury, and S. Higgins, "Development and Testing of Friction Stir Welding (FSW) as a Joining Method for Primary Aircraft Structure," *Proceedings of 4th International FSW Symposium*, (Publ: TWI Ltd. Park City, Utah, 2003), paper S04A-P3.

2. B.J. Dracup, and W.J. Arbegast, "Friction Stir Welding as a Rivet Replacement Technology," *SAE Paper 1999-01-3432*, (Publ: Society of Automotive Engineers Inc., 1999).

3. L. Cederqvist, and A.P. Reynolds, "Factors Affecting the Properties of Friction Stir Welded Aluminum Lap Joints," *Welding Journal*, 80 (12) (2001), 281s-287s.

4. O.K. Mishina, and A. Norlin, "Lap Joints Produced by FSW on Flat Aluminum Alloy EN AW-6082 Profiles," *Proceedings of 4th International FSW Symposium*, (Publ: TWI Ltd. Park City, Utah, 2003) paper S07-P1.

5. G.M.D. Cantin, S.A. David, W.M. Thomas, E. Lara-Curzio, and S.S. Babu, "Friction Skew-Stir Welding of Lap Joints in 5083-O Aluminum," *Science and Technology of Welding and Joining*, 10 (3) (2005), 268-280.

FRICTION STIR LAP WELDS OF AA6111 ALUMINUM ALLOY

M. Yadava[1], R.S. Mishra[1], Y. L. Chen[2], X. Q. Gayden[2] and G. J. Grant[3]

[1]Center for Friction Stir Processing, Materials Science and Engineering, University of Missouri, Rolla, MO 65401, USA
[2]GM R&D Center, Warren, MI 48090, USA
[3]Pacific Northwest National Labs, Richland, WA 99356, USA

Keywords: Lap Joints, AA6111, Lap Shear Tests, Effective Thickness, Hooking Defect

Abstract

Lap joints of 1 mm thick AA6111 aluminum sheets were made by friction stir welding, using robotic and conventional machines. Welds were made for advancing as well as retreating side loading. Thinning in welds was quantified. Lap shear test of welds was conducted in as-welded and paint-baked conditions. Paint bake treatment improved the weld strength; but the improvement varied with process parameters. Advancing side loaded welds achieved higher strength than the retreating side loaded welds. Fracture was found to occur on the loaded side of the weld and along the thinning defect.

Introduction

Friction stir welding (FSW) has emerged as a promising solid state joining technique [1,2]. It has been extensively studied for joining of aluminum alloys for aerospace and automotive applications [3]. In FSW, the tool, which consists of a shoulder with protruding pin, is rotated and traversed along the seam. Friction between tool and work material heats and plasticizes the work material, which gets forged and extruded behind the traveling pin, thus leaving a joint [2]. The weld nugget in FSW consists of dynamically recrystallized equiaxed fine grains with better properties as compared to fusion welds [3].

Friction stir welds do not require extensive surface treatments prior to welding. Features on the pin, like threads or steps, assist in breaking and distributing the surface oxide film. In butt welds, this oxide film does not create any significant defect. Whereas, in spot and lap welds, due to characteristic material flow in FSW, this oxide film gets deposited across the weld in a continuous pattern. This is known as thinning or hooking defect, and it leads to reduction in strength of lap joints. Thinning defect and the resultant strength in lap welds are affected by process parameters. Increase in rotation rate and decrease in the travel speed of the tool has been observed to increase thinning in lap joints [4]. Increase in the pin penetration beyond the faying surface also increases the interface pull up [4,5]. As shown later in the present work, extent of thinning is different on advancing and retreating sides and so is their dependence on process

203

variables. Due to this asymmetry, welds have different strength when loaded on advancing side or retreating side [4].

Most of the earlier work on FSW lap joints had been done on plates or sheets thicker than 4 mm [4-7]. In the present work, friction stir lap welds of 1 mm thin sheets of AA6111 were studied. The aim of this work is to study the effect of process variables on the strength and heat treatment response of FSW lap welds. Heat index (HI) has been used to represent thermal input during welding and is defined as [8]:

$$HI = 10^{-4}(\text{rotation speed in rpm})^2/(\text{traverse speed in inch per minute}).$$

Experimental Procedure

AA6111 aluminum alloy sheets of 1 mm thickness in T4 temper (approximately 3-4 years naturally aged) were cut in to 3.5" X 7" coupons. Before welding, surface of these coupons were cleaned with acetone to remove oil and grease. Some of the coupons, which had significant oxide spots, were manually polished with 600 grit silicon carbide paper. Three sets of FSW runs were made. First set of welds were made with robotic machine in retreating loaded configuration (Fig.1) using 1° tilt angle, 1 kN zeroing force and 1.4 mm plunge depth. Zeroing force is the forces applied by robot on the sheet surface before the actual weld to read the z- coordinate of top surface. During zeroing-in the tool makes an indent on the surface like a hardness test. With this combination of tilt angle, zeroing force and plunge depth, limited flash was observed. A densimet tool with shoulder diameter of 12 mm and a tapered pin of 1.45 mm height with steps, was used for the above robotic runs. Four sets of tool rotation rate and travel speed were used. Second set of FSW runs were made on a conventional machine in advancing loaded configuration. Same tool, tilt angle and plunge depth were used. To compare the advancing and retreating loaded configuration a third set of runs was made on conventional machine with increased plunge depth of 1.45 mm using same tool and tilt angle.

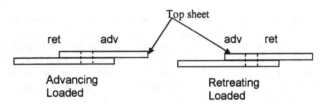

Figure 1: Advancing and retreating loading configurations for shear test of lap joints.

All welds were made on steel anvil in position control mode. Due to machine constraints, anvil or backing plates used for robotic and conventional machine welds were of different dimensions. After welding, samples were mounted for optical microscopy and percentage effective thickness (Fig. 2) was measured on both advancing as well as retreating side. Percentage effective thickness (PET) was calculated as:

PET = (minimum vertical distance from oxide film to top surface/initial top sheet thickness)*100

First set of robotic machine runs were only tested in the as-welded condition. All other welds were tested in the as-welded and welded + paint-baked conditions. The paint-bake cycle used was 175 °C for 30 min. Three samples for both as-welded and welded + paint-baked conditions were tested for each run. Before the shear tests, all samples were stored at temperature below -10°C. Weld shear specimen geometry can be found elsewhere [4]. To align the samples during shear test, 1 mm thick spacers of same material AA6111 were glued to the sample ends.

Microhardness tests were performed on the first set of welds made on robotic machine. Vickers microhardness measurements were taken at 0.5 mm below the top surface with 0.5 mm interval. Duration between FSW runs and microhardness tests was approximately 1 week. In this time period samples were kept at room temperature.

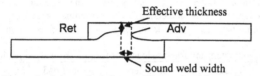

Figure 2: Schematic defect profile on advancing and retreating side.

Results and Discussion

Thinning Defect

Process parameters for all the runs are listed in Table I. Profile of the faying interface is shown schematically in Fig. 2. These features can be observed in the macrographs of weld cross sections (Fig. 3). The defect on advancing side is lifted up in the top sheet and then bent away from the nugget, whereas on the retreating side the defect is pulled into the nugget after getting pulled up. PET, as defined in the previous section, is plotted against heat index for all sets of FSW runs (Fig. 5).

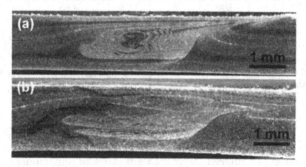

Figure 3: Macro images of welds made with (a) Robotic machine (rpm 2000, travel speed 10 ipm) (b) Conventional machine (rpm 1100, travel speed 3 ipm). Advancing (left) and retreating (right) side interfaces can be seen in the images as white curved lines.

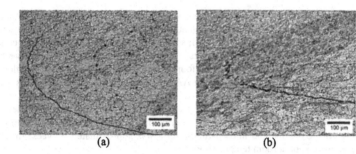

(a)	(b)

Figure 4: Advancing side defect profile in lap welds showing the difference in vertical lift of interface (nugget is on the left side of the defects): a) robotic machine welds for rpm 2000, travel speed 10 ipm and b) conventional machine welds for rpm 1100, travel speed 3 ipm.

Table I. Process parameters and fracture loads for friction stir welded lap joints.

	Run No.	Rotation Rate (rpm)	Traverse Speed (ipm)	Heat Index	Fracture Load As Welded (kN)	Fracture Load FSW+ Paint Baked (kN)
Robotic machine runs with 1.4 mm plunge depth	1	1500	15	15	4.91	-
	2	1500	10	22.5	3.12	-
	3	1500	7.5	30	2.71	-
	4	2000	10	40	2.45	-
Conventional machine runs with 1.4 mm plunge depth	1	1000	3	33.3	3.67	4.76
	2	1000	5	20	4.10	4.56
	3	1000	10	10	4.99	5.52
	4	900	5	16.2	4.01	4.18
	5	800	5	12.8	4.66	5.17
	6	700	5	9.8	4.23	5.20
Conventional machine runs with 1.45 mm plunge depth	1	1000	3	33.3	A5.30,R5.02	A*5.71,R*5.59
	2	1000	5	20	A5.00,R4.98	A5.26,R5.32
	3	1000	10	10	A3.98,R4.46	A4.53,R6.06
	4	1100	3	40.3	A4.71,R5.24	A5.70,R6.24
Parent Material	Tensile test				7.50	8.38

A*:Advancing loaded, R*: Retreating side loaded

As observed by others [4,5,10], PET for robotic welds decreased with increase in thermal input represented by heat index. Conventional machine welds showed a different pattern. PET variation with HI of robotic runs with 1.40 mm plunge and conventional machine runs with plunge depth of 1.45 mm were similar to the reported trends. In both cases, the advancing and retreating side PET decreased with increase in HI. Conventional machine welds had higher PET and less variation with HI than robotic welds.

The difference between PET values of advancing and retreating side is higher for the conventional machine welds, which were made at lower tool rotation rates, as compared to the robotic machine welds. The advancing side PET is more than retreating side PET for all HI values. PET results for 1.4 mm plunge conventional machine runs were quite different from the other two sets of runs (Fig. 5b). In these welds, though advancing side PET decreased with HI, retreating side PET showed opposite trend.

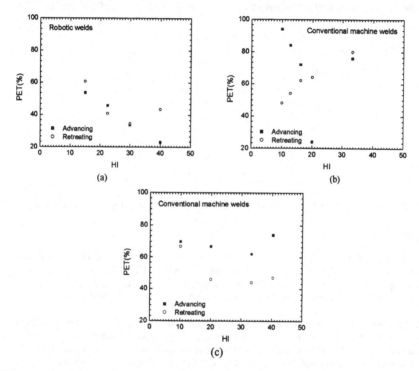

(a)

(b)

(c)

Figure 5: Percentage effective thickness (PET) variation with heat index: (a) robotic welds with plunge depth 1.40 mm (b) conventional machine run with 1.40 mm plunge (c) conventional machine run with 1.45 mm plunge.

The hooking defect profile in the friction stir lap welds is caused by the material flow during FSW, which in turn depends on the process parameters. As demonstrated by tracer techniques [10], material in the nugget undergoes vertical movement which causes the faying surface to move up. Depending on the process parameters and the resultant temperature, the material rotating under the shoulder can sweep the lifted interface from retreating side towards the advancing side, creating the retreating side defect. On the other hand, the hooking defect on the advancing side is formed because the interface moves away from the nugget as can be observed in Fig. 3.

Sound weld width (SWW) (Fig. 2) was measured for robotic welds and is plotted against heat index in Fig. 6. Sound weld width was found to decrease with increase in HI. The advancing side hook was always along the nugget boundary. Therefore, the decrease in SWW was only due to the drag of retreating side interface into the nugget. Higher HI leads to higher temperature in the weld during FSW and hence lower flow stress of the material around the tool. Therefore, the interface can be swept farther towards advancing side by the trailing side of the shoulder.

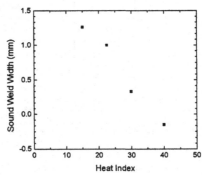

Figure 6: Plot showing sound weld width variation with heat index for robotic welds .

Lap Shear Test

Fracture load of the welds are listed in Table I. Maximum fracture load of 70.7% of parent material was recorded for conventional machine weld with a HI of 30.3 when loaded on the advancing side. Fracture load increased to a maximum of 74.5% in the paint baked condition for the weld made with highest HI (40.3) and tested in retreating side loaded configuration.

Shear test results of robotic machine welded specimens are shown in Fig. 8. Shear strength of welds decreased with increase in heat index. Shear strength of conventional machine runs with 1.4 mm plunge depth also decreased with increase in HI as shown in Fig 9(a). Influence of tool rotation rate and traverse speed on as welded and paint baked weld strength are shown in Figs. 9(b) and (c). At constant tool rotation rate, increase in traverse speed leads to higher weld strength (Fig. 9b). Higher traverse speed during welding is known to increase both the heating and cooling rates. Effect of tool rotation rate is complex and the minimum strength was recorded for the tool rotation rate of 900 rpm for a constant tool traverse rate. Although, the paint baked strength was higher for all the process parameters, the level of enhancement varied. This is not only related to the metallurgical changes in the microstructure, but also on the thinning defect.

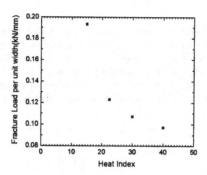

Figure 8: Fracture load per unit width vs. heat index plot for robotic welds with plunge 1.40 mm.

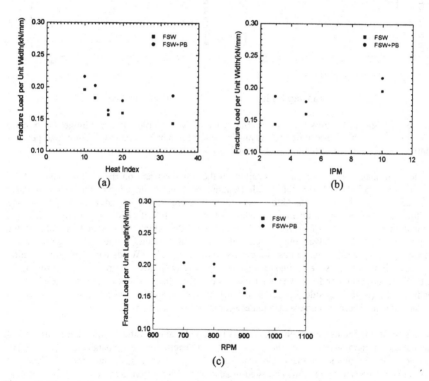

Figure 9: Shear test results for conventional machine welds with 1.40 mm plunge: (a) fracture load per unit width vs. heat index (b) fracture load per unit width vs. travel speed (constant rpm : 1000) (c) fracture load per unit width vs. rotation speed (constant travel speed: 5 ipm).

Shear test results of welds for conventional machine runs with 1.45 mm plunge depth are shown in Fig. 10. Results show the influence of loading direction and heat treatment. As-welded strength in the advancing loaded configuration achieved maxima at HI of 33.3 but in the retreating side loaded configuration it reached a plateau at HI of 20. A significant difference was observed after the paint bake heat treatment. The strength in advancing loaded configuration showed maxima at HI of 33.3 and retreating configuration showed a minima at HI of 20.

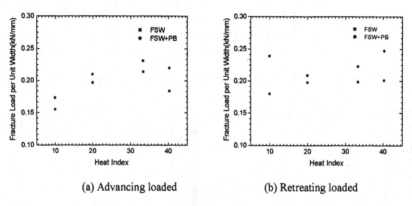

(a) Advancing loaded   (b) Retreating loaded

Figure 10: Shear test results for conventional machine welds with 1.45 mm plunge: (a) fracture load per unit width vs. heat index for advancing loaded condition, and (b) fracture load per unit width vs. heat index for retreating loaded condition.

In the weld shear test all the samples failed on the top sheet and on the loading side. All the robotic welds failed on the retreating side (loading side) and at the edge of the weld. Similarly first set of conventional machine welds, which were tested in advancing side loaded configuration, failed on advancing side edge of the weld. Effect of loading configuration was evident in the third set of conventional machine welds, which were tested in both of the configurations. All the advancing loaded welds failed at the edge of the weld nugget but the retreating loaded welds' failure position changed with heat index. Welds with 1000 rpm rotation rate and 5 ipm traverse speed failed in a distinctive way when tested in retreating configuration in both the as-welded and welded+paint baked condition. In these welds, failure started on the loading side (retreating side), progressed through the nugget and final failure occurred on advancing side. These specimens also underwent maximum rotation.

Weld strength of friction stir welded lap joints is a combination of the thinning defect as well as the matrix strength variation in the weld. Matrix strength for heat treatable alloys in FSW depends primarily on the thermal cycle, whereas thinning defect is the direct outcome of material flow in FSW. FSW lap joints, when tested in advancing loaded configuration, may fail either at the notch of the hooking defect or outside the notch, depending on the matrix strength and the stress at that location. Since on the advancing side, the hooking defect is always out of the nugget, samples loaded on this side are expected to fail out of the nugget. The same was observed in the present study, where advancing side loaded welds failed on the advancing side near the edge of the weld and never in the nugget. On the other hand, the interface on the retreating side is swept across the nugget and provides a path for the failure through the nugget. Some of the retreating side loaded welds, indeed failed through the nugget.

Microhardness Test

AA6111 aluminum alloy being a heat treatable alloy, undergoes precipitate coarsening, dissolution and reprecipitation depending on the thermal cycle. Previous work on other heat treatable aluminum alloys show that precipitate size distribution and volume fraction depend not only on the process parameters but also on the location from the seam [11]. This is indicated by the microhardness plots for robotic welds at different HI in Fig. 11, where the hardness profiles across the welds are shown to vary with HI. The highest hardness of 89 HV was observed in the HI-30 specimen at 6 mm away from the weld center on the advancing side of the weld. Hardness minima location in the weld moved closer to the center with an increase in heat index up to 30. For HI of 40, lowest hardness position is farther away from center than that for HI of 30.

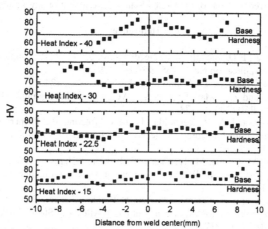

Figure 11: Vickers microhardness plot for robotic runs, showing variation of hardness across weld at 0.5 mm below the top surface.

Relative role of thinning and thermal cycle on the weld strength is not clear at this point of study. As shown in Fig. 10(a), the strength of advancing loaded welds increased with HI, even though PET decreased with heat index (Fig. 5(c)). This means that PET only is not the strength determining factor. Weld strength variation with heat index was also affected by the ~4% variation in plunge depth, which can be seen by comparing the weld strengths of the two sets of runs made on conventional machine with same loading configuration (Fig. 9(a) and 10(a)). Contact area between the tool and work material increases with increase in plunge, leading to increase in heat generation. Apart from thermal effects, higher plunge brings the shoulder closer to interface, thus also affecting thinning. A combination of these effects might have lead to this observed difference in strength of the two set of runs.

Conclusions

Percentage effective thickness was higher in the welds made at lower rotation rate on conventional FSW machine. A change in plunge depth has been found to affect the strength variation with heat index. Thinning in the welds was also observed to increase with heat index. Effect of paint bake treatment on weld strength depended on the process parameters: tool rotation rate and traverse speed. Low heat index runs, in general, had better strength in as-welded and paint baked conditions. Heat treatment response of lower and higher heat index welds was better than those of intermediate ones.

Acknowledgments

This work was performed under the NSF-IUCRC for Friction Stir Processing and the support of NSF, Boeing, GM, Pacific Northwest National Laboratory and Friction Stir Link for the UMR site is acknowledged. This report was prepared as an account of work sponsored by an agency of the United States Government. The views and opinions of authors expressed herein do not necessarily state or reflect those of the United States Government or any agency thereof.

References

1. W.M. Thomas, E.D. Nicholas, J.C. Needham, M.G. Murch, P. Templesmith, C.J. Dawes, G.B. Patent Application No. 9125978.8(Dec. 1991)
2. C. Dawes, W. Thomas, TWI Bulletin 6, Nov./Dec. 1995, 124
3. R.S. Mishra and Z.Y. Ma, "Friction Stir Welding and Processing", Materials Science & Engineering, R50 (2005), 1-78.
4. L. Cederqvist and A. P. Reynolds "Factors Affecting the Properties of Friction Stir Welded Aluminum Lap Joints", The Welding Journal Research Supplement, 80 (2001), 12, 281-287.
5. A. Elrefaey, M. Gouda, M. Takahashi, K. Ikeuchi, "Characterization of Aluminum/steel lap Joint by Friction Stir Welding", Journal of Materials Engineering and Performance, 14 (2005), 1, 10-17.
6. K. Kimapong and T. Watanabe, "Effect of Welding Process Parameters on Mechanical Property of FSW Lap Joint between Aluminum Alloy and Steel", Materials Transactions, 46(2005), 10, 2211-2217.
7. M. Merklein, A Giera, M. Geiger, "Deep drawing of Friction Stir welded Thin sheet Aluminum Steel Tailored Hybrids", Steel Research Intl., 76(2005), 2/3, 250-256.
8. W.J Arbegast, "Modeling Friction Stir Joining As A Metal Working Process", Hot Deformation of Aluminum Alloys III, 2003, 313-327.
9. K. Matsumoto, S. Sasabe, "Lap Joints of Aluminum Alloys by Friction Stir Welding", Third International Symposium on Friction Stir Welding; Kobe; Japan; 27-28 Sept. 2001, 11, 2002
10. K. Colligan, "Material Flow Behavior during Friction Stir Welding of Aluminum", The Welding Journal Research Supplement, 78(1999), 12, 229S-237S.
11. V. Dixit, R. S. Mishra, R. J. Lederich and R. Talwar, "Influence of Process Parameters on Microstructural Evolution and Mechanical Properties in Friction Stirred Al-2024 (T3) Alloy", submitted for publication, 2006.

FRICTION STIR WELDING OF AN AEROSPACE MAGNESIUM ALLOY

X. Cao, M. Jahazi and M. Guerin

Aerospace Manufacturing Technology Center, Institute for Aerospace Research, National Research Council Canada, 5145 Decelles Avenue, Montreal, Quebec, H3T 2B2, Canada

Keywords: Friction stir welding, Magnesium alloy, Porosity, Microstructure, Hardness.

Abstract

Hot rolled AZ31B-H24 sheets with a thickness of 4.95 mm were joined using an MTS ISTIR friction stir welding machine. The cavity defects in the butt joints were studied and two mechanisms are observed. The pores are thought to be due to volume deficiency, and/or poor material flow and mixing. Excessive metal loss may cause subsurface pores which usually occur at the upper half of the welding nugget at the advancing side. Poor material flow and mixing may form internal pores which are usually elongated and curved, with the concave side in the welding direction. The grain size in the weld nugget was observed to be larger than that in the thermal-mechanically affected zone or heat-affected zone due to recrystallization, causing a drop in Vickers microindentation hardness.

Introduction

Friction stir welding (FSW) is an emerging technology to join magnesium alloys, a key lightest structural material. The basic concept of FSW is remarkably simple. A tool pin is inserted into the area to be joined and stirs the material to be joined while the heat generated by the contact between the tool shoulder and the work-piece surface softens the material and allows its movement. A fully consolidated joint is produced by the movement of material from the front of the pin to the back of the pin. To date, however, the main research and development efforts in friction stir welding have been concentrated on aluminum alloys [1,2]. The application of FSW to magnesium alloys presents many attractive advantages as it can significantly reduce weld defects such as oxide inclusions, porosities, cracks, and distortions, commonly encountered in fusion welded joints. Only very limited work has been conducted on magnesium alloys. Clearly, comprehensive investigation into FSW of magnesium alloys is essential for the wider utilization of both magnesium alloys and the new welding technique. In this work, the effect of shoulder plunge depth on weld quality was studied.

Experimental Method

The experimental alloy was hot rolled AZ31B-H24 magnesium alloy sheet with dimensions of 1200 × 1200 × 4.95 mm. The alloy has nominal composition of Al 2.5-3.5 wt.%, Zn 0.7-1.3 wt.%, Mn 0.2-1.0 wt. % and the balance Mg. The 400 × 90 × 4.95 mm specimens were cut from the as-received sheet with the end surfaces machined along the specimen length. After the faying surfaces were cleaned, the specimens were fixed on a support plate that was positioned on the anvil (backing plate) of an ISTIR MTS FSW equipment. The welding tool had a scrolled shoulder with a diameter of 19.05 mm and a ¼-20 left hand threaded pin with a diameter of 6.35 mm, a thread spacing of 1.27 mm and a pitch of 0.8 thread/mm. The length of the threaded pin was 4.45 mm. The pin was made of H13 steel. The welding direction was perpendicular to the roll direction of the work-piece. The butt joints were welded at tool rotational rate of 1750 rpm

213

clockwise and welding speed of 6 mm/s. The pin length (i.e. distance between the pin bottom and the tool shoulder) was 4.7 mm. The penetration depth was controlled through shoulder plunge depth from 0.07 to 0.25 mm. Here shoulder plunge depth is the distance that the tool shoulder has penetrated into the top surface of the workpiece, measured normal to the weld panel surface.

For each FSWed butt-joint, at least three metallurgical specimens were cut. These specimens were mounted using cold-setting resin, ground and polished to produce a mirror-like finish. To reveal the metal flow patterns, some specimens were etched in nital [5 mL HNO_3 (conc.), 100 mL ethanol (95%) or methanol (95%)]. Other specimens were etched in acetic picral [10 mL acetic acid (99%), 4.2 g picric acid, 10 mL H_2O, 70 mL ethanol (95%)] for about 6 seconds to reveal grain structures. The macrostructures and microstructures were examined using an Olympus Inverted System Metallurgical Microscope GX71 equipped with Olympus digital camera and AnalySIS Five digital image software. The shoulder plunge depth was measured and the average values were reported. The grain size was obtained according to ASTM standard E112 using linear intercept method. The Vickers microindentation hardness was measured at the mid-thickness of some butt joints using Struers Duramin A-300 hardness tester at a load of 100 g force, a dwell period of 15 seconds and an interval of 0.3 mm.

Results and Discussion

Defects

Figure 1 shows the top and back surface morphologies of the FSWed butt joints. At shoulder plunge depth of 0.07 and 0.08 mm, little flash was produced on the weld surface. When shoulder plunge depth is 0.18 mm, flash was observed along the entire length of the retreating side (RS) and little flash appeared at the advancing side (AS). Heavy and uniform flash was produced at larger shoulder plunge depth (0.24-0.25 mm). Therefore, the flash formed at the RS increases with increased shoulder plunge depth. The flash was ejected out due to the softening of the metal and deep shoulder plunge. At low shoulder plunge depth, some surface flaws such as surface depressions and notches was observed as shown in Figure 1a,b. The joint morphologies on the back surface were well formed with no indication of lack of penetration. In addition, little distortion was observed on the welded joints.

At shoulder plunge depth of 0.07 mm, subsurface cavity defects were widely observed. Figure 2 shows a huge subsurface porosity with a length up to approximately 3.5 mm. This porosity almost reached the top weld surface. Figure 3 shows some porosity where several pieces of metal are observed within the cavity indicating that debonding and collapse of metal appear during the formation of the porosity. Long cavity defect with saw-tooth shape was also observed as shown in Figure 4. The appearance of these cavity defects is further confirmed at shoulder plunge depth of 0.08 mm as demonstrated in Figures 5 and 6. In Figure 5e, some small microporosities are also visible within the weld nugget (WN). When shoulder plunge depth varies from 0.18 to 0.25 mm, no cavity defects were observed on all the transverse sections. To further confirm whether there are any cavities within the weld joints, two horizontal planes at distances of 4.01 and 3.35 mm from the base of the work-piece were examined. Some porosity is still observed at shoulder plunge depth of 0.25 mm as shown in Figure 7.

Various terms have been used to describe such volume defect as cavity, porosity, void, running voids, buried voids, wormhole, "tunnel", tunnel discontinuity, etc. The formation of cavity defects has not been systematically investigated but they are probably formed by two main mechanisms (i) volume deficiency, and (ii) poor materials flow and mixing. The first rule to avoid the formation of cavity defects is that volume loss of metal excluding that for joint

214

thinning should be no more than the volume gain obtained during FSW. The metal volume loss may occur due to flash. The original porosity in cast materials also contributes to some metal loss. If tool shoulder is poorly designed, it will fail to confine the metal and may cause metal loss. The volume gain of metal is mainly from the top surface of the work-piece due to the shoulder plunge. If the metal volume loss excluding that for joint thinning is more than the gain obtained during FSW, the cavity may be formed. As shown in Figures 2-6 most pores are nearly close to the top surface of the work-piece (i.e. subsurface porosity) and they are nearly located at the upper half of the WN at the AS indicating that the formation of these pores may be due to the volume loss of metal. This is further supported by the flow lines appearing around the porosity as shown in Figure 6. In this work, the pin has left hand threads and rotates clockwise. Therefore, the material in the WN moves upwards on the AS. This is clearly demonstrated in the bent flow lines towards the left upward direction at the AS (Figure 6a-d), indicating that metal may be expelled out from the AS and forms flashes at the RS due to the shoulder rotation. In addition, this region is also the farthest to be fed by the plasticized material from the RS. Therefore, these defects are most probably formed at the shoulder region near the half WN on the AS. By contrast, little subsurface porosity is observed in the upper half of the WN near the top surface on the RS since the material in this region move downwards due to the clockwise movement of the left hand threaded pin and the constraint of the shoulder (Figures 6). The methods to avoid subsurface porosity are usually to (i) increase shoulder plunge depth, (ii) increase forge force [3,4], (iii) use work-piece with low levels of porosity and (iv) employ well-designed welding tools.

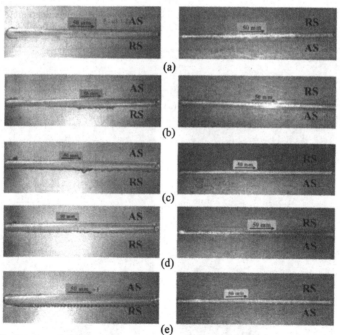

Figure 1. Surface morphologies of weld joints at shoulder plunge depth of (a) 0.07, (b) 0.08, (c) 0.18, (d) 0.24 and (e) 0.25 mm. The left photographs are the front views and the right back views. Welding direction is from the left to the right.

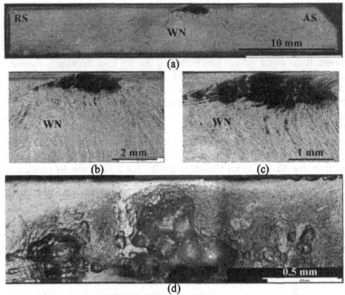

Figure 2. (a) Overview of transverse section obtained at shoulder plunge depth of 0.07 mm. Micrographs (b-d) showing subsurface porosity at higher magnification.

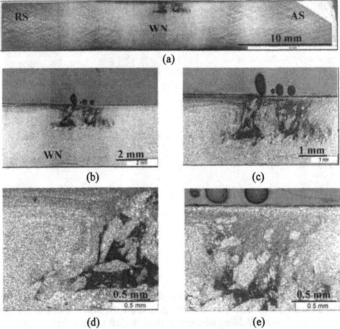

Figure 3. (a) Overview of transverse section obtained at shoulder plunge depth of 0.07 mm. Micrographs (b-e) showing the subsurface pores at higher magnification.

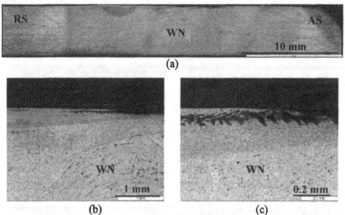

Figure 4. (a) Overview of transverse section obtained at shoulder plunge depth of 0.07 mm. Micrographs (b-c) showing subsurface defects at higher magnification.

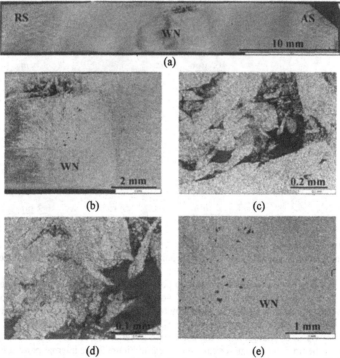

Figure 5. (a) Overview of transverse section obtained at shoulder plunge depth of 0.08 mm. Micrographs (b-d) showing the subsurface porosity at higher magnification and (e) indicating microporosity in the welding nugget.

217

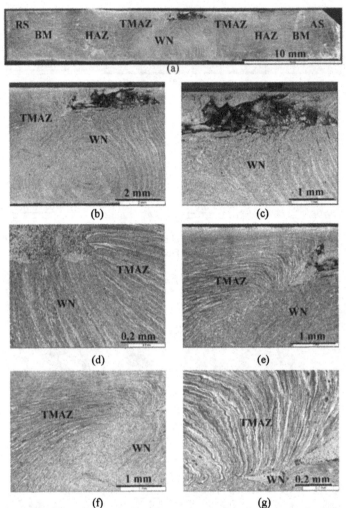

Figure 6. (a) Overview of transverse section obtained at shoulder plunge depth of 0.08 mm. Micrographs (b-g) showing the subsurface porosity or flow lines at higher magnification.

Another possibility of pore formation is due to poor material flow or mixing as typically shown in Figure 7. For instance, the formation of vortex may cause pores. It was reported that inadequate stirring and mixing can be caused by too fast joining speed, or wrong combination of welding speeds and pin rotational speed [3,5,6]. When the traversing and the tool rotational speeds are not vigorous enough to stir the plasticized material in front of the probe to completely fill the rear of the trailing edge, pores can be formed. In Figure 7, the internal pores are elongated and curved, with the concave side in the welding direction indicating that their formation is directly related to the material flow and mixing. The poor stirring and mixing can also be caused by low heat input [6], particularly for dissimilar joints. At low heat input (i.e. low temperature) material mixing is difficult, leading to the formation of discontinuity and ultimately initiate pores. Clearly, the optimized process condition may avoid poor stirring and mixing, and thereby decrease the formation of these pores.

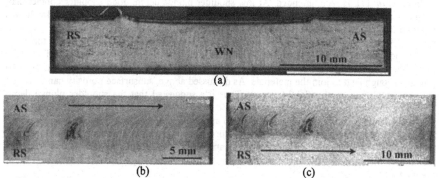

(a)

(b) (c)

Figure 7. (a) Overview of transverse section obtained at shoulder plunge depth of 0.25 mm. Micrographs indicating pores at horizontal planes from the joint base (b) 4.01 and (c) 3.35 mm (Arrows indicate welding direction.).

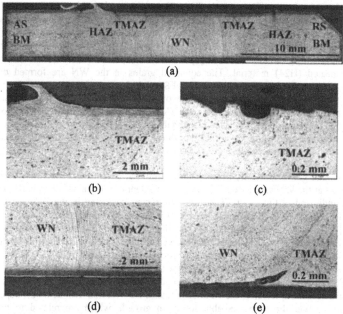

(a)

(b) (c)

(d) (e)

Figure 8. (a) Overview of transverse section obtained at shoulder plunge depth of 0.24 mm. Micrographs indicating (b-c) surface and (d-e) root flaws.

In addition to cavity defects as discussed above, other flaws observed in this work are surface grooves or notches and heavy flashes as shown in Figure 8. The surface grooves or notches are small but they may influence the fatigue properties of the weld joints. The large mass of flash may cause thinning of the work-piece (i.e. decrease in joint thickness) and thereby lower the load-bearing area. For example, a shoulder plunge depth of 0.25 mm for 4.95-mm sheet means a

219

decrease of approximately 5% in thickness. Therefore, too deep shoulder plunge is not desired. A typical value around 0.2 mm, which is in agreement with that reported by Johnson [4] and Schmidt *et al.* [7], is recommended. At low shoulder plunge depth, good shoulder contact may not be well realized with the top surface of the work-piece, as illustrated in Figure 1a. It was reported that, the welding heat is mainly produced by the friction between the shoulder and weld surface [7]. In a case analysis by Schmidt *et al.* [7], the shoulder friction with the top weld surface occupied about 86% of the total heat generated, with the other contributions of 11% for the probe side and 3% for the probe tip [7]. Therefore, good shoulder plunge is critical to produce enough heat to join the materials. As discussed above, subsurface porosity can be found at shoulder plunge depth of 0.7-0.8 mm. Thus it's suggested that the shoulder plunge depth should be controlled above 0.1 mm. In this work, the pin-to-back distance (penetration ligament, or pin clearance) was up to 0.18 mm (i.e. up to approximately 4% thickness of the work-piece). The lack of penetration was not observed (Figures 2-6) and the back surface of the joints was well formed (Figure 1) indicating that a pin clearance up to 0.2 mm is feasible. Lower values of pin-to-back distance may cause contact between the pin and the support plate, or even fracture of the pin tool.

Microstructure and Hardness

Figure 9 shows typical microstructures obtained at shoulder plunge depth of 0.25 mm. The WN and thermal-mechanically affected zone (TMAZ) mainly consist of equiaxed grains. The heat-affected zone (HAZ) consists mainly of equiaxed grains near the TMAZ side and elongated grains close to the BM side. The base metal has both equiaxed and pancake grains with various sizes. The heterogeneity in grain structure may originate from deformation and incomplete recrystalization during the hot extrusion [8] since the base metal AZ31B-H24 is hot rolled and partially annealed (H24) material. The equiaxed grains in the WN are formed due to the occurrence of dynamic recrystalization during FSW [9]. It was reported that in the TMAZ the grains are severely deformed, rotated, and elongated, but usually do not recrystallize for aluminum alloy [1]. In this work, however, the TMAZ was mainly composed of equiaxed grains indicating that recrystalization has taken place in this zone. The grain structures at the top surface under the shoulder are equal or similar to those in the nugget (Figure 9) indicating that recrystallization has also taken place in this region.

It was found that grains gradually become larger from the BM through the HAZ, to the TMAZ and culminate at the WN as shown in Figure 9. The extent of the grain growth in the TMAZ and the WN is indicated in Figure 10. The grain coarsening occurring in the WN and the TMAZ during FSW of AZ31B-H24 alloy has been related to the annealing effect induced by welding heat [8,10]. By contrast, Nagasawa *et al.* [11] and Katoh *et al.* [12] observed grain refinement in the WN while Woo *et al.* [13] did not observe any significant difference in the grain size between the WN and BM during friction stir processing of annealed AZ31B-O alloy. Although the results obtained in the present work agree with those reported by Lee *et al.* [8], Lim *et al.* [10] and Cao *et al.* [2], further work is still needed to clarify the effect of processing parameters on grain size evolution. Clearly, the sizes of grains in the WN and TMAZ might depend on the recrystallization and the time available for grain growth, which are related to the process conditions.

As shown in Figure 10, the hardness decreases gradually from the BM through the HAZ, the TMAZ and then to the WN where the lowest hardness appears. The average hardness in the WN is approximately 79% of that in the BM. The decrease of hardness may be correlated to the grain growth [8].

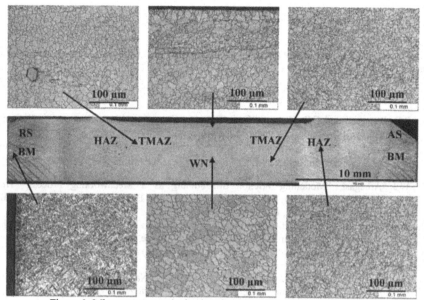

Figure 9. Microstructures obtained at shoulder plunge depth of 0.25 mm.

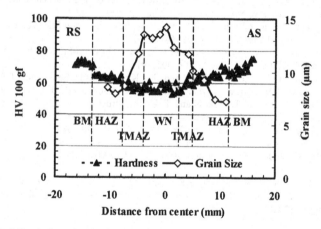

Figure 10. Microindentation hardness and grain size at shoulder plunge depth of 0.25 mm.

Conclusions

Two types of cavity defects, i.e. subsurface and internal pores, are observed in the friction stir welded magnesium alloy joints. The subsurface cavities usually occur at the upper half of the welding nugget at the advancing side, due to volume deficiency caused by excessive metal loss during the welding. The internal pores are usually elongated and curved, with the concave side in welding direction and are formed probably due to poor material flow and mixing. Grain growth was observed in the FSWed AZ31B-H24 alloy butt joints, causing the decrease in Vickers microindentation hardness.

Acknowledgments

Great thanks are due to R. Mehta, a co-op student from McGill University, for the metallurgical analysis of the weld joints.

References

1. P.B. Prangnell and C.P. Heason, "Grain Structure Formation during Friction Stir Welding Observed by Stop Action Technique," *Acta Materialia*, 53 (2005), 3179-3192.
2. X. Cao, M. Jahazi, and M. Mehta, "Friction Stir Welding of AZ31B-H24 Magnesium Alloy butt joints" (Paper presented at the 6th Int. Symp. on Friction Stir Welding, Montreal, Quebec, Canada, 10-12 Oct. 2006).
3. W.J. Arbegast, "Modelling Friction Stir Joining as a Metalworking Process," *Hot Deformation of Aluminum Alloys III*, ed. Z. Jin, A. Beaudoin, T.A. Bieler and B. Radhakrishnan (Warrendale, PA: TMS 2003), 313-327.
4. R. Johnson, "Forces in Friction Stir Welding of Aluminum Alloys" (Paper presented at the 3rd Int. Symp. of Friction Stir Welding and Processing, Kobe, Japan, 27-28 Sept. 2001).
5. Y.S. Sato and H. Kokawa, "Friction Stir Welding (FSW) Process," *Welding Int.* 17 (11) (2003), 852-855.
6. Y.G. Kim, H. Fujii, T. Tsumura, T. Komazaki, and K. Nakata, "Three Defects Types in Friction Stir Welding of Aluminum Die Casting Alloy," *Mater. Sci. Eng. A* 415 (2006), 250-254.
7. H. Schmidt, J. Hattel and J. Wert, "An Analytical Model for the Heat Generation in Friction Stir Welding," *Modelling Simul. Mater. Sci. Eng.* 12 (2004), 143-157.
8. W.B. Lee, Y.M. Yeon and S.B. Jung, "Joint Properties of Friction Stir Welded AZ31B-H24 Magnesium Alloy," *Mater. Sci. Tech.* 19 (6) (2003), 85-90.
9. J.A. Esparza, W.C. Davis, E.A. Trillo, and L.E. Murr, "Friction-Stir Welding of Magnesium Alloy AZ31B," *J. Mate. Sci. Letters*, 21 (2002), 917-20.
10. S. Lim, S. Kim, C.G. Lee, C.D. Yim, and S.J. Kim, "Tensile Behaviour of Friction-Stir-Welded AZ31-H24 Mg Alloy," *Metall. Mater. Trans. A* 36A (6) (2005), 1609-1612.
11. T. Nagasawa, M. Otsuka, T. Yokota and T. Ueki, "Structure and Mechanical Properties of Friction Stir Weld Joints of Magnesium Alloy AZ31," *Magnesium Technology 2000*, ed. H.I. Kaplan, J. Hryn and B. Clow (Warrendale, PA: TMS 2000), 383-387.
12. K. Katoh, H. Tokisue and T. Kitahara, "Microstructures and Mechanical Properties of Friction Stir Welded AZ31 Magnesium Alloy," *J. Light Metal Welding and Construction* 42 (3) (2004), 130-39.
13. W. Woo, H. Choo, D.W. Brown, P.K. Liaw, and Z. Feng, "Texture Variation and its Influence on the Tensile Behaviour of a Friction-Stir Processes Magnesium Alloy," *Script Materialia*, 54 (2006), 1859-1864.

FRICTION STIR JOINING OF THERMOPLASTICS

S.Mattapelli[1], W.Arbegast[2], and R.Winter[1]

[1]Department of Chemical and Biological Engineering
[2]Advanced Materials Processing and Joining Center
South Dakota School of Mines & Technology; Rapid City, SD, 57701, USA

Keywords: Thermoplastics; Friction stir joining; polypropylene

Abstract

Over the last two years we have investigated the joining of thermoplastic polypropylene using the Friction Stir Joining (FSJ) technique. Our work has focused on optimizing the FSJ process parameters, specifically material travel speed, material pre-heating temperature, material pre-heating time, pin tool rotation speed, and pin tool thread pitch. The FS joint was characterized using optical microscopy, scanning electron microscopy, differential scanning calorimetry (DSC) and mechanical property analysis including tensile strength and percent elongation. As a result of this research a process map was developed which shows that higher quality joints are obtain at high pin tool rotation speeds and low material travel speed. It was observed that the tensile strength of the friction stirred joints was ~90% of the parent material. DSC shows that the degree of variation of the crystallinity at FS joint interface is increased and that the average crystallinty throughout the FSJ sample remains constant. Work is currently ongoing to spatially resolve crystalline density, flow patterns and the temperature distribution through the FSJ part.

Introduction

Over the past few years, the process of composite materials joining has undergone significant advances. In spite of the development of some novel joining processes for the production of high quality plastic parts, considerable effort is going into inventing better joining processes for structural and high performance applications. A good joining method should meet several functional criteria such as: reproducibility of joint efficiency, suitability for small and large bonding areas, ability to join materials in various geometries, minimal surface preparation, minimal use of specialty equipment, potential for production applications and retention of joint integrity in a variety of environments and load configurations [1]. One such new method which meets these criteria is friction stir joining (FSJ) which also has low machine costs when applied to polymeric systems [2]. The Advanced Materials Joining and Processing Center has extensive experience in joining metals using friction stir joining. It is thus a natural extension of this expertise to explore optimizing the friction stir joining of polymeric materials. The main goal of the work herein reported is to understand the FSJ technique when applied to thermoplastic materials and the effect of FSJ process parameters on the joint quality. In order to achieve this goal the following objectives define our research effort: designing of the FSJ experimental set-up, development of the FSJ technique, mapping process parameters to joint quality and characterizing the joint.

Background

FSJ was first invented for joining of metals and was first successful demonstrated with aluminum alloys in 1991 at The Welding Institute (TWI) in England [2]. Recognizing that engineering plastics have a higher strength to weight ratio when compared to many metals, has resulted in many existing metal parts being redesigned and replaced by their plastic analog. This had driven researcher in recent years to extend the FSJ process for the joining of polymers [1]. Friction stir joining (FSJ) of polymers is a new solid state technique to join plastic parts without going through the melt phase. The first successful FSJ process for polymers was reported in 1999 by the Brigham Young University researcher group [3]. There after the BYU research group reported on the 'effect of FSJ process on microstructure of polymers'. The joining of polymers by this process is similar to the conventional polymer extrusion in which a high deformation occurs through a die cavity.

Machine set up and Experimental procedure

In the interest of developing the FSJ technique at SDSM&T, the design and fabrication of the FSJ experimental set-up was required. After the success of producing the FSJ joints on thermoplastic materials using a simple milling machine at BYU, a similar milling machine set-up was developed at SDSM&T with several modifications [2]. The milling machine used at SDSM&T was a Sharp mill which is shown in Figure 1. The modifications to the mill to create a functional FSJ system included: a clamp around the quill to hold the shoe stationary with the help of turnbuckles, an anvil on the mill table acts as a support for placing the thermoplastic sheet during FSJ process, and a temperature controller for controlling the set point temperature provided for the shoe. The additional modifications beyond the first generation system developed by BYU included temperature sensors and meters for obtaining the temperatures of the pin-tool and the part of the shoe where heat is not provided. These are shown in Figure 1a and 1b.

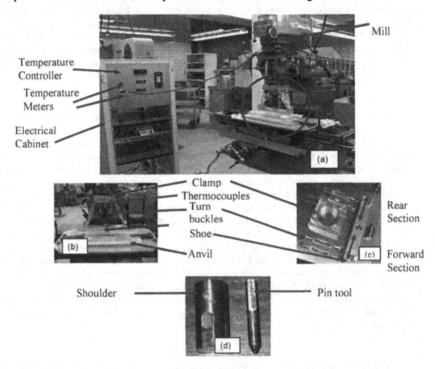

Figure 1. (a) Sharp Milling Machine with FSJ Experimental setup (right) and Electrical Cabinet (left) , (b) FSJ Experimental setup, (c) Accessories of FSJ experimental set up , and (d) Shoulder and Pin tool.

The tooling consists of stationary aluminum shoe, rotating pin tool, shoulder, heater, steel clamping system, and thermocouples. The heated shoe is used to soften the polymer material and also for trapping the material stirred by the pin tool within the joint region. The forward section of the shoe is heated using an electrical cartridge heater inserted into pre-drilled holes in the shoe. The pin tools are made of H13 tool steel which is used for its resistance to degradation during the FSJ process. The pin tool tip of 6 mm in length is tapered to a diameter of 3 mm from a shank diameter of 9.5 mm. Three pin tools with having thread pitches of coarse (4.2 mm), medium (2.1) mm and fine (1 mm) are used in this study.

Preliminary trials were first done to develop the FSJ technique and there by to achieve acceptable FSJ joints using the polypropylene (PP) as the "model" thermoplastic material. Initial studies have been performed on a single sheet of polypropylene in order to: verify the operation of the newly designed FSJ experimental set-up and also as a precursor to conduct the experiments for developing the FSJ technique in making quality FSJ joints. Polypropylene was chosen for its known suitability for the FSJ process [2], cost, and availability.

The FSJ process parameter set investigated includes: the rotation speed of the pin tool, table speed with which the material sheet travels during the process, pre-heating temperature, penetration ligament, and shoe pressure. Table 1 summarizes the processing variables used in this study and Table 2 provides the parameter ranges.

Table I. Processing variables for the FSJ of thermoplastic materials.

Process Variables			
Pre-Heat	Joining	Post Heat	Comments
Pre-Heat Time	Pin Tool Diameter	Post Heat Temperature	1. Shoe pressure can be specified by setting the gap between the shoe and the anvil.
Pre-Heat Temperature	Pin Tool Thread Pitch (PTTP)	Post Heat Cool Down Rate	
Pre-Heat Shoe Pressure	Pin Length	Post Heat Shoe Pressure	2. Penetration ligament is difference between pin length and gap between shoe and the anvil.
	Penetration (pin tool) Ligament		
	Rotation Speed of the pin tool		3. Pin length is extension past bottom of shoe.
	Travel Speed of the table		
	Shoe Pressure		4. PTTP is the distance between 2 threads of the pin tool.

Friction Stir Joining of polymers is accomplished in a series of steps. First the part (in this study the PP sheet) is to be held tightly on to the anvil using iron rods and clamping system as shown in the Figure 2. Then the region, at which the polymeric parts are to be joined, is pre-heated by the forward section of the shoe. The material to be processed is continuously brought into contact with the pre-heated shoe to achieve the desired temperature.

Table II. FSJ process parameters and their value or ranges.

Parameters	Range	Optimized Process Parameters
Rotation speed of the pin tool	900 - 3000 rpm	3000 rpm
Travel speed of the table	38.1-241.3 mm/min (1.5 – 9.5 inch/min)	38.1 mm/min (1.5 inch/min)
Pre-heat temperature	120 - 175°C	120°C
Pre-heat time	30 – 500 seconds	200 seconds
Pin tool thread pitch (PTTP)	1, 2.1 & 4.2 mm	4.2 mm
Constant pin tool (penetration) ligament	0.5 mm	0.5 mm
Constant shoe pressure	5.6 mm (0.22 inches)	5.6 mm (0.22 inches)

The shoe temperature is dictated by the thermal properties of the thermoplastic material and the pre-heat time. The pin tool is rotated with a suitable rotation speed to achieve sufficient frictional energy and the rotating pin tool is then plunged into the joint region, to further elevate the temperature in the region of the pin tool. The thermal process efficiency is dependent on the partitioning of the flow of energy between that which goes into the material and that which goes into the pin tool and is conducted away from the joint. Through the motion of the anvil the pin tool, in relation to the part, advances along the joint. Material in front of the pin tool is moved and deposited behind the pin tool. The process proceeds until the pin tool reaches the edge of the joint as shown in Figure 3.

Figure 2. Clamping system of the thermoplastic material on one side.

During the FSJ process material extruded by the pin tool is trapped under the shoe if sufficient shoe pressure is applied. The non-heated rear section of the shoe then comes in contact with the newly formed FS joint resulting in hardening of the FS joint and surrounding material while simultaneously smoothening the top surface of the FS joint. At the point where it is desired to terminate the joint the pin tool rotation is stopped and the shoe is raised up from the joint edge. It is important that before raising the shoe from the joint edge, that the joint should be cooled under the shoe pressure because the extruded material will eject out of the joint if it is not hardened under sufficient shoe pressure.

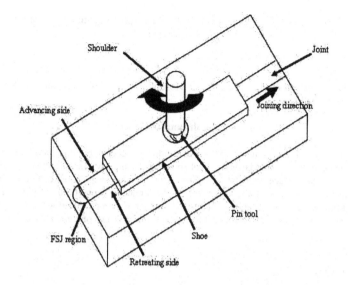

Figure 3. Schematic drawing of FSJ of thermoplastics system.

Process parameters were varied to explore the region and appropriate ranges to obtain visually good joints. Initial values are used based on the optimized reported parameter values of previous work [BYU] using polypropylene [2]. With the polypropylene used in this study these optimized values did not yield a good visual joint, which required further optimization of FSJ process parameters.

Characterization of mechanical properties, fracture surfaces and percent crystallinity of the friction stir joints was performed to evaluate the performance of the FSJ technique. The tensile strength of the untreated parent material and FSJ material are measured and compared assess what if any effect FSJ has on the material and in particular the joint strength. Morphological analysis is also performed on the fracture surfaces of both the parent material and the FSJ material to assess the effect of the FSJ process on the fracture surfaces morphology. The macroscopic and microscopic ductility of the fracture surfaces were observed using optical microscopy (OM) and scanning electron microscopy (SEM).

Since it was observed that the ductility of the material changed, work was initiated to assess the possible causes for this change. Possible factors responsible for the transition from ductility to brittleness include: degree of crystallinity, stress concentration, contamination, temperature, poor fusion, degradation, strain rate, and chemical contact [5]. It was hypothesized that a change in the degree of crystallinity was a possible a major contributor to this transition. A study to investigate this hypothesis was initiated where percent crystallinty was evaluated in different regions of the FS joint and surrounding material using differential scanning calorimetry. Since the FSJ process subjects the material to a varied thermal history DSC analyses were performed at the center of the friction stir joint, the interface of the friction stir joint, and the non-processed material. Additionally the heat effected material under the shoe, outside of the shoe region and untreated PP material were analyzed for percent crystallinity.

Percent crystallinity (%Xc) is calculated using the Equation (1). The reference heat of melting ($H_{100\%}$) of 100% crystalline PP is 190J/g [8].

$$X_c = \frac{H_c}{H_{100\%}} \quad \text{---------- (1)}$$

Where H_c = Actual heat of melting; $H_{100\%}$= Heat of melting of 100% crystalline PP = 190 J/g; % Xc= % crystallinity.

Results and Discussions

The designed system was capable of producing acceptable joints based on visual quality. At the lower rotation speeds of pin tool, poor visual quality was seen with a rough surface appearance and surface indentation (joint surface is lower in level compared to untreated material surface level) was also observed. At the higher rotation speeds of pin tool and slower forward travel speed of the table, good visual quality FSJ was produced with out any top surface indentation (see Figure 4). See Table II for the optimized values of a good visual joint.

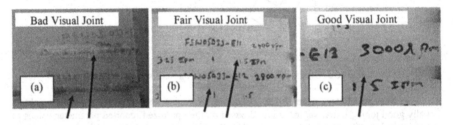

Figure 4. Visual Appearance of FSJ in 6.35 mm (¼ inch) Polypropylene Sheet Showing (a)Bad, (b) Fair, and (c) Good, Visual Joint Quality using Coarse PTTP of 4.2 mm, 120°C Pre-Heat Shoe Temperature, and 200 seconds Pre-Heat Time.

At lower rotation speeds, joints with rough surface appearance, surface indentation and lack of material within the joint after FSJ were created and were characterized as having 'bad visual quality'. This qualitative analysis was utilized to expedite process parameter survey. An FS joint of "fair visual quality" is characterized by having a reduction in surface roughness, a decrease in the surface indentation and reduction in the lack of material within the joint after FSJ. Joints of "good visual quality" occurred at higher rotation speeds and are characterized as joints with smooth surface appearance, no surface indentation and good flow of material. More over for a "good visual joint", it is very hard to differentiate between the joint and parent material. Figure 5 shows the processing map of the FSJ of 6.35 mm (1/4 inch) polypropylene sheet with coarse PTTP 4.2 mm.

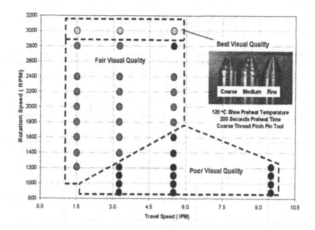

Figure 5. Processing Map for the FSJ of 6.35 mm (¼ inch) Polypropylene Sheet with Coarse PTTP of 4.2 mm, 120°C Pre-Heat Shoe Temperature, and 200 seconds Pre-Heat Time.

Mechanical Properties Testing

Using a MTS (Material Testing System) 858 model servo hydraulic testing system, the mechanical properties of the FS joint analyzed in this study included tensile strength and percentage elongation. Tensile tests are conducted on the untreated PP samples as baseline values for comparison to the FSJ samples mechanical properties. Tensile test sample geometries were determined based on ASTM standard D 638, Standard Test Method for Tensile Properties of Plastics.

The tensile strength of FSJ PP using all 3 types of PTTP is ~90% compared to tensile strength of untreated PP (see Table III). For untreated PP the percent elongation was found to be 180%. In PP samples after FSJ process, the elongation was reduced substantially and was estimated to be 10%. Untreated PP material showed ductile nature with high elongation where as FSJ PP material showed brittleness with much reduced elongation (see Figure 6). It was observed that a majority of fractures occurred at interface of the FS joint and non-processed material on retreating side. Microscope analysis was employed to analyze the tensile test fracture surfaces of the FSJ treated materials for localized ductility.

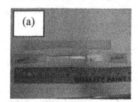 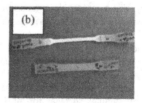

Figure 6. (a) Tensile Specimen of untreated parent PP after and before tensile test (b) Tensile tested specimen of untreated PP with elongation (top) and FSJ PP without elongation (bottom).

Table III. Average Tensile Strength and % Elongation of the Pure Polypropylene and FSJ Polypropylene with 3 Different PTTP.

Material	Pin tool thread pitch (mm)	Average tensile strength (MPa)	% FSJ material tensile strength w.r.t. base material tensile strength	% Elongation
Pure PP		34.9	100	180
	4.2	30.9	89	9.3
	2.1	32.2	92	12
Friction Stir Joined PP	1	30.8	88	10

Fracture Surface Analysis

Optical microscopy was used assess whether micro-ductility or macro-ductility was present on the fracture surfaces of the untreated PP and FSJ PP material. Figure 7 shows the fracture surfaces of untreated PP showing broken elongated fibers.

Figure 7. Image of the untreated fractured PP samples.

The percent elongation and elongated fibers of the fractured untreated parent PP sample surface demonstrated the ductile nature of this material. The fractured FSJ PP sample surfaces for all 3 types of PTTP showed brittleness as demonstrated by a dramatic decrease in the percent elongation of the FSJ treated material. The fractured surface is rough without fibrous structures (see Figure 8a). The macroscopic optical microscopy observations of brittleness guided the use of SEM for observing the fracture surfaces microscopically.

Microscopic analysis of the fracture surfaces were accomplished using a Jeol 840 Scanning Electron Microscope (SEM). The magnification of SEM used in this research provides 10 to 300,000 X. The samples are coated with a thin film of gold as polymers are low conductive materials. The samples are cut from the mechanically tested friction stir joints. In SEM also fractured FSJ PP samples surfaces for all 3 types of PTTP showed brittle nature without any fibrous structure. Typical brittle fractures are classified as type I and type II (see Figure 8b). Type I brittle fractures represent roughness of the material with plateaus. Type II brittle fractures represent smoothness of the material. Both types of fractures do not show any fibrous surfaces. Macro- and microscopic examinations conclusively showed the change in failure from ductile to brittle as a result of the FSJ process.

230

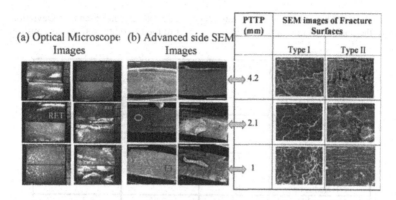

(a) Optical Microscope Images	(b) Advanced side SEM Images	PTTP (mm)	SEM images of Fracture Surfaces	
			Type I	Type II
		4.2		
		2.1		
		1		

Figure 8a and b. Fracture surface images of FSJ PP in OM and SEM. Type I fractures represent -- ○ ; Type II fractures represent -- □ .

<u>Differential Scanning Calorimetry</u>

The transition from ductile to brittle behavior was hypothesized to be strongly dependent on the local crystallinity. To test this hypothesis untreated and FSJ materials at specific points within the samples were analyzed using differential scanning calorimetry (DSC). In this study DSC was used to determine the percent crystallinity change in the polypropylene material from the region of friction stir joint to the region of material where there is no heat effect during the friction stir region process.

The samples for these experiments are extracted from the top surface of the polypropylene sheet at different selected regions (Table IV) to a depth of approximately 0.041 inch. Using a sharp blade PP material is extracted from selected regions. The sample mass should be in between 2-10 mg in order to avoid temperature gradient within the sample during the DSC experiment. In order to avoid thermal lag between the sample and DSC sensors, the heating rate should be 5-10 °C/min [4], here 10°C/min is chosen as it is the accepted "standard" heating rate in thermal analysis as specified by TA Instruments. The melting range of polypropylene is 164-177°C. The temperature range chosen for this study is 25°C to 200°C, as we need only the heat of melting for calculation of percent crystallinity. However to insure that the melting event is captured the maximum temperature should generally be approximately 30°C above the anticipated melting temperature. Table IV shows the selected materials for thermal analysis. At each region 5 samples were analysed.

Statistical analysis was then conducted on the DSC results. The averages from the 5 positions were compared to each other to determine if results were statistically different. To accomplish this, the Gosset's t-Test was employed [6]. The number of samples, the average crystallinity from each position and standard deviations are the only required data from experiments to apply the Gosser t-Test. The t-test analysis shows that the percent crystallinities from each position are statistically the same. However it is noted that the variation in crystallinty as indicated by the standard deviation does vary significantly for the "interface" samples (See Table IV). This higher variability in crystallinty correlates with the position at which a majority of fracture events occurred, which was at the interface of the FS Joint and the non-processed material on retreating side.

Table IV. Thermal analysis--average percent crystallinity and Standard Deviation with respect to the position of the samples analyzed. Center of the FS joint position was considered as origin.

Sample region	Position (inches)	Average Percent Crystallinity	SD
Center of the FS joint	0	30.03	2.67
Interface of the FS joint and non-processed material	0.2	32.95	14.27
With in the shoe	0.63	29.59	4.07
Outside the heat effected zone	1.19	32.63	2.52
Untreated PP	4	32.05	1.31

Conclusions

A FSJ system was designed and fabricated which was capable of producing good quality FS joints based on visual quality and mechanical properties. The process parameter evaluation showed that good visual quality joints are produced at higher rotation speeds and lower travel speeds. Joint integrity, as indicated by tensile strength, was found to be acceptable. Tensile strength of the friction stir joints of PP was ~ 90% of the untreated parent material. However ductility as indicated by percent elongation at failure decreased dramatically. Crystallinity studies reveal that the loss in elongation cannot be attributed to the changes in crystallinity. Further investigation of other factors responsible for the ductile to brittle transition is needed [5].

Acknowledgements

This research is supported by the Army Research Laboratory and was accomplished under Cooperative Agreement Number DAAD19-02-2-0011 with the assistance of the staff of the SDSM&T Advanced Materials Processing and Joining Center (AMP).

References

1. Nelson et al., United States Patent 6,811,632 B32: Nov. 2, 2004.
2. S. R. Seth, "Effects of Friction Stir Welding on Polymer Microstructure" (M.S. Thesis, Brigham Young University, April 2004).
3. Strand, Seth R et al., "Effects of Friction Stir Welding on Polymer Microstructure", (61st Annual Technical Conference ANTEC, 1, 2003), 1078-1082.
4. J.Scheirs, "Compositional and Failure Analysis of Materials – A Practical Approach", (New York, NY: John Wiley and Sons, Ltd., 2000), 37-39, 55, 108.
5. Jeffrey A. Jansen, "Ductile-To-Brittle Transition of Plastic Materials", Advanced Materials & Processes, February (2006), 39-42.
6. J.P.Homan, "Experimental Methods for Engineers,"(New York, NY: McGraw -Hill, Inc., 6th edition, 1994), 92-98.
7. A. Vander Wal et al., "Fracture of Polypropylene: 2. The Effect of Crystallinity", Polymer, 39, (1988), 5477-5481.
8. Mettler Toledo, "Collected Applications Thermal Analysis – Thermoplastics", (New Jersey, NJ: Mettler-Toledo, Inc.), 40.

FRICTION STIR WELDING OF BULK METALLIC GLASSES – VITRELOY106A

Rakesh C. Suravarapu[1], Stanley M. Howard[1], William J. Arbegast[2], Katharine M. Flores[3]

[1]Department of Materials and Metallurgical Engineering
[2]Advanced Materials Processing and Joining Laboratory
South Dakota School of Mines and Technology
[3]Department of Material Science and Engineering
Ohio State University

Keywords: Bulk Metallic Glasses, Friction stir welding, Devitrification

Abstract

The objective of this work was to determine if a processing window could be found for the friction stir welding of the bulk metallic glass Vitreloy 106A that preserved the material's amorphous nature. In this investigation a sample of Vitreloy 106A, was friction stir welded with selected processing parameters including preheating to 662 °F (350 °C). X-ray diffraction and differential scanning calorimeter were used to determine the extent of devitrification induced into the metal. The extent of devitrification was plotted as a function of processing parameters and observed that the extent of devitrification decreased with the increase in the average specific heat, pseudo heat index I (RPM/IPM * Plunge Force) and pseudo heat index II (RPM/IPM * Forge Force). This paper describes the effect of processing parameters on the friction stir welded bulk metallic glass.

Introduction

Amorphous metals being strong, elastic, and very tough as compared to crystalline metals have wide-ranging potential applications. However, a significant, limiting factor for their expanded use is the unfulfilled need for a means of joining them by a welding process. The primary problem in joining amorphous metals by conventional fusion welding methods is the loss of the amorphous state, or so called devitrification. A possible means of obviating the occurrence of devitrification during welding is to use friction stir welding (FSW). The bulk metallic glass used in this project was Vitreloy106A with composition $Zr_{58.5}Nb_{2.8}Cu_{15.6}Ni_{12.8}Al_{10.3}$. X-ray diffraction (XRD) and differential scanning calorimeter (DSC) are the most commonly used methods to determine the extent of amorphous structure; however, transmission electron microscopy provides the most definitive information but at much greater effort and cost.

Materials in which disordered amorphous structure is produced directly from the liquid state by rapid cooling are called *metallic glasses*. Metallic glasses are a class of metallic alloys in which, disordered, short-range atomic structure is formed by rapid melt cooling to prevent crystallization [1]. These materials exhibit extraordinary mechanical [2] and magnetic properties. They resist breaking when stretched, return to their original shape with little permanent strain, and are difficult to shatter [3]. They are very strong, highly elastic, very tough, corrosion resistant, springy, and wear resistant. They are opaque with a very shiny gray surface. However, the surface is so smooth that paint does not adhere to it.

The time-temperature-transformation (TTT) curve for Vitreloy 106A in Figure 1 shows the time to reach crystallization at a given temperature [4]. The circles and squares in the TTT diagram denote the devitrification limits on cooling from the liquidus state to lower temperatures. Diamonds denote the devitrification limits while heating from the amorphous state. Successful friction stir welding would require a processing window that remains within the amorphous region (left of the nose of the C curve) of the TTT diagram.

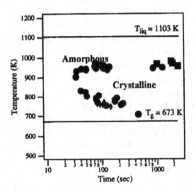

Figure 1. Time temperature transformation curve for Vitreloy106A [4]

The percentage crystallinity was computed from the XRD spectrum based on the relative peak intensity, I_R. This normalized intensity is the ratio of the measured intensity at a $2\theta = 37$, where the highly crystallized alloy has a prominent diffraction peak, divided by the measured intensity at $2\theta = 20$, where no diffraction peak was observed.

$$I_R = \frac{I_{37}}{I_{20}} \tag{1}$$

Uncertainty in the XRD data and even minor lattice strain variations were mitigated by taking the sum of intensities over a range of 2θ values that included approximately 50 digitized channels. The width, in 2θ units, of each channel was approximately 0.02. The crystalline fraction, x_C, was computed using the equation

$$x_C = \frac{I_R - I_{RC}}{I_{RA} - I_{RC}}$$ (2)

where
    I_{RC} = the relative intensity of the completely crystallized sample
    I_{RA} = the relative intensity of the completely amorphous standard sample (A).

The percent crystallinity was calculated from the measured heat for glass crystallization.

$$\text{Percent Crystallization} = \left[\frac{\Delta H_{Am} - \Delta H_u}{\Delta H_{Am} - \Delta H_{Cry}} \right] * 100\%$$ (3)

where ΔH_u – heat of crystallization for (unknown) sample under analysis
       ΔH_{Cry} – heat of crystallization for completely crystalline sample
       ΔH_{Am} – heat of crystallization for amorphous sample

Experimental

Three samples of Vitreloy106A were provided for this investigation. Sample A was used as the reference material while samples B and C were FSW processed. Initially, sample B was used to determine if the material could be plasticized by making simple friction stir plunges. This also provided some understanding of likely processing parameters. This initial work is termed Campaign I.

A second series of friction stir welds, called Campaign II, were made at selected processing parameters on sample C. Campaign III consisted of one linear weld on the surface of sample C preheated to 650 °F (343 °C) to decrease tool wear and possibly lessen the time above Tg at the weld start. Campaigns I and II were both conducted on material at room temperature. Table 1 summarizes the campaign conditions. Each weld was performed at selected processing parameters shown in Table 2.

Campaign III consisted of one linear weld made on the sample preheated to 650 °F (343 °C). Table 3 shows the processing parameters used in Campaign III and Figure 2 shows the locations of the weld relative to Campaign II welds (Welds 1 to 4) and the locations of the six thermocouples spot welded to the surface of the sample on either side of the Campaign III weld.

235

Table 1. Summary of the campaign conditions

Description	Sample	Dimensions, in. (1 in.= 2.54 cm)	Preheated	Process
Reference	A	0.12 x 0.6 x 0.39	N/A	N/A
Campaign I	B	0.12 x 0.6 x 0.39	No	Plunge
Campaign II	C	0.12 x 1.78 x 1.84	No	Four linear welds
Campaign III	C	0.12 x 1.78 x 1.84	Yes	Preheated linear weld

Table 2. Processing parameters used in Campaign II

Pin Tool	Weld #	Weld Length (in.)	Weld Speed (in./min)	Spindle Speed (RPM)	Forge Depth (in.x10^{-3})	Plunge Rate (in./min/s)	Forge Force (lbf)	Atmo-sphere
PCBN	1	1.00	1.0	500	5	0.130	882	Air
PCBN	2	0.65	1.0	500	15	0.130	950	Air
W-Re	3	0.35	0.5	300	10	0.130	2070	Ar
W-Re	4	0.85	0.5	800	10 - 16	0.065	647	Ar

1 in. = 2.54 cm, 1 lbf = 4.45 N

Table 3. Processing parameters used in Campaign III

Pin Tool	Weld No.	Weld Length (in.)	Weld Speed (in./min)	Spindle Speed (RPM)	Forge Depth (in.)	Plunge Rate (in./min/sec)	Forge Force (lbf)
W-Re	5	1.0	2.0	800	0.010	0.065	2000

1 in. = 2.54 cm, 1 lbf = 4.45 N

The preheating plate was placed over an electrically heated coil and the sample of Vitreloy106A was clamped to the preheating plate. Fiber fax was used to insulate the sides of the preheating plate so as to moderate possible thermal gradients in the plate. The preheating plate controller thermocouple was spot welded to the sample so as to control the sample temperature rather than the preheating plate. The sample was heated to and held at 662 °F (350 °C), friction stir welded, and rapidly quenched below the glass transition 752 °F (400 °C) using the liquid nitrogen.

Samples were cut perpendicular to the processing direction using a diamond cut off saw. The 0.1 cm thick, oil-cooled saw blade rotated at 3500 RPM and progressed at a 0.0127 cm/min. feed rate. X-Ray diffraction and DSC analyses were performed on the surface and the cross sections of the samples from all campaigns and compared to the corresponding reference samples consisting of the initial amorphous parent material and a completely crystallized (devitrified) sample. The devitrified sample was prepared by

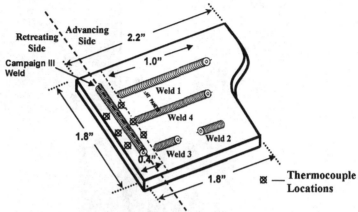

Figure 2. Thermocouple locations in the amorphous region of the
sample of Vitreloy106A for Campaign III (preheated weld)

heating a piece of sample A to 1202 °F (650 °C) in a tube furnace under flowing argon for
40 min.

Results

The results consist of XRD and DSC analyses. XRD analyses were performed on both
the surface and cut cross sections. DSC work is necessarily a bulk analysis. Figure 3
shows photographs of the welds resulting from the three campaigns. The combined
graphs shown in Figures 4 to 6 summarize the surface and cross section XRD and the
DSC results. The plotted DSC results are before adjustment for sensible heat. The
computed percent crystallinity according to Eq. [2] indicated no devitrification for the
Campaign I plunge welds on Sample B. The percent crystallinities computed on the
weld locations on Sample C are summarized in Table 4. The linear fit of percent
crystallinity versus: RPM, IPM, *average specific energy* [5], and *pseudo heat index* [5]
are shown in Table 5.

There were no thermal data for Campaign I. The thermal data for Campaign II were
inadequate for determining thermal history, specifically the sample's progress on the TTT
diagram. The temperature record for Campaign III showed that the estimated cooling rate
behind the weld to be on the order of 8 °C/s. The TTT diagram in Figure 1 shows the
required initial cooling rate from the worst-case required cooling rate from the curve nose
to be approximately the same. Consequently, the cooling rate in the Campaign III weld
was very close to being rapid enough to maintain the amorphous state.

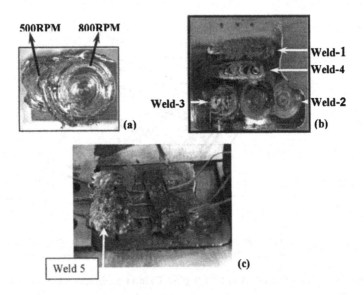

Figure 3. Photographs of (a) Campaign I (Sample B), (b)
Campaign II (Sample C), and (c) Campaign III (Sample C)

Discussion

The weld results show that plasticization of Vitreloy106A is possible by FSW. The plunges into Sample B in Campaign I not only support this conclusion but also demonstrated that friction stir plunges do not cause devitrification. Friction stir welded bulk metallic glass (Sample C) showed different behavior with different processing parameters. The results in Table 4 showed significant, but not total, devitrification from 35 to 60 percent on the surface of the welded region. These values agree well with the cross sectioned sample results in Table 4, which range from 24 to 53 percent crystallization. The results of all Campaign II and III XRD and DSC determinations of percent crystallinity are given in Table 4. These results are remarkably consistent within each weld with the exception of Weld 3, but that is mostly likely explained by the W-Re pin tool failure during the weld. This was likely caused by the low rotation speed of the pin tool (300 RPM).

Results obtained from Campaign II welds indicated that too much plunge and travel time was required in room-temperature material to maintain the amorphous state. Campaign II results shown in Table 5 indicate that the correlation between percent crystallinity and the various process variables is not specifically linear. There is a generally decreasing trend in percent crystallinity with pseudo heat index (I) and (II). However, there is no clear trend with the other parameters. This points to an amorphous-favoring processing window comprised of hotter welds, which would require rapid quenching. The preheating plate was built to achieve this condition. Campaign III, designed to determine the efficacy of this technique, and was somewhat successful. The material was more plasticized and pin tool wear was lessened but the resulting 50 percent crystallization was less than a complete success. However, the very short weld distances imposed by the sample size may have prevented a more favorable outcome.

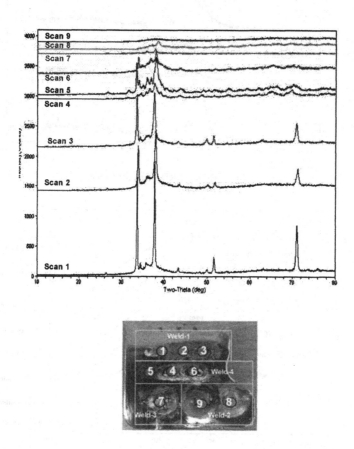

Figure 4. Locations and scans of surface XRD scans for Campaign II

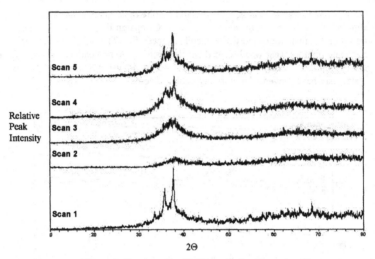

Figure 5. Cross section XRD scans for Campaign II

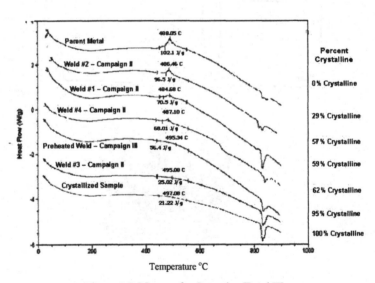

Figure 6. DSC scans for Campaign II and III

Table 4. Comparison of crystallinity determinations for Campaigns II and III

XRD Location	Crystallinity, (%)		
	Surface XRD	Cross section XRD	DSC
Crystallized Bulk Metallic Glass Sample	100	100	78
Weld 4 (Campaign II)	60	53	56
Weld 1 (Campaign II)	53	41	55
Weld 3 (Campaign II)	51	28	82
Weld 5 (Campaign III - Pre Heated)	51	61	77
Weld 2 (Campaign II)	35	24	25
Amorphous Bulk Metallic Glass -Parent Metal	0	0	0

Table 5. Fraction crystallinity correlation to selected processing parameters

Independent Variable, x	Fraction Crystallinity	R^2
Average Specific Energy (Watts)	0.587-0.0014x	0.37
Forge Force (lbf)	0.4735-0.0001x	0.04
Pseudo Heat Index I (lbf/min)	0.5145-0.0108x	0.32
Pseudo Heat Index II (lbf/min)	0.5773-0.2111x	0.38
Average Power (Watts)	0.5593-0.0039x	0.04
RPM	1.4872-0.0016x	0.67

1 in. = 2.54 cm, 1 lbf = 4.45 N

Conclusion

Vitreloy106 A can be plasticized by friction stir plunging while maintaining the amorphous condition. In fact, there is some evidence that the processing further disrupts even long-range ordering. Friction stir welds approximately 2 in. long on the small plate of Vitreloy106A material available resulted in percent crystallinity ranging from 24 to 61 percent. Preheating the sample resulted in approximately 50 percent crystallinity but superior plasticization. Keeping the pin tool shoulder small is believed to allow faster quenching of the processed material. The process window appears to contain 800 RPM. A travel speed of 2 IPM appears to be a reasonable starting point. Preheating with quenching is recommended primarily for the weld quality and pin tool life.

Acknowledgement

Army Research Laboratory (ARL) provided funding for this project under contract # DAAD 19-02-2-0011.

References

1. Mark Telford, "The Case for Bulk Metallic Glasses", *Materials Today*, vol. 7, March (2004), 36-43.
2. C. J. Gilbert, V. Schroeder, and R. O. Ritchie, "Fracture and Fatigue in a Zr-Based Bulk Metallic Glass", *Bulk Metallic Glasses*, ed. A. Inoue, W. L. Johnson, and C. T. Liu, (Pittsburgh, PA, Materials Research Society Symposium Proceedings, vol. 554, 1999)
3. A. Masuh, R. Busch, W. L. Johnson, "Rheometry and Crystallization of Bulk Metallic Glass Forming Alloys at High Temperatures", (Trans Tech Publications, ISMANAM 1997 - Materials Science Forum, Barcelona, Spain, Switzerland:, 1998), 779-84.
4. C.C. Hays, J. Schroers, W.L. Johnson, T. J. Rathz, R.W. Hyers, J.R. Rogers, and M.B. Robinson, "Vitrefication and Determination of the crystallization time scales of the bulk-metallic-glass-forming liquid $Zr58.5Nb2.8Cu15.6Ni12.8Al10.3$", *Applied Physics Letters*, 79, (Sept)(2001).
5. William J. Arbegast, "Using Process Forces as Statistical Process ContrbTool for Friction Stir Welds", (Warrendale, PA, TMS, Friction Stir Welding Symposium Proceedings, 134th Annual Meeting and Exhibition, San Francisco, 2005).

FRICTION STIR WELDING OF HSLA-65 STEEL

P.S. Pao, R.W. Fonda, H.N. Jones, C.R. Feng, and D.W. Moon

Naval Research Laboratory, Washington, DC 20375, USA

Keywords: Friction stir welding, Mechanical properties, Fatigue crack growth, Microstructure, Steel.

Abstract

The microstructure, mechanical properties, and fatigue crack growth kinetics of friction stir welded HSLA-65 steel were investigated. TEM studies reveal the formation of Widmanstatten structure in the weld region, which results in a higher tensile strength and microhardness as compared to the base plate. A yield strength minimum is observed in the outer regions of the HAZ. Because of the presence of compressive residual stresses, the fatigue crack growth rates are significantly lower and the fatigue crack growth threshold significantly higher in the weld nugget and HAZ than those in the base plate. The tensile and fatigue crack growth properties are discussed in terms of the observed microstructure in various regions of the weld.

Introduction

HSLA-65 is a high strength, low alloy steel, and is processed by controlled rolling to produce a small ferrite size with minimum yield strength of 448 MPa (65 ksi). This steel has low carbon and low alloy content and thus affords good weldability and toughness. The high strength would also allow substantial structural weight reduction and cost savings through the use of thinner plates. HSLA-65 has been used in bridge structures, pipelines, and is being considered for use in ship hulls and other non-primary structures.

Friction stir welding (FSW), which was invented by TWI in 1991, has been quickly maturing as an alternative to conventional gas metal arc (GMA) and submerged arc welding (SAW) [1]. FSW is a solid-state joining process in which the material is welded by the conjoint action of plastic deformation and friction heat. Because during FSW the steel stays below its melting point, the amount of fumes that contain hexavalent chromium is greatly reduced. In addition, because of FSW's low heat input, weld distortion and the subsequent post-weld flame straightening is expected to be substantially less than that of conventional fusion welding.

Friction stir welding produces significantly different weld region microstructures when compared to the base plate [2-14]. The changes in microstructure can potentially affect mechanical and fatigue properties of the material. Previous studies on FSW HSLA-65 have shown the microstructural variation from a fine equiaxed ferrite grain structure in the base plate to Widmanstatten type ferrite laths in the weld nugget [2-3]. However, most of the mechanical property studies are limited to microhardness mapping and transverse tension tests. Microhardness mappings of the FSW HSLA-65 weld region indeed confirm the weld nugget region exhibits higher hardness than that of the base plate [2-3]. In addition, softer heat-affected zones (HAZ) are also identified outside the weld nugget region. These microhardness measurements, though easy to perform, cannot directly evaluate the yield strength, tensile

243

strength, and ductility variations across various regions of the weld. Finally, the fatigue crack growth properties through the FSW and HAZ of HSLA-65 have not been investigated.

In this paper, the microstructures, longitudinal and transverse tensile properties, and fatigue crack growth kinetics of weld region and HAZ in the FSW HSLA-65 were studied and were compared to those of base metal.

Experimental Procedures

The material used in this study was 6.35 mm thick HSLA-65 plate. Single-pass FSW butt joints were made at a rotational speed of 600 rpm and a weld speed of 5 mm/s. A microhardness map was obtained across the width of the weld zone. Optical and transmission electron microscopy (TEM) metallographic examinations were made at various weld locations.

For longitudinal tensile tests, a set of rectangular specimens, with a nominal cross section of 2.1 x 1.4 mm, were cut from the top and bottom halves across the weld region. These longitudinal specimens are parallel to the weld direction. Though longitudinal tensile specimen may have different microstructures at different locations, the microstructure in each specimen is expected to be uniform.

For fatigue crack growth studies, 5.8 mm-thick, 50.8 mm-wide compact tension (CT) specimens were used. The notch direction and hence the crack growth direction are parallel to the welding direction. 5% side grooves were introduced on both specimen surfaces to provide additional constraint. CT fatigue specimens were made with notch located in the center of the FSW weld and in the base metal. Based on longitudinal tensile test results, CT specimens with notches located in the outer region of HAZ, which was 14.2 mm from the weld center on the advancing side, were machined. All fatigue crack growth tests were carried out in ambient air (20°C and 42% relative humidity) at a cyclic frequency of 10 Hz and two stress ratios, R, of 0.1 and 0.5. A compliance technique was used to continuously monitor the fatigue crack growth.

Results and Discussion

Microstructural Analysis

Figure 1 shows the microhardness scan from just above the mid-plane and the microhardness map across the transverse cross section of the FSW HSLA-65 weld. As shown in Fig. 1, there is a significant overmatch in the FSW weld region. The thermo-mechanically affected zone (TMAZ) and heat affected zone (HAZ) are not discernible from the microhardness scan and microhardness map presented in Fig. 1.

The low magnification macro image of the HSLA-65 FSW weld is shown in Fig. 2. A concentric "onion ring" structure often observed in friction stir weld is clearly visible in the weld nugget (WN). Just outside this weld nugget, the distorted banding identifies this region as TMAZ. The TMAZ experiences significant heating during welding, which permits the material to be deformed by the welding process. The TMAZ on the retreating side is wider than that on the advancing side, likely the result of the greater material flow around the retreating side of the weld. Outside the TMAZ is the HAZ, where the material is modified by the friction heat but does not mechanically deform. The HAZ can be further divided into two zones, the inner HAZ (iHAZ) and the outer HAZ (oHAZ). The iHAZ is immediately adjacent to TMAZ and has a similar etching appearance as TMAZ but without the distorted banding. Outside the iHAZ are oHAZ regions, which etch barely with light etching as shown in Fig. 2.

244

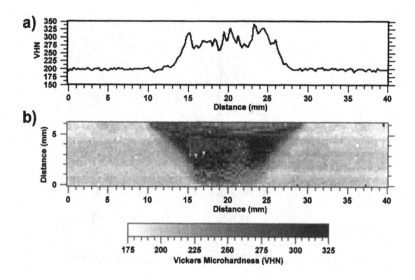

Fig. 1. (a) Microhardness scan from just above the mid-plane and (b) microhardness map across the HSLA-65 weld.

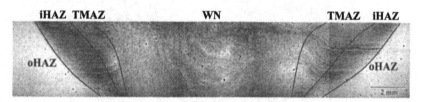

Fig. 2. Montage of optical micrographs showing the center 4 mm thickness of the HSLA-65 friction stir weld, with lines delimiting the regions discussed in the text.

Detailed optical microscopy was performed to determine the microstructural variations across the HSLA-65 FSW weld. As shown in Fig. 3, the base plate consists predominantly of equiaxed ferrite grains with a relatively small grain size, but does not contain the large regions of pearlite typical of medium-carbon, microalloyed steels. Transmission electron microscopy (TEM) of the base plate, shown in Fig. 4, confirms the optical observations of fine, equiaxed ferrite grains. In addition, TEM examinations reveal numerous particles, which appear to be cementite, and considerable dislocation structures.

The grain structure in the outer portions of the HAZ (oHAZ) is similar to that of base metal. Previous study of the HAZ of a similar steel suggested that the ferrite grains in this region contained fewer but coarser precipitates and had a lower dislocation density [2]. These observations indicate that the temperature excursion in the HAZ during friction stir welding, though still below the A_1 temperature, was sufficient to dissolve fine particles and coarsen larger ones, as well as annealing the dislocation structures in the ferrite grains. Though detailed

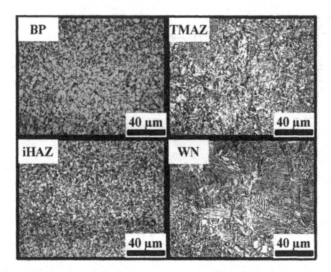

Fig. 3. Optical micrographs from the base plate (BP), thermo-mechanically affected zone (TMAZ), inner heat affected zone (iHAZ), and weld nugget (WN) near the plate mid-thickness.

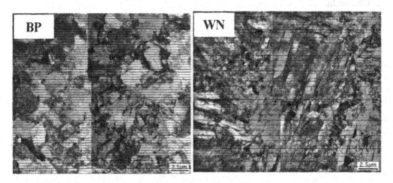

Fig. 4. Montage of TEM images of the HSLA-65 base plate (BP), showing the fine equiaxed ferrite grains and precipitates, and the weld nugget (WN), showing MAC at the interlath boundaries between the long laths of Widmanstatten ferrite.

TEM analysis was not performed, a similar fewer but coarser particle distribution and a lower dislocation density in the oHAZ of the present study are anticipated. The inner regions of HAZ (iHAZ), as shown in Fig. 3, have significantly finer equiaxed grains when compared to those in the base plate and oHAZ. In addition, TEM analyses indicate the presence of retained austenite in the microstructure, suggesting the temperatures in iHAZ regions are above the A_1 temperature, where the ferrite begins transformation to the high-temperature austenite phase.

The microstructure in the weld nugget is distinct and different from those observed in the base plate and HAZ. Instead of polygonal ferrite, the weld nugget microstructure, as shown in Fig. 3, consists of Widmanstatten ferrite plates within a large prior austenite grain structure. TEM examination of the weld nugget reveals, as shown in Fig. 4, long, thin layers of martensite/austenite component at the interlath ferrite boundaries and extensive dislocation structures.

The microstructure in TMAZ also consists of Widmanstatten type ferrite plates and is similar to that observed in the weld nugget region, except the prior austenite grains are significantly smaller.

Mechanical Properties

Transverse tension tests were conducted to determine the weakest part of the weld. The transverse tension specimens had a 12.8 x 6.2 mm gage cross-section and 52 mm gage length. The transverse yield and tensile strength are 523 MPa and 610 MPa, respectively. The transverse specimens all failed in the base plate located just outside of the oHAZ on the advancing side.

Because the transverse tension tests failed to detect the weakest part of the weld, a series of tension tests were conducted in the longitudinal direction. Across the FSW weldment, thin slices of tension specimens were sectioned parallel to the welding direction. Figure 5 shows the longitudinal tension specimen locations relative to the macro image of the weld. Figure 6 shows the variations of yield and tensile strength across the weld in the longitudinal direction. As shown in Fig. 6, in the center nugget region, the yield and tensile strengths are highest. The strengths progressively decrease away from the center nugget region. The yield strength reaches a minimum in the oHAZ, which is about 15% lower than the base plate yield strength. The high yield strengths in the weld nugget and TMAZ are due to the presence of Widmanstatten ferrite plates. The annealing of dislocation structures and the coarsening of particles during friction stir welding probably cause the low yield strength in the oHAZ. The tensile strength is essentially the same outside the iHAZ and does not exhibit a dip in the oHAZ. Except in the weld nugget region, where the strengths are higher in the top half of the weld, the strengths are comparable for specimens taken from either halves of the weld. The anomalously low strength near location #10 is due to the presence of an entrained oxide layer at that location.

Besides the variation in yield strength, the plastic flow behavior in various regions of the weldment is also different. As shown in Fig. 7, the base plate material (BP) exhibits distinctly sharp upper and lower yield behavior associated with the Luders extension with almost no work hardening. The sharp yield behavior progressively disappears and the work hardening increases as the longitudinal specimen location moves toward the center of the weld. As shown in Fig. 7, the weld nugget (WN) exhibits a smooth yield behavior with a significant work hardening.

Fatigue Crack Growth

The fatigue crack growth rates as a function of stress intensity factor range, ΔK, for starting cracks located in the weld center, HAZ, and base plate of HSLA-65 steel are shown in Fig. 8. As indicated before, the fatigue crack growth rate data in HAZ was obtained from specimens with starting notches located at 14.2 mm from the weld center on the advancing side. As shown in Fig. 8, the fatigue crack growth rates at both stress ratios in the weld and HAZ are significantly lower than that in the base plate at the comparable stress intensities. Furthermore, the fatigue crack growth thresholds in the weld and HAZ specimens are considerably higher than that in the base plate specimen. At $R = 0.1$, the fatigue crack growth rates and fatigue

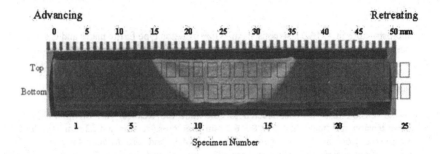

Fig. 5. Longitudinal tension test gage section locations relative to the macro image of the weld.

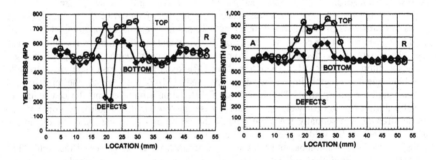

Fig. 6. Variations in longitudinal yield and tensile strength across the weld.

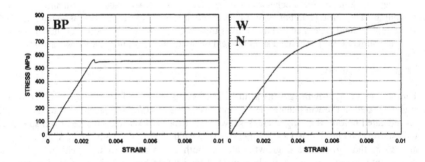

Fig. 7. Stress-strain curves for longitudinal tension tests of the base plate (BP) and weld nugget (WN) regions.

248

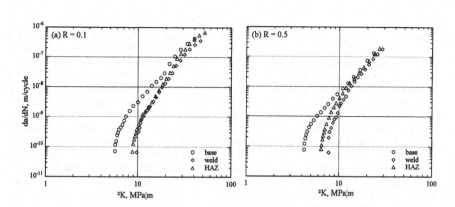

Fig. 8. Fatigue crack growth kinetics in air at stress ratio of (a) 0.1 and (b) 0.5 through the base plate, HAZ (advancing side), and weld nugget.

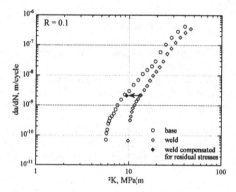

Fig. 9. Fatigue crack growth rate in the weld before (open diamond) and after (solid diamond) compensated for compressive residual stresses.

crack growth thresholds in the weld and HAZ are similar. At a higher stress ratio of R = 0.5, fatigue crack growth rate in HAZ is only slightly higher than that in the weld. These results are similar to those in high strength 2000- and 7000-series aluminum alloy FSW welds in which lower fatigue crack growth rates and higher fatigue crack thresholds are also reported [6-8]. The difference in fatigue crack growth rates among base plate, weld nugget, and HAZ progressively diminish as the stress intensity increases.

The observed different fatigue crack growth behavior in the FSW weld, the HAZ, and the base plate of HSLA-65 steel are not the result of differences in crack tip microstructures and/or strengths in these regions. The fatigue crack growth behavior in the FSW weld and the HAZ are similar, yet the microstructures and the yield strength in these two regions are very different. On the other hand, the grain structure and the yield strengths in the HAZ and in the base plate, though are not identical, are close. However, the fatigue crack growth resistance in HAZ is significantly better than that in the base metal.

The significantly lower fatigue crack growth rates and higher fatigue crack growth thresholds in the weld and HAZ regions are believed to be caused by the presence of compressive residual stresses in these regions. The existence of significant residual compressive stresses in the FSW welds has been reported in previous studies on high strength aluminum alloys [6]. The compressive stresses in the weld nugget and HAZ regions can "close" the crack and reduce the crack tip stress intensity and thus lower the fatigue crack growth rates and increase the fatigue crack growth thresholds. To test this assumption, a series of CT specimens with starting cracks located in the weld nugget and in the base plate were fatigue cracked to a predetermined crack length. Compliance measurements were then performed on these fatigue cracked specimens to determine the onset load of crack closure (the deviation of compliance slope). Comparing the closure loads from the weld nugget and base plate regions, compressive residual loads in the weld nugget can be estimated. This compressive load is then used to offset the fatigue load and adjust crack tip stress intensity in the fatigue crack growth rate curve. After accounting for the compressive residual stress effect, as shown in Fig. 9, the fatigue crack growth rate in the weld region is essentially the same as that obtained in the base plate. The same compressive stress effect is expected in HAZ. Thus, the apparently lower fatigue crack growth rates and higher fatigue crack growth thresholds in the weld and HAZ are caused by the presence of compressive residual stress in these regions.

Conclusions

1. The weld nugget is characterized by the presence of Widmanstatten ferrite plates as opposed to the equiaxed ferrite grains in the base plate.
2. The yield strength is highest in the weld nugget and is lowest in the outer regions of HAZ.
3. The base plate exhibits distinct upper and lower yield behavior with minimum work hardening while the weld nugget region shows a smooth yield behavior with significant work hardening capability.
4. The weld nugget and HAZ exhibit significantly lower fatigue crack growth rates and higher fatigue crack growth thresholds, when compared to the base plate. The superior fatigue crack growth resistance in the weld nugget and HAZ can be attributed to the presence of compressive stresses in these regions.

Acknowledgements

The authors would like to acknowledge financial support from the Naval Research Laboratory under the auspices of the Office of Naval Research.

References

1. W.M. Thomas, E.D. Nicholas, J.C. Needham, M.G. Nurch, P. Temple-Smith, and C.J. Dawes, "Friction Stir Butt Welding," International Patent Application No. PCT/GB92/02203, GB Patent Application No. 9125978.8 (1991), and U.S. Patent No. 5,460,317 (1995).
2. M. Posada, J. DeLoach, A.P. Reynolds, R.W. Fonda, and J.P. Halpin, "Evaluation of Friction Stir Welded HSLA-65," in Proceedings of the Fourth International Symposium on Friction Stir Welding, TWI Ltd., S10A-P3 (2003).
3. P.J. Konkol, J.A. Mathers, R. Johnson, and J.R. Pickens, "Friction Stir Welding of HSLA-65 Steel for Shipbuilding,"*J. Ship Production*, 19 (2003), 159-164.
4. T.J. Lienert, W.L. Stellwag, Jr., B.B. Grimmett, and R.W. Warke, "Friction Stir Welding Studies on Mild Steel – Process Results, Microstructures, and Mechanical Properties," *Welding J.*, 82 (2003), 1S-9S.

5. C.G. Rhodes, M.W. Mahoney, W.H. Bingel, R.A. Spurling, and C.C. Bampton, "Effects of Friction Stir Welding on Microstructure of 7075 Aluminum," *Scripta Materialia*, 36 (1997), 69-75.

6. K.V. Jata, K.K. Sankaran, and J.J. Ruschau, "Friction-Stir Welding Effects on Microstructure and Fatigue of Aluminum Alloy 7050-T7451," *Met Mat Trans A*, 31A (2000), 2181-2192.

7. P.S. Pao, S.J. Gill, C.R. Feng, and K.K. Sankaran, "Corrosion-fatigue crack growth in friction stir welded Al 7050," *Scripta Materialia* 45 (2001), 605-612.

8. P.S. Pao et al., *Aluminum 2001* (Warrendale, PA: TMS, 2002), 265.

9. J-Q Su, T.W. Nelson, R. Mishra, and M.W. Mahoney, "Microstructural Investigation Friction Stir Welded 7050-T651 Aluminum," *Acta Materialia*, 51 (2003), 713-729.

10. R.W. Fonda, and J.F. Bingert, "Microstructural Evolution in the Heat-Affected Zone of a Friction Stir Weld," *Met Mat Trans A*, 35A (2004), 1487-1499.

11. J. Corral, E.A. Trillo, Y. Li, and L.E. Murr, "Corrosion of Friction-Stir Welded Aluminum Alloys 2024 and 2195," *J Mat Sci Letters*, 19 (2000), 2117-2122.

12. Y.S. Sato, H. Kokawa, M Enomoto, and S. Jogan, " Microstructural Evolution of 6063 Aluminum During Friction-Stir Welding," *Met Mat Tran A*, 30A (1999), 2429-2437.

13. C.J. Dawes, and W.M. Thomas, "Friction Stir Process Welds Aluminum Alloys," *Welding J.*, 75 (1996), 41-45.

14. B. Heinz, B. Skrotzki, and G. Eggeler, "Microstructural and Mechanical Characterization of Friction Stir Welded Al-Alloy," *Mat Sci Forum*, 331-337 (2000), 1757-1762.

CHARACTERIZATION OF DUAL PHASE STEEL FRICTION-STIR

WELD FOR TAILOR-WELDED BLANK APPLICATIONS

Seung Hwan C. Park[1], Satoshi Hirano[1], Kazutaka Okamoto[2],
Wei Gan[3], Robert H. Wagoner[3], Kwansoo Chung[4], Chongmin Kim[5]

[1]*Materials Research Laboratory, Hitachi Ltd., Omika 7-1-1, Hitachi 319-1292, Japan*
[2]*Automotive Products Research Laboratory, Hitachi America Ltd., 34500 Grand River Ave,
Farmington Hills, MI 48335, USA*
[3]*Department of Materials Science and Engineering, The Ohio State University, 2041 College
Road, Columbus, Oh 43210, USA*
[4]*Department of Materials Science and Engineering, Intelligent Textile System Research
Center, Seoul National University, San 56-1 Silim-dong, Gwanak-gu, Seoul, 151-742 Korea*
[5]*Materials and Processes Laboratories, General Motors R&D Center, 30500 Mound Road,
MC 480106224, Warren, MI 48090, USA*

Keywords: Friction stir welding, Dual phase steel, Tailor welded blank, Microstructure,
Mechanical properties

Abstract

Friction stir welding (FSW) was applied to dual phase (DP) steel sheet for tailor-welded
blank (TWB) applications. The mechanical properties and the microstructural evolution in the
weld were investigated. The mechanical behavior of the friction stir (FS) welds was
characterized through hardness and tensile tests. It was found that the hardness of the stir
zone (SZ) is higher than the base material (BM). The roughly same ultimate tensile strength
(UTS) and half elongation of the BM have been shown through the transverse tensile test. On
the other hand, the proof stress and UTS of the longitudinal tensile specimen having only the
SZ were much higher than those of the BM. The microstructural observation revealed that
martensite was formed in the SZ, with rapid cooling and high strain during FSW. The
martensite formation in the SZ is the cause for the higher hardness, proof stress and UTS, and
lower elongation in the FS weld.

Introduction

DP steels, which are comprised of soft ferrite and, depending on strength, between 20 and
70% volume fraction of hard phases, normally martensite, generally show higher strength
than 350 MPa yield strength and 600 MPa tensile strength. The soft ferrite phase is generally
continuous, giving these steels excellent ductility. When these steels deform, however, strain
is concentrated in the lower strength ferrite phase, creating the unique high work hardening
rate exhibited by these steels. The work hardening rate along with excellent elongation
combine to give DP steels much higher UTS than conventional steels of similar yield
strength. This excellent combination of mechanical properties leads to the increasing
applications for automotive components requiring high strength. For the applying DP steel in
automobile industry, TWB allowing the welding and subsequent forming of sheets is an

effective method for white body and panel parts, yielding greater cost and weight savings.

FSW [1-9], which is a solid-state joining process providing a relatively low distortion, good dimensional stability and good mechanical properties, is one of candidate welding methods for TWB applications. These promising mechanical properties are generally characterized by frictional heating and intense material flow arising from the rotating welding tool. FSW is also an environmental-friendly process, since the welding temperature is kept below the materials melting point so there are no fumes or sputter generated in this welding process. FSW has been commercially applied to various metal alloys such as aluminum, magnesium and copper for numerous industrial sectors such as transportation, marine, aerospace, IT devices etc. However, FSW applications of steels are still very challenging and therefore it is valuable to develop new applications of this technology.

As part of an FSW feasibility study for DP steels, linear FSW was applied to DP steel sheets. This work is also aim to an application study for TWB and sheet metal assembly. During this study, microstructure and mechanical properties were observed in the joints and BM.

Experimental Procedures

Materials and welding process

DP 600 sheet steels with 1.6 and 2.0 mm thickness were used in this study. The steel sheets with the dimension of $1000mm^L$ x $200mm^W$ x 1.6 and $2.0mm^t$ were prepared. The chemical composition of the BM is shown in Table 1. The BM microstructure of DP steel observed by optical microscopy (OM) is shown in Fig. 1. DP steel has Zn coated layer on the substrate. The thickness of the Zn coated layer is about 45 μm. Small amount of martensite was observed in the BM. The ferrite grain size measured by the mean linear intercept method was approximately 10 μm.

Table 1 Chemical composition of DP steel. (wt%)

	C	Mn	Si	Cr	Mo	Fe
DP steel	0.07	1.83	0.01	0.20	0.18	Bal.

Fig. 1 Base material microstructures of DP steel.

The similar gauge (SG) FSW of 2.0/2.0 mm-thick sheets and dissimilar gauge (DG) FSW of 1.6/2.0 mm-thick sheets were carried out. For the DG weld, 1.6 mm-thick sheet was placed on the retreating side. FSW of DP steel for TWB was tried using a polycrystalline cubic boron nitride (pcBN) tool that was made in MegaStir. All of the welding trials were carried out on a 3D FSW machine with 7.5 kW spindle drive motor power. The steel sheets were tightly clamped by clamping fixtures and then FSW was performed on the plate. Inert Ar gas shroud was utilized for shielding which prevented surface oxidation.

Microstructural observation

Following FSW, the microstructures were observed in the joint. The cross section perpendicular to the welding direction was observed by OM. The specimen for OM was etched in a solution of 3 % nitric acid + 97 % ethanol for about 30 s. The detailed microstructure of the weld was observed by OM. The workpiece temperature in the optimized FSW parameters during FSW was measured using alumel-chromel thermocouples embedded in steel sheets as shown in Fig. 2. The measured regions are 2, 5, 8mm from the weld center as indicated in Fig. 2.

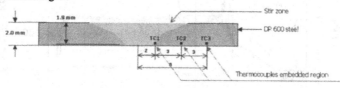

Fig. 2 Temperature measurement using alumel-chromel thermocouples

Mechanical properties

Mechanical properties for the weld were evaluated by bend, Vickers hardness and tensile tests. The penetration welding was examined by root and face bend tests after welding. The diameter of bending roll was 8 mm. The Vickers hardness profiles of the weld were measured on the cross section perpendicular to the welding direction with a load of 500 g force and 15 seconds of dwell time. The stress-strain response of the materials was obtained by tensile test at 1.6×10^{-3} s^{-1}. Both full-sized and sub-sized tensile samples were tested. The gage sections for the full-sized specimens measured 63.5 mm long by 12.7 mm wide, while the sub-sized samples have gage sections of 25.4 by 6.4 mm. The tensile tests were repeated and scatter was found to be within 10 MPa. The average stress-strain curves were calculated.

Results and Discussion

Welding

The applied FSW parameters were 80 mm/min and 1000 rpm for SG weld, and 100 mm/min and 800 rpm for DG weld. The tool shoulder glowed a bright orange color during the weld since the shoulder temperature reached upto 1200°C. Fig. 3 shows the top surface of the weld. Smooth and glossy surface ripple appears on the stir-on surface since the drastic oxidation was prevented by using inert Ar gas, but the Zn adhesion surface often is formed due to the interfusion of Zn coated layer on the surface during FSW.

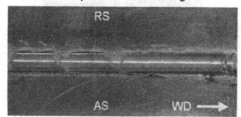

Fig. 3 Top surface overview of friction stirred DP steel weld

255

Fig. 4 shows the vertical welding force during the process. The maximum Z force is about 11 kN. The total power consumption that is including the frictional term of main spindle, about 1 kW, was roughly 4 kW. Therefore, the actual power needed for the DP steel FSW is 3 kW in the present study. From these data collected from the preparation of the test pieces, the rigid but energy-saving FSW equipment with 7.5 kW power spindle motor is enough for linear FSW of steel sheets.

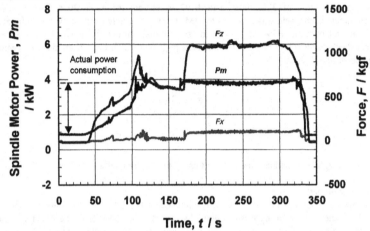

Fig. 4 Welding force and power spindle motor consumption

Microstructural features in the weld

Macrostructure and optical micrographs of SG and DG welds are shown in Figs. 5 and 6, respectively. The both weld shows good consolidation with no evidence of defects. In the SZ, the duplex microstructure of the martensite and ferrite has been formed as shown in Figs. 5 and 6. The thermo mechanically affected zone (TMAZ) was observed outside the SZ where the material is plastically deformed in general. However, the TMAZ showed a similar microstructure with the SZ. The heat affected zone (HAZ) was also observed outside the TMAZ. The two types of HAZ were observed in the DP steel FS weld. The HAZ close to the TMAZ showed the fine grain structure of ferrite with small amount of martensite while the HAZ near the BM has the slightly coarse ferrite structure.

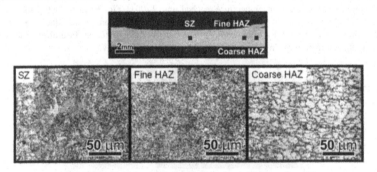

Fig. 5 Optical micrographs of similar gauge friction stir weld

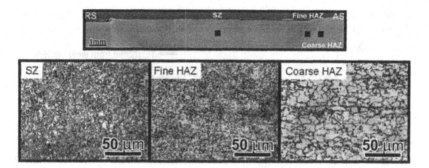

Fig. 6 Optical micrographs of dissimilar gauge friction stir weld

Thermal profiles of the workpiece on the advancing side of SG weld during the process are shown in Fig. 7. The maximum temperature of the workpiece was around 1000°C at 2 mm away from the weld center. This temperature is lower than that expected by tool shoulder color, which seems to be originated from the thermal gradient between the top surface and bottom. The thermocouples were embedded into the location of 2, 5 and 8 mm from the weld center, which are roughly corresponding to the SZ or TMAZ, fine HAZ and course HAZ. The 2 mm away from the weld center was heated to 1000°C during FSW. When the DP steel is heated over 1000°C, austenite would be formed in the SZ. This austenite would be transformed to the martensite and retained austenite during cooling process. The rapid cooling rate during FSW could accelerate the martensite formation in the SZ. In addition, large amount of strain applied during FSW could lead to rapid transformation of martensite in the SZ. These two reasons would probably result in the martensite formation with large amount. The TMAZ showed the roughly same microstructure with the SZ. This is maybe attributed to the heating around 1000°C in the TMAZ, which resulted in the formation of austenite during FSW. This is the reason why the distinction between the SZ and TMAZ is difficult in the steel FS weld. The HAZ with fine grain structure (5 mm from the weld center) has been heated up around 800°C, which roughly corresponds to the temperature between A1 and A3 transformation temperature of carbon steel. The small size of austenite would be formed along the grain boundary of the ferrite, which probably resulted in the fine grain structure of ferrite. The coarse HAZ (8 mm from the weld center) has been heated up over 500°C for 10s, which probably results in grain growth in this region during the process. However, the more detailed microstructural evaluation, for example, orientation imaging microscopy (OIM) map, is needed to clarify the microstructural evolution during FSW.

Mechanical properties of the joint

Root bend test was used as an important tool to understand about the ductility and toughness of the FS weld. The as-welded specimens passed the 180° bend test as shown in Fig. 8. The face bend tested specimens are also shown Fig. 8.

The hardness distributions on the cross section perpendicular to the welding direction are shown in Fig 9. The hardness profiles were measured along the centerline of the plate thickness. The average hardness of the BM was about 180 Hv. The hardness of the SZ was higher than that of the BM. Hardness-increased region roughly corresponds to the SZ. The maximum hardness was measured to about 316 Hv in the SZ of SG weld. Basically, there was no large difference in hardness distribution in SG and DG welds, which is maybe

257

attributed to slight difference in the welding conditions between the both welds. It is thought that the increased hardness in the SZ is mainly due to the martensite formation.

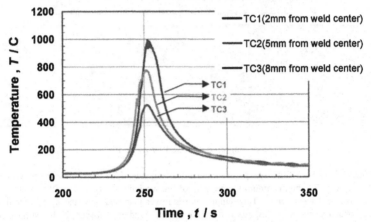

Fig. 7 Temperature hysteresis on the advancing side of similar gauge weld during linear FSW

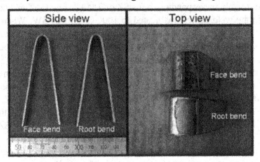

Fig. 8 Root and face bend test of DP steel friction stir welds

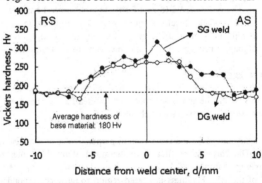

Fig. 9 Vickers hardness distributions of both similar and dissimilar gauge welds

Fig. 10 shows the tensile properties of the BM. The yield stress of the DP steel as received is close to that of the Al 5083-H18 alloy, but its UTS is 50% higher. The steel alloy is as ductile as the Al 6111-T4 material with an elongation of slightly over 30% strain. The rate jump tests yielded an average rate sensitivity index, m, of 0.001 for DP steel. This m-value is essentially zero, and will be neglected in further analysis. The final width and thickness measurements of the parallel-sided tensile specimens were used to determine the plastic anisotropy parameter, r, for each direction. Averages of r were then used to determine the average plastic anisotropy ratio, r-bar, and delta-r. For 2 mm thick DP steel sheet, r-bar and delta-r were 1.09 and –0.01, respectively.

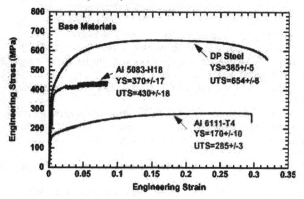

Fig. 10 Tensile properties of the BM.

Fig. 11 shows the transverse and longitudinal tensile properties of the weld. The DP-steel weld has higher UTS along the longitudinal direction than its base, but the tensile stress does not change after welding along the transverse direction. Because DP steel is stronger in the welded region than its BM due to the martensite formation, the tensile deformation concentrates in the softer BM when tested along the transverse direction. That limits the tensile strength and elongation of the welded specimen.

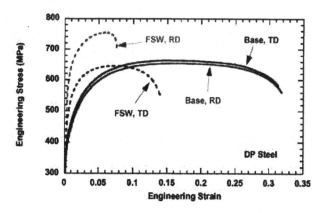

Fig. 11 Tensile properties of DP steel friction stir welds

Summary

FSW using pcBN tool was applied to DP steel sheets. The SZ was heated to over 1000°C during FSW, which resulted in the martensite formation in the SZ. The harness and UTS largely increased in the SZ due to the martensite microstructure. Bend test suggested that the SZ had toughness and ductility enough for forming in despite of the martensite formation. The HAZ showed two grain structures originated from the temperature gradient. The tool wear was negligible throughout the present work.

References

[1] C.J. Dawes and W.M. Thomas: *Welding J.*, 1996, vol. 75 (3), p. 41.

[2] W.M. Thomas and E.D. Nicholas: *Mater. Design*, 1997, vol. 18(4-6), p. 269.

[3] S.H.C. Park, Y.S. Sato, H. Kokawa, K. Okamoto, S. Hirano, and M. Inagaki: *Scripta Mater.*, 2004, vol. 51, p. 101.

[4] W.M. Thomas, P.L. Threadgill, and E.D. Nicholas: *Sci. Tech. Weding Join.*, 1999, vol. 4 (6), p. 365.

[5] C.D. Sorensen, T.W. Nelson, and S.M. Packer: *Proceedings of the 3rd International Symposium on Friction Stir Welding*, TWI, Kobe, Japan, 2001, CD-ROM.

[6] T.J. Lienert, W.L. Jr. Stellwag, B.B. Grimmett, R.W. Warke: *Welding J.*, 2003, vol. 82 (1), p. 1s.

[7] A.P. Reynolds, W. Tang, T. Gnaupel-Herold, and H. Prask: *Scripta Mater.*, 2003, vol. 48, p. 1289.

[8] K. Okamoto, S. Hirano, M. Inagaki, S.H.C. Park, Y.S. Sato, H. Kokawa, T.W. Nelson, C.D. Sorensen: *Proceedings of the 4th International Symposium on Friction Stir Welding*, TWI, Park City, Utah, 2003, CD-ROM.

[9] S.H.C. Park, Y.S. Sato, H. Kokawa, K. Okamoto, S. Hirano, and M. Inagaki: *Scripta Mater.*, 2003, vol. 49, p. 1175.

MICROSTRUCTURE AND PROPERTIES OF FRICTION STIR WELDED 304 STAINLESS STEEL USING W-BASED ALLOY TOOL

Yutaka S. Sato, Masahiro Muraguchi, Hiroyuki Kokawa

Department of Materials Processing, Graduate School of Engineering, Tohoku University; 6-6-02 Aramaki-aza-Aoba, Aoba-ku, Sendai 980-8579, Japan

Keywords: Friction stir welding, 304 stainless steel, W-based alloy tool, microstructure, mechanical properties, corrosion property

Abstract

Microstructure, and mechanical and corrosion properties were examined in 304 stainless steel friction-stir-welded using a tool made of a W-based alloy, and then these were compared with those of a weld produced by polycrystalline cubic boron nitride (PCBN) tool. Severe wear of the W-based alloy tool occurred during friction stir welding (FSW). The weld contained some tunnel-type defects in the stir zone. The microstructure in the bottom and advancing sides of the stir zone was relatively more sensitive for etching than the other regions, and it was identified as the duplex microstructure consisting of austenite and ferrite phases. W was enriched in the ferrite phase. The W-based alloy tool made the weld having roughly the same mechanical properties as the base material. The weld produced by the W-based alloy tool had the better corrosion resistance than that by the PCBN tool, although the corrosion resistance was slightly lower in the advancing side of the stir zone and heat affected zone (HAZ) than in the base material.

Introduction

Austenitic stainless steels are widely used in nuclear power plant applications requiring high temperature components such as heat exchangers and chemical reactors, because of their good mechanical properties at high temperatures and excellent corrosion properties. However, fusion welding of austenitic stainless steels often causes stress corrosion cracking and weld decay due to sensitization in the heat affected zone (HAZ). Friction stir welding (FSW) is a solid-state joining process [1], which would result in the formation of fine grain structure having no segregation and reduce the distortion and degree of sensitization in the austenitic stainless steels.

Polycrystalline cubic boron nitride (PCBN) and W-based alloy are known as the possible candidates of the FSW tool material for high melting temperature materials [2]. Some papers on FSW of austenitic stainless steels using these tool materials [3-8] have been reported. Park and co-authors [4-7] examined microstructure and several properties of a friction stir (FS) weld of 304 stainless steel produced by PCBN tool. They have shown that FSW produces the stir zone having the finer grain structure than the base material. The FS weld had roughly the same mechanical properties as the base material, and it exhibited the much lower degree of the sensitization in the HAZ than the gas tungsten arc (GTA) weld. However, small sigma phases [5] and Cr-rich borides [9] were rapidly formed in the advancing side of the stir zone having the recrystallized grain structure and deteriorated the corrosion properties in this region significantly. On the other hand, examination of 304L stainless steel FS-welded by a W-based alloy tool was performed by Reynolds et al. [3]. The FS weld had the finer grains causing the higher mechanical properties in the stir zone. The longitudinal residual stresses close to the base material yield strength in the weld were reported. Results on the 304 stainless steel weld produced by PCBN tool suggest that the microstructure and properties of the weld would depend on the tool materials, but the details of them have not been systematically examined in the 304-series stainless steel weld produced by W-based alloy tool.

The present studies used a W-based alloy tool for FSW of 304 stainless steel. Microstructure, mechanical and corrosion properties of the weld were examined, and then these were compared with those of a 304 steel weld produced by PCBN tool. The objective of this study is to clarify difference in microstructure and properties between the welds produced by the W-based alloy and PCBN tools.

Experimental Procedures

The base material used in this study was a 6mm-thick 304 austenitic stainless steel, whose chemical composition (wt%) was 18.10Cr - 8.56Ni - 0.59Si - 1.08Mn - 0.040C - 0.032P - 0.003S. This was the same material plate as that used in previous studies on FSW using PCBN tool [4-6]. The FSW tool was made of a W-based alloy. A stir-in-plate FS weld was produced on the base material plate at a travel speed of 1.33 mm/s and a rotational speed of 550 rpm using the W-based alloy tool with a 25 mm shoulder diameter and a 4.75mm pin length. The pin tapered from 9 mm at the shoulder to 6 mm at the pin tip. The welding parameters including the tool dimension were totally same as those used in the previous studies on FSW using PCBN tool [4-6].

Microstructure in the FS weld was observed by optical microscopy, scanning electron microscopy (SEM) and orientation imaging microscopy (OIM). Sample for optical microscopy and SEM were electrolytically etched in a 10 wt% oxalic acid solution at 30 V for 20 s. Chemical composition analysis was conducted in a HITACHI S4700 SEM equipped with an energy-dispersive X-ray spectroscopy (EDS) system. Crystallographic data collection by OIM was used for phase identification in a HITACHI S4700 SEM, operating at 25 kV under step size of 1 μm. Crystallographic data were expressed by phase map with grain boundaries. In the phase map, phases with the fcc and bcc structures were colored by light and dark gray, respectively.

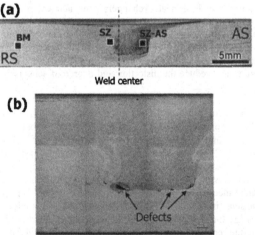

Figure 1. (a) Macroscopic overview of the cross section perpendicular to the welding direction and (b) magnified image of the stir zone.

262

Vickers hardness was measured on the cross section perpendicular to the welding direction, using a Vickers indenter with a 9.8 N load for 15 s. Transverse tensile specimens were cut perpendicular to the welding direction. Tensile test was carried out at room temperature at a crosshead speed of 0.05 mm/s.

Corrosion property of the FS weld was qualitatively examined by a ferric sulfate-sulfuric acid test [10] for 72 h. Two cross sections perpendicular to the welding direction, the top surface and the bottom surface of the weld were in contact with the test solution during testing. Moreover, double-loop electrochemical potentiokinetic reactivation (DL-EPR) test [11] was also performed with small pieces of specimens cut from the various regions to quantitatively evaluate distribution of the corrosion resistance in the weld.

Results and Discussion

<u>Microstructure</u>

A FS weld could be produced in 304 stainless steel using the W-based alloy tool. However, severe wear of the W-based alloy tool was obviously found after FSW.

Macroscopic overview of the cross section perpendicular to the welding direction is shown in Fig. 1(a). The stir zone is formed around the weld center. Border between the stir zone and thermo-mechanically affected zone (TMAZ) is distinctly observed at the advancing side, while the border at the retreating side is faint. Some tunnel-type defects are also found at the bottom and advancing sides of the stir zone, as shown in Fig. 1(b). Formation of the defects would be due to the un-optimized welding parameters for the W-based alloy tool, because the present study used the suitable welding parameters for PCBN tool, i.e. these parameters made the stir zone with no defects in 304 stainless steel using PCBN tool [4-6].

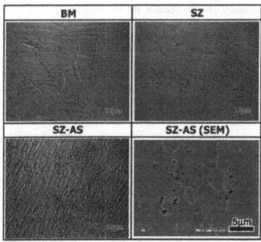

Figure 2. Optical micrographs of the BM (base material), SZ (stir zone) and SZ-AS (advancing side of the stir zone) shown in Figure 1(a). SEM image of the SZ-AS is also included.

Optical micrographs of the BM (base material), SZ (stir zone) and SZ-AS (advancing side of the stir zone) shown in Fig. 1(a) are presented in Fig. 2. The BM has an annealed grain structure including a high density of twins. The most part of the stir zone, which is represented by the SZ in Fig. 2, had the slightly smaller grains having a lower density of twins than the base material, but a difference in grain size between the BM and SZ is not large. On the other hand, microstructure of the SZ-AS is relatively more sensitive for etching than the other regions. Reynolds et al. [3] have shown the similar microstructure in the 304L stainless steel weld produced by the W-based alloy tool. Inspection by SEM revealed that this microstructure in the SZ-AS had the fine microstructure consisting of two phases, as shown in Fig. 2.

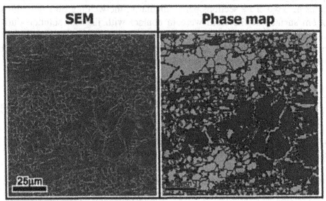

Figure 3. Example of SEM image and phase map obtained by OIM of the sensitive microstructure for etching.

SEM image and phase map obtained by OIM in the other region having the sensitive microstructure for etching are shown in Fig. 3. The sensitive microstructure for etching consists of two phases with the fcc and bcc structures. Interfaces between the two phases are hardly corroded during etching in the 10wt% oxalic acid solution. The fcc phases could be identified as the austenite phases, because the initial microstructure of 304 stainless steel consists of the austenite phase having the fcc structure. On the other hand, there are some candidates for the bcc phases, because both the ferrite phase, which is formed at high temperatures in 304 stainless steel, and the W-based alloy itself, which may remain when the tool wears during FSW, have the bcc structure. EDS analysis revealed that both the phases mainly consisted of Fe, Cr and Ni, and that the bcc phases contained the higher Cr and lower Ni than the austenite ones. Moreover, the W content of about 5 at% was detected in only the bcc phases. From these results, the bcc phases can be identified as ferrite phases containing the W. Careful inspection did not found any debris of the W-based alloy tool in the stir zone.

It has been reported that FSW of 304 stainless steel using PCBN tool [5] makes the different microstructural features from the present result (FSW using W-based alloy tool). The stir zone of the PCBN weld has tiny sigma phases [5] and Cr-rich borides [9] in the advancing side, and ferrite phases were not found. Since the lower heat-input parameters often leave a small amount of ferrite phases in the PCBN weld [7], it is suggested that sigma phases are formed through rapid decomposition of ferrite phases associated with dynamic recrystallization [5]. On the other hand, the FSW using W-based alloy tool made the stir zone having a large amount of ferrite phases containing the W of about 5 at%. This result suggests that maximum temperature during FSW is much higher in the FSW using W-based alloy tool, because ferrite phases are more stable at higher temperatures in 304 stainless steel. Since any debris of the W-based alloy tool was not found in the stir zone, the tool-wear products, i.e. W-based alloy itself, would be dissolved into the 304 stainless steel during stirring. It is likely that the W concentrates into ferrite phases at higher temperatures because the W is a ferrite stabilizer [12]. There is a possibility that the W solutionization to ferrite phase raises the stability of ferrite phase. This may be a reason why the ferrite phases did not decompose to sigma phases in the stir zone of the weld produced by W-based alloy tool.

Hardness profile across the stir zone of the weld is shown in Fig. 4. The weld exhibits roughly the constant hardness profile due to no large variation in grain size in the weld. Transverse tensile properties of the weld are summarized in Fig. 5. The weld had the slightly lower ultimate tensile strength and elongation than the base material, and failed from the defects existing in the stir zone.

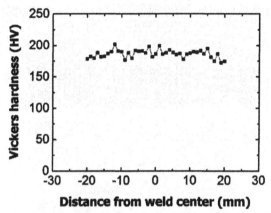

Figure 4. Hardness profile across the stir zone of the weld.

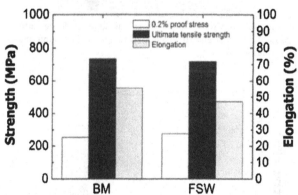

Figure 5. Transverse tensile properties of the base material (BM) and the weld (FSW).

Cross section of the weld after ferric sulfate-sulfuric acid test is shown in Fig. 6(a). Magnified image of the advancing side of the stir zone shown by open square in Fig. 6(a) are presented in Fig. 6(b). The bottom and advancing sides of the stir zone are preferentially corroded. The similar result has been reported in the PCBN weld [6]. SEM observation revealed that the corrosion sites in these regions were ferrite phases. Corrosion in the HAZ was not severe in the weld. Current ratios obtained by DL-EPR test for the BM, HAZ, SZ and SZ-AS shown in Fig. 6(a) are summarized in Fig. 7. The current ratios in the SZ-AS of the weld produced by PCBN tool and the HAZ of the GTA weld [6] are also added in this figure. The higher current ratio means the larger degree of the Cr depletion in the DL-EPR test. In the weld produced by W-based alloy tool, the current ratios of the HAZ and SZ-AS are slightly higher than that of the base material. The lowest current ratio was obtained in the

265

SZ. The HAZ exhibits the highest current ratio, but this value (0.36%) was much smaller than that of the HAZ of the GTA weld. This result shows that the corrosion resistance in the HAZ of the FS weld is much superior to that in the HAZ of the GTA weld. The SZ-AS in the FS weld produced by PCBN tool exhibited the much higher current ratio [6], which was caused by formation of the Cr depleted zone around the sigma phases and Cr-rich borides [9], while the SZ-AS produced by W-based alloy tool maintains the current ratio close to that of the base material. This result suggests that no precipitates leading to the Cr depleted zone exist in the SZ-AS of the FS weld produced by W-based alloy tool, which is supported by SEM image in Fig. 3. The present study suggests that the W-based alloy tool can produce the weld having the better corrosion resistance than the PCBN tool.

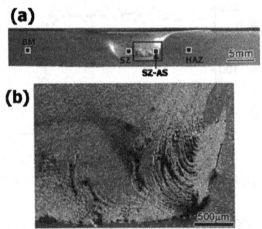

Figure 6. (a) Cross section of the weld after ferric sulfate-sulfuric acid test and (b) magnified image of the region shown by square in (a).

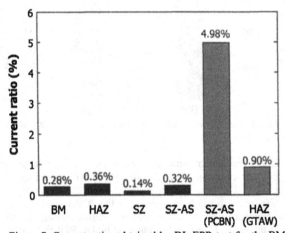

Figure 7. Current ratios obtained by DL-EPR test for the BM, HAZ, SZ and SZ-AS produced by the W-based alloy tool, the SZ-AS produced by PCBN tool, and the HAZ of the GTA weld.

266

Summary

In the present study, 304 stainless steel was FS-welded using W-based alloy tool, and then microstructure, mechanical and corrosion properties of the weld were examined. The W-based alloy tool severely wore during FSW. Duplex microstructure consisting of ferrite and austenite phases was formed in the bottom and advancing sides of the stir zone, and the W debris arising from the tool wear was dissolved into ferrite phases. The weld produced by the W-based alloy tool had roughly the same mechanical properties as the base material, and exhibited the higher corrosion resistance than the weld produced by PCBN tool.

Acknowledgements

The authors are grateful to Mr. A. Honda, Mr. M. Michiuchi, Mr. C.J. Sterling and Mr. R.J. Steel for technical assistance and acknowledge Prof. T.W. Nelson, Prof. C.D. Sorensen, Dr. S.H.C. Park and Prof. Z.J. Wang for their helpful discussions. They wish to thank Prof. T.W. Nelson, Brigham Young University and MegaStir Technologies for making the W-based alloy tool and providing the friction stir weld. Financial support from the Japanese Ministry of Education, Science, Sports and Culture with a Grant-in-Aid for Encouragements for Young Researchers and Education and a Grant-in-Aid for the 21st Century COE program in International Center of Research and Education for Materials at Tohoku University is gratefully acknowledged.

References

1. W.M. Thomas, E.D. Nicholas, J.C. Needham, M.G. Nurch, P. Temple-Smith, and C.J. Dawes, "Friction Stir Butt Welding," International Application No. PCT/GB92/02203.

2. C.D. Sorensen, "Progress in friction stir welding of high temperature materials" (Proceedings of the 14th International Offshore and Polar Engineers Conference (ISOPE 2004), Tulon, France, 23-28 May 2004), vol. 4, 8-14.

3. A.P. Reynolds, W. Tang, T. Gnaupel-Herold, and H. Prask, "Structure, properties, and residual stress of 304L stainless steel friction stir welds," *Scripta Materialia*, 48 (2003), 1289-1294.

4. K. Okamoto, S. Hirano, and M. Inagaki, S.H.C. Park, Y.S. Sato, H. Kokawa, T.W. Nelson, and C.D. Sorensen, "Metallurgical and mechanical properties of friction stir welded stainless steels" (Proceedings of the 4th International Friction Stir Welding Symposium, Park City, UT, USA, 14-16 May 2003), CD-ROM.

5. S.H.C. Park, Y.S. Sato, H. Kokawa, K. Okamoto, S. Hirano, and M. Inagaki, "Rapid formation of the sigma phase in 304 stainless steel during friction stir welding," *Scripta Materialia*, 49 (2003), 1175-1180.

6. S.H.C. Park, Y.S. Sato, H. Kokawa, K. Okamoto, S. Hirano, and M. Inagaki "Corrosion resistance of friction stir welded 304 stainless steel," *Scripta Materialia*, 51 (2004), 101-105.

7. S.H.C. Park, Y.S. Sato, H. Kokawa, K. Okamoto, S. Hirano, and M. Inagaki, "Microstructural characterization of stir zone containing residual ferrite in friction stir welded 304 austenitic stainless steel," *Science and Technology of Welding and Joining*, 10 (2005), 550-556.

8. Y.S. Sato, T.W. Nelson, and C.J. Sterling, "Recrystallization in type 304L stainless steel during friction stirring," *Acta Materialia*, 53 (2005), 637-645.

9. S.H.C. Park, "Microstructures and properties of friction stir welded stainless steels" (Ph.D. thesis, Tohoku University, 2005)

10. J.B. Lee, "Modification of the ASTM standard ferric sulfate-sulfuric acid test and copper-copper sulfate-sulfuric acid test for determining the degree of sensitization of ferritic stainless steels," *Corrosion*, 39 (1983), 469-474.

11. A.P. Majidi and M.A. Streicher, "The double loop reactivation method for detecting sensitization in AISI 304 stainless steel," *Corrosion*, 40 (1984), 584-593.

12. F.C. Hull, "Delta ferrite and martensite formation in stainless steels," *Welding Journal*, 52 (1973), 193s-203s.

Friction Stir Welding of X-65 Steel

Tracy W. Nelson, Sterling J Anderson, and David J. Segrera

Brigham Young University
Department of Mechanical Engineering
435 CTB
Provo, UT 84602

Friction Stir Welding, HSLA Steels, Welding, Joining, X-65 steel

Abstract

Friction Stir Welding (FSW) has been successfully applied to API X-65 grade line pipe steel using Polycrystalline Cubic Boron Nitride (PCBN) tooling. Thermal cycles throughout the weld correlate strongly with heat input, post weld microstructure, microhardness, and tensile properties. Transverse weld tensile samples consistently fail in the base metal well outside of the HAZ. Longitudinal all-weld tensile test indicate significant property variation through the thickness of the weld. These improved properties relative to traditional arc welding offer significant benefits. This paper presents in detail the results described above.

Introduction

The need for a high-strength, ferritic steel exhibiting desirable weldability, malleability, and corrosion resistance properties has led to the development of high strength, low alloy (HSLA) steels. This class of steels utilizes small amounts of alloying material and very little carbon to make it more corrosion resistant than normal carbon-containing steels. As HSLA has gained popularity in automotive, naval, and other industries, the inherent need for better welding processes has arisen.

Most fusion welding techniques used for steel degrade mechanical properties. They do this by: 1) causing dramatic phase changes in the microstructure of the existing metal, and 2) changing the metal composition at the joint by the addition of a filler metal.

Originally patented in 1991 by The Welding Institute [1], friction stir welding (FSW) has proven to be a suitable technology in joining aluminum and copper alloys. Since inceptions, FSW has found many applications in marine, automotive, rail transpiration industries, and launch vehicles such as the space shuttle and delta rockets. However, applications to date have been limited to lower melting temperature alloys like aluminum and copper.

Given the benefits of FSW in aluminum and copper alloys, similar benefits and cost savings should also be achievable if applied to high temperature metals like steels and stainless steels. One of the major costs in the fabrication of larger structures is post-weld straightening. In panel lines, more man hours are spent straightening welded panels than actual weld time [2-3]. Similarly, in high alloy materials like stainless steels and nickel-base super alloys, post-weld repair of weld defects is a major cost in welded fabrication.

Much of the preliminary work in FSW of steels successfully demonstrated feasibility. Thomas and co-workers [4-5] first reported FSW of ferritic steel in 1999. They reported excellent results with regards to weld quality, reduced distortion, and resulting mechanical properties. Thomas [4-5], Lienert [6-7], Reynolds [8], Sterling [9-10] and Steel [11] also reported excellent weld quality and post weld mechanical properties in ferritic steels and stainless steels..

269

The objective of this investigation was to explore the possible range of parameters over which X-65 steels could be successfully FSW, and characterize the as-weld properties and microstructure at several different parameters.

Approach

Initial parameters studies were undertaken to develop a set of parameters over which X-65 steels could be successfully welded. A polycrystalline cubic boron nitride (PCBN) [12-13] FSW tool with a concave shoulder and step spiral pin tool was used for this investigation. A picture of the tool used is shown in Figure 1.

Figure 1. Picture of the PCBN Step Spiral tool

All welds in this investigation were performed as bead-on-plate welds. The X-65 steel was sectioned into 6.35 mm (0.25 in) thick plates that were 178mm wide by 914 mm long (7 x 36 in). All oxide/millscale was removed prior to welding and the surface of each plate was cleaned with methanol. The underside of each plate was coated with boron nitride to prevent adhesion to the anvil. Plates were rigidly clamped at 102 mm (4 in.) intervals to an anvil and a head tilt angle 2.5° was used.

The ranges of RPM and IPM investigated were 350-550 and 50.8-177.8 mm/min (2-7 IPM), respectively. Once the process window was established specific parameters were selected for further mechanical and microstructural analysis. The three parameters investigated were: 76.2 mm/min (3 IPM) – 350 RPM, 177.8 mm/min (7 IPM) – 350 RPM, and 177.8 mm/min (7 IPM) – 550 RPM.

Heat input for each weld was calculated by multiplying spindle torque and spindle speed, and dividing by travel speed. Weld properties were correlated with heat input to give a uniform standard of comparison.

Thermal couples were strategically placed in the heat affected zone (HAZ) of each weld. Six thermocouples were placed on either side of the weld centerline, beginning at the edges of the pin, and spaced at one millimeter intervals outward from the pin centerline at the midplane of the FSW tool pin. To avoid interference or heat transfer inconsistencies between thermocouple holes, 138 mm (5.45 in.) of weld travel separated thermocouple pairs. Data from these readings were recorded, plotted and analyzed for peak temperatures and cooling rates at each thermocouple location. Post weld analysis was performed to account for tool deflection during the weld.

Transverse samples were removed from each parameter set. These were mounted, polished, etched, and examined via optical microscopy (OM) for weld quality and microstructure. Each sample was etched with a modified Winsteard reagent to delineate flow lines, grain boundaries, and phases present in the weld nugget and HAZ. This etch consisted of a 35-second application of 4% Picral followed by an 8-second application of 2% Nital. Optical microscopy images were examined to determine grain structure and phase morphology in each weld.

Vickers microhardness was performed on each of the three weld parameters. Microhardness tests were performed with an automatic microhardness tester using a Vickers diamond indenter, a 300 gram load, and an 8 second dwell. A 500X magnification lens was used for optical measurement of indentations. A grid of points spaced at 600 micrometer intervals across the surface was tested and the results were plotted with Matlab v 7.0 to show relative hardness areas.

Both transverse and longitudinal all-weld metal tensile properties were evaluated. Transverse tensile specimens were removed from each weld and milled according to ASTM E8 Standard. Each specimen was pulled using an MTS tensile testing machine. Data were recorded using TestStar and analyzed.

Three longitudinal samples were water jet cut from each parameter set. Each weld sample was milled square and three tensile specimens were EDM cut though the thickness of each sample (see Figure 2). Each specimen was lightly hand ground on 200 grit SiC wet sand paper to remove the EDM HAZ from all sides of the sample. Specimens were tensile tested using an Instron.

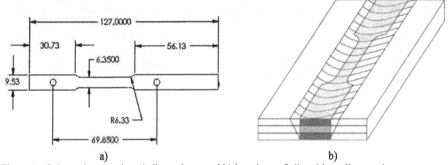

a)   b)

Figure 2. Schematics showing 1) dimensions, and b) locations of all-weld tensile samples.

Results and Discussion

Significant differences in weld thermal cycles, post weld microstructure and tensile properties were observed over the range of parameters investigated. All welds conducted over the range of process window parameters showed complete consolidation and excellent surface finish. Machine load data showed that Z- and X- loads increase with decreasing heat input. Y-loads depended almost solely on spindle speed, increasing with increasing RPM.
Table 1 shows machine loads for the three parameters of interest.

Table 1. Machine loads for the step spiral tool at three parameters of interest.

Parameter (mm/min – rpm)	Heat Input (kJ/mm)	Z-Load (N)	X-Load (N)	Y-Load (N)
76.2 - 350	4.3	48000	3050	1570
177.8 - 350	1.75	51000	13940	1612
177.8 - 550	2.4	49000	8380	2000

Peak temperatures and cooling rates were calculated from adjusted thermocouple data. The cooling rates reported are the average rates of cooling from 800°C, or the peak temperature if lower than 800°C, to 500°C. Similarly, heat input was also calculated for each of the parameter sets. These results are shown in Table 2.

271

Peak temperatures in the HAZ are higher for welds with greater heat input. Cooling rates appear more dependent on travelspeeds, increasing as linear travel speed increases.

Table 2. Peak temperatures and cooling rates achieved at various weld locations.

SS Tool														
Distance from Weld Center			3.53		4.53		5.53		6.53		7.53		8.53	
	Heat Input (kJ/mm)	Side of Weld	Adv	Ret	Adv	Ret	Adv	Ret	Adv	Ret	Adv	Ret	Adv	Ret
76.2 mm/min 350 RPM	4.3	Peak Temp (°C)	862	1105	841	998	819	890	798	783	777	675	758	567
		Cooling Rate (°C/s)	25	33	28	33	24	17	19	20	19	15	28	9
177.8 mm/min 350 RPM	1.75	Peak Temp (°C)	933	666	822	631	712	595	602	560	491	524	381	488
		Cooling Rate (°C/s)	50	60	44	22	44	19	13	17	15	9	.**	.**
177.8 mm/min 550 RPM	2.4	Peak Temp (°C)	1002	999	941	921	880	842	819	763	758	685	697	606
		Cooling Rate (°C/s)	49	52	.*	45	39	42	44	40	41	34	40	26

* = Thermocouple Failure
** = Didn't reach 500 C

Post-weld microstructural characterization of FSW X-65 indicates a strong dependence on location within the weld. All weld microstructures showed refined grain size relative to the base metal. Figure 3 contains micrographs of center, advancing side edge of pin and the near HAZ from each of the three different weld parameters investigated. Microstructures at the center of the weld nuggets consisted of mostly upper bainite with the exception of the lowest heat input weld (b). This particular weld consists of polygonal ferrite which is consistent with lower peak temperatures resulting in only partial transformation. The microstructure in the advancing side pin edges (which correspond to the regions of highest hardness) consists of coarse upper bainite and carbide for the two higher heat in put welds (a and c). For the lower heat input weld, the microstructure is more refined, e.g. bainite platelets are narrower, with finer carbides. Similar trends were observed in the near HAZ.

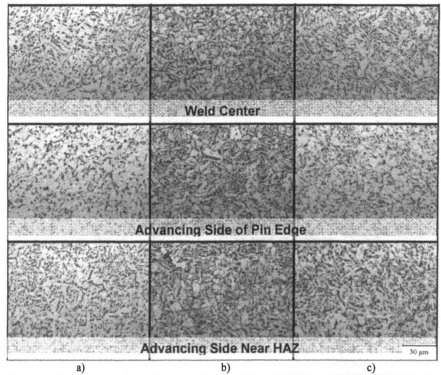

| a) | b) | c) |

Figure 3. Micrographs at the weld center, advancing side edge of pin and the near HAZ of welds made at: a) 76.2 mm/min and 350 RPM, b) 177.8 mm/min and 350 RPM, and c) 177.8 mm/min and 550 RPM.

Correlating peak temperatures from Table 2 with weld phases in Figure 4 helps explain some of the microstructural characteristic in FSW X-65. None of the temperatures recorded for the SS tool in the HAZ exceeded 1100°C. Given this maximum temperature limit, combined with a finite cooling rate, FSWs in X-65 do not exhibit a coarse grain region in the HAZ typical of a fusion welding process. Generally, the coarse grain region suffers the greatest loss in toughness. As such, FSW should retain much higher toughness that fusion welding processes in the weld nugget. In addition, overall grain size refinement in all regions of the weld and HAZ should produce excellent tensile strengths and fracture toughness throughout the weld and HAZ.

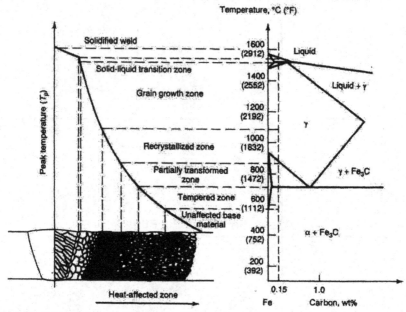

Figure 4. Schematic showing the relationship between peak temperatures and various sub zones that can form in the HAZ of low carbon steels. (ASM Handbook Vol. 6)

Microhardness Mapping

Microhardness tests were completed on one test specimen from each parameter set. Figure 5 shows hardness profiles of the weld cross sections along with the maximum, minimum, and average hardness values for each tool/parameter.

Figure 5 shows interesting trends in hardness relative to weld parameters. All three parameters show small regions of high hardness within the advancing side of the weld nugget. Comparing the maximum to average hardness, the two high heat input welds (76.2 mm/min at 350 rpm and 177.8 mm/min at 550 rpm) show a much greater increase (>100 DPH) in hardness compared to the low heat input weld (60 DPH) in this region. Microstructural comparison of these regions indicates higher peak temperatures were attained in the high heat input welds relative to the low heat input weld as evident by the larger prior austenite grain size.

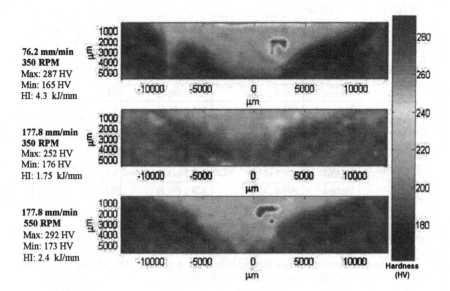

76.2 mm/min
350 RPM
Max: 287 HV
Min: 165 HV
HI: 4.3 kJ/mm

177.8 mm/min
350 RPM
Max: 252 HV
Min: 176 HV
HI: 1.75 kJ/mm

177.8 mm/min
550 RPM
Max: 292 HV
Min: 173 HV
HI: 2.4 kJ/mm

Figure 5. Microhardness maps of welds conducted using a SS tool.

<u>Tensile Properties</u>

Both transverse and all-weld longitudinal tensile properties were evaluated. All transverse tensile specimens failed in the base metal for all three parameters investigated. Tensile, yield, and elongation exceeded the base metal. These results give no specific data relative to the actual properties of the weld or HAZ. Table 3 shows the results from the transverse weld tests and Table 4 those from the all-weld tests.

Table 3. Results from transverse tensile tests.

	Yield Strength MPa (ksi)	Ultimate Tensile MPa (ksi)	Strain to Failure (%)
72.6 mm/min 350 RPM	440.8 (63.9)	533.4 (77.4)	19.9
177.8 mm/min 350 RPM	465.2 (67.5)	536.6 (77.8)	20
177.8 mm/min 550 RPM	467.0 (67.7)	538.3 (78.1)	19.7
Minimum X-65 Specification	448 (65)	538 (78)	19

Longitudinal all-weld metal tensile tests were performed to better understand how the properties of the weld metal varied though the plate thickness. Although weld properties varied considerably from the weld top to weld bottom, all-weld-metal tensile and yield strengths were significantly higher than the base metal and transverse weld tensile properties while ductility was comparable.

Table 4. Results from all-weld tensile tests.

		Yield Strength MPa (psi)	Ultimate Tensile MPa (psi)	Strain to Failure (%)
177.8 mm/min 350 RPM	Top	555.4 (80.5)	655.6 (95.1)	17.7
	Middle	531.1 (77.0)	650.3 (94.3)	19.2
	Bottom	493.3 (71.5)	621.9 (90.2)	21.4
177.8 mm/min 550 RPM	Top	562.6 (81.6)	666.2 (96.6)	17.9
	Middle	547.8 (79.4)	658.8 (95.6)	17.6
	Bottom	510.4 (74.0)	632.8 (91.8)	23.1
76.2 mm/min 350 RPM	Top	557.9 (80.9)	663.0 (96.2)	17.5
	Middle	533.8 (77.4)	642.8 (93.2)	17.7
	Bottom	505.8 (73.4)	628.1 (91.1)	20.6

Longitudinal all-weld metal tensile and yield strength exceeded the base metal at all weld parameters. Longitudinal tensile and yield strengths were on average 100 MPa (14.5 kpsi) greater than the transverse properties. Elongations were roughly 2% lower at the top and middle of the welds, while the bottom regions exceeded base metal minimums. These strength increases are likely due to the grain refinement in the weld and HAZ.

The increase in tensile properties, as evident from the longitudinal all-weld samples, is significant. These strength increases are sufficient to have an impact on fracture toughness. The authors are confident that with further parameter studies, the differences between weld metal and base metal can be reduced.

Conclusions

From the results presented above, the following conclusions can be made:
1. X- and Z-loads decrease with increasing heat input.
2. Peak temperatures in the HAZ increase with increasing heat input.
3. Cooling rates increase with increasing linear travel speed.
4. All post weld microstructures show refined grain size relative to base metal and are strongly dependent on location within the weld.
5. Weld microhardness at different locations increases with increasing heat input.
6. Weld transverse tensile properties exceed base metal minimums.
7. All-weld metal longitudinal tensile properties exceed base metal minimums.

Acknowledgements

This material is based upon work supported by the National Science Foundation under Grant No. EEC-0437358. The authors would also like to express their gratitude to JFE Steel Corporation (Japan) for providing the steel for this investigation, and the NSF Center for Friction Stir Processing for support of this work. The authors also acknowledge the contributions of Scott Wood, Trevor Downs and Scott McEuen in completing this work.

References

1. Thomas, W.M., Nicholas, E.D., Needham, J.C., Murch, M.G., Temple-Smith, P., Dawes, C.J. (1991), *International Patent Application* No. PCT/GB92/02203 and GB application No. 9125978.8.

2. Cahill, P.D., (2003), Personal communication: FSW in Panel Line Production for Ship Building, April.

3. Midling, O.T., (1999), Personal communication: FSW in Panel Line Production for Ship Decking and Substructures, August.

4. Thomas, W.M. (1999). "Friction Stir Welding of Ferritic Steels-A Feasibility Study," *First International Symposium on Friction stir Welding*, ISFSW, Thousand Oaks, CA, USA, June 14-16.

5. Thomas W.M., et al, "Feasibility of Friction Stir Welding Steel," *Science and Technology of Welding and Joining*, vol. 4, no. 6, pp. 365-372, 1999.

6. Lienert T.J., Gould, J.E. (1999), "Friction Stir Welding of Mild Steel", *First International Symposium on Friction Stir Welding*, Thousand Oaks, CA, USA, June 14-16.

7. Lienert, T.J., Stellwag, W.L., Grimmett, B.B. and Warke, R.W., (2003), "Friction Stir Welding Studies on Mild Steel," *Welding Journal*, vol. 82, no. 1, pp. 1s-9s.

8. Reynolds, A.P., Tang, W., Gnaupel-Herold, T., and Prask, H., (2003) "Structure, Properties and Residual Stress of 304L Stainless Steel Friction Stir Welds", Scripta Mat., 48, pp. 1289-1294.

9. Sterling, C.J., Nelson, T.W., Sorensen, C.D., Steel, R.J. and Packer, S.M., (2003). "Friction Sir Welding of Quenched and Tempered C-Mn Steel," *Friction Stir Welding and Processing II*, TMS, San Diego, CA, March.

10. Sterling, C.J., Nelson, T.W., Sorensen, C.D. and Posada, M., (2003), "Effects of Friction Stir Processing on the Microstructure and Mechanical Properties of Fusion Welded 304L Stainless Steel," *THERMEC 2003*, Leganés, Madrid, Spain, July 2003.

11. Steel, R.J., Pettersson, C., Nelson, T.W., Sorensen, C.D., Sato, Y.S., Sterling, C.J., and Packer, S.M., (2003), "Friction Stir Welding of SAF 2507 (UNS S32750) Super Duplex Stainless Steel," *Stainless Steel World 2003*, Maastricht, Netherlands, November.

12. Packer, S.M., Steel, R.J., (2003) "Tool and Equipment Requirements for Friction Stir Welding Ferrous and Other High Melting Temperature Alloys," *4th International Symposium on Friction Stir Welding*, Park City, UT, May 2003.

13. Sorensen, C.D., Nelson, T.W. and Packer, S.M., (2001) "Tool Material Testing for FSW of High-Temperature Alloys," *3rd International Symposium of Friction Stir Welding*, Kobe, Japan, September.

Tool Mushrooming in Friction Stir Welding of L80 Steel

Wei Gan, Z. Tim Li, Shuchi Khurana

Edison Welding Institute, 1250 Arthur E. Adams Dr., Columbus, OH 43221, USA

Keywords: Friction stir welding, Finite element simulation, Deformation, Mushrooming

Abstract

In this study, friction stir welding of L80 steel was investigated with experiments and modeling. Severe tool deformation and some wear were observed when a commercial pure tungsten was used. Finite element analysis (FEA) was used to predict the mushrooming of the tool. The temperature histories in the weld region and tool forces were measured and supplied to a finite element model. The simulated pin deformation matched the experimental observation. Using this model and optimization techniques, the required yield strength of the pin material is estimated to be 400 MPa at 1000°C to avoid mushrooming.

Introduction

Friction stir welding (FSW) has been widely applied to the aluminum alloys which are difficulty to join by fusion welding techniques. In the past few years, the interest in FSW of high melting temperature materials such as steels has increased [1-3]. However, the application of FSW in steels has been hindered by the structural performance of tool materials. Because of the high temperature and loading forces during welding, conventional tool materials may not have the needed strength. As a result, tool wear or deformation could occur. Therefore, it's important to understand the loading conditions of the pin tool during service so that the guidelines for selecting tool materials can be established.

There are three types of materials that are used for pin tool: tool steels, refractory metals, and super abrasive materials. Tool steels have too much wear when the base material is strong [5-6]. Refractory alloys have better strength than tool steels at high temperature, but tool wear can still occur. Super abrasive materials, such as polycrystalline diamond (PCD) and polycrystalline cubic boron nitride (PCBN), have excellent high temperature strength and are chemically stable with most metal. However, they are very difficult to machine and can be made only in small pieces [7], which limits the application of super abrasive materials in the manufacturing of FSW tools.

To apply FSW to steels, the desired tool material should have high fracture toughness, excellent yield strength at high temperature, stable microstructure, and be inert to workpiece. The objective of this work was to determine the strength requirement of the tool material at high temperature to successfully weld L80 steel. This is achieved through a combination of experimental and numerical approaches as presented in this article.

Experiments

High-strength pipe steel L80 was selected for weld trails in this study. Potassium doped commercial pure tungsten (CPW) was used to make the pin tool. Table 1 shows its typical composition. The material was delivered in stress relieved condition as forged rod. The stress strain behavior of the pin tool material at room temperature was provided by the vendor although its strength near 1000°C is unknown. It was estimated in simulation to fit the observed pin deformation.

Table 1. Chemical Composition of CPW (in ppm unless otherwise specified)

W	K	C	Mo	P	Fe
Balance (~99.98%)	50	15	20	20	20

FSW experiments were conducted on a Nova-Tech® Friction Stir Welder at EWI. Several sets of welding conditions were attempted to optimize the parameters. It was found that a spindle speed of 170 rpm and travel speed of 4 ipm can produce void free welds with a smooth surface finish. They were, therefore, selected as welding input parameters.

Transient temperatures at different locations were monitored by thermocouples during welding. These thermocouples were located in the stir zone and heat-affected zone (HAZ) as illustrated in Figures 1. The measured peak temperatures were used in FEA model to simulate the temperature distribution near the stir zone.

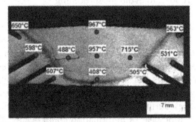

Figure 1. Thermal couple positions and measured peak temperatures

The applied tool forces in vertical and horizontal directions were recorded to be 20 kips and 6 kips respectively. The FSW pin tool was deformed under these welding conditions. Figure 2 compares the pin geometries before and after welding. Mushrooming of the pin was apparent. It was observed that the mushrooming consisted of both material loss and plastic deformation. An analytical model is needed to quantify the percentages and where they occurred.

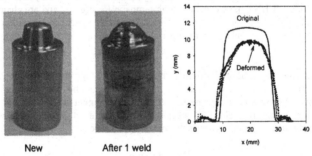

Figure 2. Pin profiles before and after welding.

280

Numerical Analysis

Since the pin tool is rotating fast during welding, the stress and temperature field near the pin tool tends to be axisymmetric in an average sense. Therefore, a 2D axisymmetric finite element model was constructed as shown in Figure 3a. It was assumed that the temperature at the interface between the pin and the base material was about 1050°C, slightly higher than that of the SZ of 957°C (Figure 1). The temperature near the holder was assigned to be 100°C, the same as the boiling point of water. This is a good approximation considering that the holder was water cooled. The temperature of the bottom surface of the plate was fitted so that the predicted temperatures in the HAZ matched the experiments. A value of 300°C was adopted.

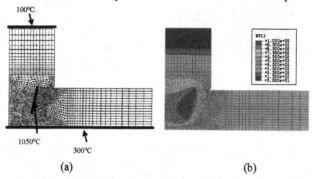

(a)   (b)

Figure 3. Thermal analysis: a) the thermal model, b) simulated temperature distribution

The predicted temperatures in the SZ and HAZ agree well with the measurements, Figures 4a and b. The temperature difference between experiments and numerical simulations were within 13%. It is a good match considering the scatter of the thermocouple measurement itself is about this magnitude. Therefore, the modeled temperature distribution was used in the subsequent mechanical analysis to analyze the mushrooming behavior of the pin tool.

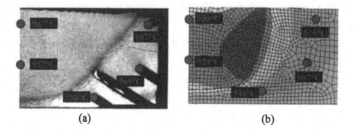

(a)   (b)

Figure 4. Comparison of temperature distribution in: a) experiment, b) the finite element model

After the thermal simulation of heat conduction the temperature distributions in both pin tool and base plate were mapped to the stress analysis model, Figure 5a. The bottom surface of the plate was supported by an anvil and its movement in the y direction was fixed in the model. The axisymmetric condition was imposed on the left edge of the model to fix the horizontal

281

displacement. A pressure load of 112 Mpa was applied on the top surface of the pin tool. This value was calculated by dividing the push-down force (20kips) by the cross section area of the pin tool. As previously mentioned, the strength of the CPW material at 1000°C was not directly available. It was calibrated in this model so that the amount of pin shortening matched with experiments. A value of 100 MPa was therefore derived.

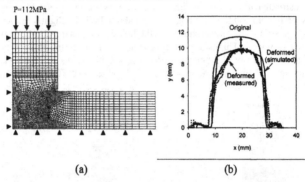

(a)   (b)

Figure 5. Mechanical analysis: a) the model, b) simulated pin deformation

It was found that the pin volume decreased by 7% after welding by comparing the original volume with the measured value after deformation. Plastic deformation should not cause material loss because it's a mass conservative process. The volume change from elastic strains is so small that it can be ignored in the analysis. Therefore, this material loss is only produced by the wear process during welding. Figure 5b suggests that material loss happened mostly at the side of the pin. That part of the pin experienced both high stress and temperature conditions. At the same time, the relative velocity between the pin and the plate material was also high. The combination of those factors caused the severe wear at the location.

Mushrooming can be avoided if the pin tool material is strong enough at higher temperature. To find out the strength requirement the yield strength of the pin tool material at 1000°C ($\sigma_y^{1000°C}$) was varied in a series of simulations. The amounts of pin length reduction using the hypothetic materials were plotted in Figure 6. When $\sigma_y^{1000°C}$ was increased to 400 MPa there is negligible pin shortening. It is therefore recommended as the requirement for pin material to avoid mushrooming.

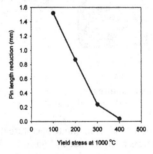

Figure 6. pin length reduction vs. $\sigma_y^{1000°C}$

282

Conclusions

Mushrooming was observed in this study during FSW of L80 steel. Temperature distributions in the SZ and HAZ were measured by thermocouples, while tool loading forces were recorded through the FSW machine. A sequentially coupled thermal mechanical finite element model was constructed to simulate the pin tool deformation under service conditions. The calibrated model was then used to determine the required $\sigma_y^{1000°C}$ to avoid mushrooming. The following conclusions have been reached from this work:

1. Both wear and plastic deformation occurred on the CPW pin tool during welding.
2. Wear and tear account for 7% material loss on the pin.
3. In order to avoid the problem, pin tool material should have a yield strength larger than 400 MPa at 1000°C.

Reference

1. Thomas WM, Threadgill PL, Nicholas ED, Feasibility of friction stir welding steel, SCIENCE AND TECHNOLOGY OF WELDING AND JOINING 4 (6): 365-372 1999.

2. Reynolds AP, Tang W, Gnaupel-Herold T, et al., Structure, properties, and residual stress of 304L stainless steel friction stir welds, SCRIPTA MATERIALIA 48 (9): 1289-1294 MAY 2003.

3. Lienert TJ, Stellwag WL, Grimmett BB, et al., Friction stir welding studies on mild steel - Process results, microstructures, and mechanical properties are reported, WELDING JOURNAL 82 (1): 1S-9S JAN 2003

4. Defalco J, Friction stir welding vs. fusion welding, WELDING JOURNAL 85 (3): 42-44 MAR 2006.

5. Prado RA, Murr LE, Shindo DJ, et al., Tool wear in the friction-stir welding of aluminum alloy 6061+20% Al2O3: a preliminary study, SCRIPTA MATERIALIA 45 (1): 75-80 JUL 13 2001.

6. Fernandez GJ, Murr LE, Characterization of tool wear and weld optimization in the friction-stir welding of cast aluminum 359+20% SiC metal-matrix composite, MATERIALS CHARACTERIZATION 52 (1): 65-75 MAR 2004.

7. Sorensen C.D., Progress in Friction Stir Welding of High Temperature Materials, The Proceedings of the 14th International Offshore and Polar Engineering Conference, Toulon, France, May 23-28, 2004.

PROPERTIES AND STRUCTURE OF FRICTION STIR WELDED ALLOY 718

Carl Sorensen, Ben Nelson, Sam Sanderson

Brigham Young University, 435 CTB, Provo, UT 84602, USA

Keywords: Friction Stir Welding, Inconel 718, Mechanical Properties

Abstract

Friction Stir Welding (FSW) has been successfully applied to annealed Alloy 718 sheet using Polycrystalline Cubic Boron Nitride (PCBN) Tooling. Bead-on-plate welds were performed using a new tool design, the Convex Scrolled Shoulder Step Spiral (CS4) pin tool. Processing windows under position control were developed. Measured tool torque, spindle speed, and feedrate were used to calculate heat input. Transverse specimens were used for microstructural analysis. Longitudinal all-weld-metal tensile specimens were tested to determine mechanical properties. The CS4 tool was shown to have an increased processing window compared with the concave shoulder tool. CS4 tools have higher tool life than traditional tools. Weld metal strengths are greater than the base metal. Weld ductility is 30%. Grains in the stir zone are approximately 1/5 the size of those in the base metal. A region of high hardness is found near the end of the pin on the advancing side in low-heat-input welds, which is thought to be primarily due to a reduction in grain size.

Introduction

Friction Stir Welding (FSW) has been widely used in aluminum as a means for joining high-strength, difficult to weld alloys. A major obstacle to welding alloys with higher melting temperatures is the availability of tools with sufficient high-temperature strength [1]. Tools made from PCBN have been used to weld a variety of high-temperature alloys, but most of the work has been done on high-strength ferritic alloys [2-9].

FSW of austenitic alloys is of interest due to the excellent corrosion resistance of these alloys. Preliminary work has demonstrated the ability to weld type 301, 304, and 316 stainless steel, and nickel alloys 200, 600, and 718, among others[10-15]. This work reports on an extended study of the process, parameters, and properties of FSW alloy 718 in the annealed state.

Alloy 718 is a high-temperature-oxidation resistant, precipitation hardening alloy. It is strengthened with gamma prime (Ni_3Nb), and has excellent strength from cryogenic temperatures up to about 700 C. It has excellent oxidation resistance up to about 1000 C. It is commonly used in jet engines and rocket motors, as well as cryogenic storage tanks. Of particular interest in FSW of alloy 718 is the potential for forming sheet parts for jet engine applications, such as nozzles and thrust reversers.

This paper describes the process loads, process window, grain size, mechanical properties, and heat input for FSW of alloy 718. Welds are shown to have excellent strength and ductility. Under low-heat-input welding conditions, a region of high hardness is formed in the weld.

Experimental Method

Bead-on-plate welds were made on a TTI RM-2 FSW machine. The tool used was a PCBN tool having a convex scrolled shoulder and a stepped spiral pin, hereafter referred to as a CS4 tool (Figure 1). This tool allows welding with the tool perpendicular to the material surface and compensates for slight surface irregularities. During welding, the tool was shielded with argon gas to prevent oxidation of the weld metal and increase material flow. The welds were made under Z-axis position control and fixed travel speed conditions. Process loads were measured by means of piezoelectric load cells mounting the machine spindle to the frame. Process torque was measured by the motor controller, with a conversion factor for the gear ratio between the motor and spindle. Tool wear was assessed by examining photographs of the tool following welding and looking for cracks or other evidence of wear or failure.

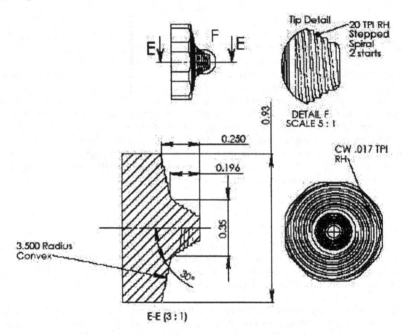

Figure 1: Pin and shoulder geometry for the CS4 tool used in this study.

Following welding, transverse specimens were removed from the weld with a waterjet cutting machine. The specimens were mounted, polished, and etched for optical microscopy. Stir zone width at the end of the pin and at the junction between the shoulder and pin were measured using image analysis software. Grain size at the top, middle, and bottom of the stir zone was measured with image analysis software and optical microscopy. Two-dimensional microhardness scans were also performed.

All-weld-metal longitudinal tensile specimens were removed from the plate by wire EDM. The cut surfaces were polished to remove the fine HAZ resulting from the EDM. The specimens were then pulled to measure yield strength, ultimate strength, and plastic elongation to failure.

286

Heat input for each weld was calculated by multiplying spindle torque and spindle speed, and dividing by travel speed. Weld properties were correlated with heat input to give a uniform standard of comparison.

Results and Discussion

Figure 2 shows the CS4 tool after 102 cm of welding as compared with a traditional concave shoulder, step spiral pin tool after 84 cm of welding. Note that there are significant fractures in both the shoulder and the pin of the traditional tool, while there are no visible fractures on the CS4 tool. This reduced propensity to fracture has been observed for CS4 tools in every alloy tested. It is believed that this is due, in large part, to a reduction in cyclic loading due to the elimination of the tool tilt.

Figure 2: Traditional PCBN tool after 84 cm of welding (left) compared with CS4 tool after 102 cm of welding (right)

The process window for FSW of alloy 718 is shown in Figure 3. As can be seen, the window for the CS4 tool is considerably larger than the window for concave shoulder tools. Of particular interest is the fact that the CS4 window is wider at 200 rpm than at 400 rpm. In many other alloys, higher spindle speeds lead to a wider process window. It appears that lower spindle speeds are more conducive to welding 718 than higher spindle speeds. Future work will involve evaluating spindles speeds lower than 200 rpm.

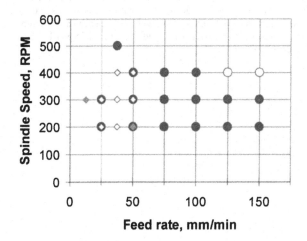

Figure 3: Process windows for the concave shoulder (diamonds) and CS4 (circles) tools Dark shapes indicate consolidated welds, light shapes indicate unconsolidated welds.

Welding direction (X-axis) loads are plotted as a function of travel speed for various spindle speeds in Figure 4. As expected, the X load increases with travel speed. Somewhat surprisingly, the X load increases with increasing spindle speed. It had been expected that higher spindle speeds, with their correspondingly higher heat inputs, would result in a softer extrusion zone and lead to reduced process forces. However, in every case, a higher spindle speed created a higher X force. It would appear that, for this alloy, welding at lower spindle speeds would be better than at higher spindle speeds. Future work will investigate the lowest possible spindle speeds to use on alloy 718.

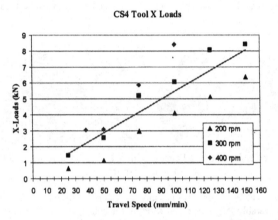

Figure 4: Welding direction (X) load as a function of travel speed for various spindle speeds. As expected, the X force increases with travel speed.

As expected, heat input varies with both spindle speed and travel speed, as shown in Figure 5. Heat input is inversely proportional to travel speed. At 50 mm/min, spindle speed has little effect on heat input, while at higher spindle speeds, heat input is approximately proportional to spindle speed. The lack of variation in heat input with spindle speed at 50 mm/min indicates that at this travel speed, torque varies strongly with spindle speed. At 75 and 100 mm/min, the torque is almost independent of spindle speed.

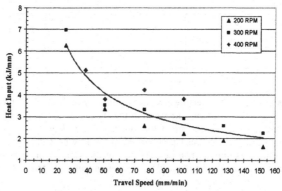

Figure 5: Heat input as a function of travel speed for various spindle speeds.

It was expected that high heat input would lead to large stir zone width, and low heat input would lead to small stir zone width. At the root of the pin (near the transition from the pin to the shoulder, this expectation was met. However, at the end of the pin, the stir zone width is essentially independent of heat input. Thus, there is little increase in the width of the weld at the back side when the heat input increases, but there is significant increase in width near the shoulder-affected region.

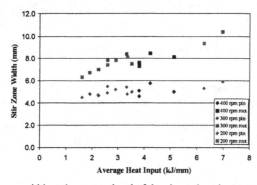

Figure 6: Stir zone width at the root and end of the pin vs. heat input at various spindle speeds.

The mechanical properties measured all-weld-metal longitudinal tensile specimens are compared with the base metal in Table 1. The yield and tensile strengths of the weld metal are significantly above those of the base metal. The yield strength of the high heat input weld is almost twice that of the base metal. The ultimate strength of the weld metal is about 20% above that of the base metal. The increase in yield strength appears not to be based primarily on grain size, as the high heat input weld had larger grains than the low heat input weld. Both of the weld metal specimens exhibited considerable ductility, with a plastic strain to failure of over 30%.

Table 1: All-weld-metal longitudinal mechanical properties

Parameters	Heat Input, kJ/mm	Yield Strength, MPa	Ultimate Strength, MPA	Plastic Strain to Failure, 25 mm gage
200 rpm, 25 mm/min	6.27	833	1101	0.317
200 rpm, 150 mm/min	1.62	746	1017	0.339
Base Metal	N/A	423	908	0.449

As can be seen in Figure 7, the specimens showed primarily uniform elongation with relatively small amounts of necking. However, the strain in the specimens appears to be inhomogeneous, with regions of high strain and low strain that appear to correlate with the tool advance per revolution. This phenomenon is most pronounced in the low heat input weld, which has the lowest yield strength. It is thought that perhaps the greater width of the dark bands in Figure 7 for the 150 mm/min weld causes the reduced yield strength compared to the 25 mm/min weld.

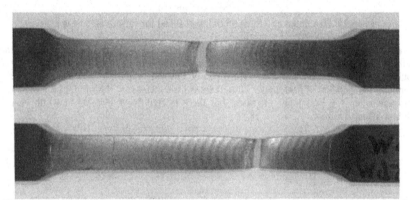

Figure 7: Tensile specimens after failure. 200 rpm, 25 mm/min (top); 200 rpm, 150 mm/min (bottom)

Grains in the stir zone of the welds were consistently smaller than in the base metal. Figure 8 shows the ratio of average grain size in the stir zone to average grain size in the base metal for a variety of weld parameters. Grain size in the stir zone is from 10% to 30% of the grain size in the base metal. Grain size reduction decreases with increasing heat input, as expected. There is no consistent pattern for grain size reduction as a function of position in the stir zone.

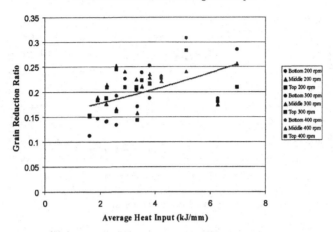

Figure 8: Grain size reduction in the stir zone as a function of heat input

Microhardness maps such as the one in Figure 9 were made for each weld. The stir zone and heat affected zone were typically harder than the base metal. In low-heat-input welds, a hard zone was typically found near the end of the pin, with the highest hardness on the advancing side. This high-hardness region corresponds to a fine-grain region of the weld, so it is thought that the grain size effect is responsible for most of the gain in hardness.

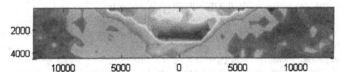

Figure 9: Microhardness map from a low heat input weld. Distances in μm.

The correlation of peak stir zone hardness with heat input is shown in Figure 10. As would be expected if grain size were the primary factor in determining hardness, the peak hardness decreases with increasing heat input. However, the yield strength increases with increasing heat input. This result is surprising, but is thought to be due to the inhomogeneity of the microstructure, as revealed in the strain localization during tensile testing. Further investigation of this phenomenon is planned.

Peak Hardness and Yield Strength vs. Average Heat Input

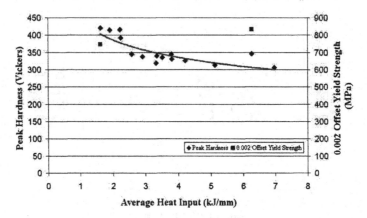

Figure 10: Peak hardness and yield strength vs. heat input.

Conclusions

As presented in the paper, the following observations have been made concerning FSW of alloy 718 in the annealed condition:

- Using as convex scrolled shoulder, step spiral pin tool, alloy 718 can be welded at up to 150 mm/min. In contrast, a traditional concave smooth shoulder, step spiral pin tool produces successful welds at a maximum of 50 mm/min.
- CS4 tools are more robust in welding austenitic alloys than traditional tools.
- Welding direction (X-axis) loads are proportional to travel speed, reaching a maximum of 8 kN at 150 mm/min.
- As expected, measured heat input is inversely proportional to travel speed. For some welding conditions, heat input is proportional to spindle speed, but for others, the effect of spindle speed is much smaller.
- Welds with lower heat input had higher peak hardness in the stir zone. However, the highest heat input weld had lower yield and ultimate strengths than the lowest heat input weld.
- All-weld-metal specimens exhibited plastic strain to failure above 30%. The high-heat-input weld had lower strain to failure than the low-heat-input weld.
- Longitudinal weld-metal tensile specimens showed inhomogeneous strain fields that correspond to the advance per revolution of the tool. It appears that there are regions of strain localization which may correspond to high strain regions of the weld.
- Grains in the stir zone are between 1/6 and 1/3 the size of grains in the base metal. The largest grain size reduction occurs in the low-heat-input welds.

Acknowledgements

The authors gratefully acknowledge the Center for Friction Stir Processing, a NSF Industry/University Cooperative Research Center, for providing the funding for this project.

References

1. Sorensen, CD, Nelson TW, and Packer, SM (2001), "Tool Material Testing for FSW of High-Temperature Alloys", *Proceedings of the Third International Symposium on Friction Stir Welding*, TWI, Kobe, Japan, September 2001, paper on CD.

2. Sorensen, CD, "Progress in Friction Stir Welding of High Temperature Materials", *Proceedings of the 14th Annual Offshore and Polar Engineering Conference*, Volume 4, pp. 8-14.

3. Thomas, WM, Threadgill, PL, and Nicolas, ED (1999). "Feasibility of Friction Stir Welding Steel", *Science and Technology of Welding and Joining*, 4:6, pp 365-372.

4. Sterling, CJ, Nelson, TW, Sorensen, CD, Steel, RJ, and Packer, SM (2003), "Friction Stir Welding of Quenched and Tempered C-Mn Steel", *Friction Stir Welding and Processing II*, TMS, pp 165-171.

5. Posada, M, DeLoach, J, Reynolds, AP, Fonda, R, and Halpin, J (2003), "Evaluation of Friction Stir Welded HSLA-65", *Proceedings of the Fourth International Symposium on Friction Stir Welding*, TWI, Park City, Utah, May 14-16, 2003, paper on CD.

6. Ozekcin, A, Jin, H, Koo, JY, Bangaru, NV, Ayer, R, and Packer, S (2004), "A Microstructural Study of Friction Stir Welded Joints of Carbon Steels", *ISOPE 2004*, May 23-28, 2004, Toulon, France.

7. Konkol, P (2003), "Characterization of Friction Stir Weldments in 500 Brinell Hardness Quenched and Tempered Steel", *Proceedings of the Fourth International Symposium on Friction Stir Welding*, TWI, Park City, Utah, May 14-16, 2003, paper on CD.

8. Johnson, R, dos Santos, J, and Magnasco, M (2003), "Mechanical Properties of Friction Stir Welded S355 C-Mn Steel Plates", *Proceedings of the Fourth International Symposium on Friction Stir Welding*, TWI, Park City, Utah, May 14-16, 2003, paper on CD.

9. Konkol, PJ, Mathers, JA, Johnson, R, and Pickens, JR (2003), "Friction Stir Welding of HSLA-65 Steel for Shipbuilding", *Journal of Ship Production*, vol 19 no 3 (August 2003), pp 159-164.

10. Okamoto, K, Hirano, S, Inagaki, M, Park, SHC, Sato, YS, Kokawa, H, Nelson, TW, and Sorensen, CD (2003), "Metallurgical and mechanical properties of friction stir welded stainless steels", *Proceedings of the Fourth International Symposium on Friction Stir Welding*, TWI, Park City, Utah, May 14-16, 2003, paper on CD.

11. Posada, M, DeLoach, J, Reynolds, AP, and Halpin, JP (2002), "Mechanical Property and Microstructural Evaluation of Friction Stir Welded AL-6XN", *Trends in Welding Research, Proceedings of the Sixth International Conference*, pp. 307-311.

12. Reynolds, AP, Posada, M, DeLoach, J, Skinner, MJ, and Lienert, TJ (2001), "FSW of Austenitic Stainless Steels", *Proceedings of the Third International Symposium on Friction Stir Welding*, TWI, Kobe, Japan, September 2001, paper on CD.

13. Reynolds, AP, Tang, W, Gnaupel-Herold, T, and Prask, H (2003), "Structure, properties, and residual stress of 304L stainless steel friction stir welds", *Scripta Materialia*, Vol 48 No. 9 (May 2003), pp 1289-1294.

14. Steel, RJ, Pettersson, CO, Sorensen, CD, Sato, Y, Sterling, CJ and Packer, SM (2003), "Friction Stir Welding of SAF 2507 (UNS S32750) Super Duplex Stainless Steel", *Proceedings of Stainless Steel World 2003*, Paper PO346.

MICROSTRUCTURAL EVOLUTION
IN TI 5-1-1-1 FRICTION STIR WELDS

R.W. Fonda, K.E. Knipling, C.R. Feng, and D.W. Moon

Naval Research Laboratory;
4555 Overlook Ave., SW; Washington, DC, 20375, USA

Keywords: Friction Stir Welding, Microstructure, Titanium

Abstract

A friction stir weld in 5-1-1-1 titanium was prepared and the end of the weld was quenched to preserve the microstructure surrounding the tool. Both the transverse cross section of the deposited weld and the plan-view cross sections through the weld end were analyzed to determine the microstructural evolution that occurs in this alloy during the welding process. This study revealed a complex microstructure separating the initial deformation of the base plate α microstructure from the regions of the weld that experienced temperatures above the β transus, developing a fine, equiaxed grain structure.

Introduction

Friction stir welding is a solid-state joining process that was developed in the early 1990's by TWI [1]. Most of the development of this process, and the commercial applications, has been on aluminum alloys, primarily because of the lower temperatures required to join those materials and thus the ready availability of tool materials to perform the welding. High strength alloys such as titanium and steel require tools that can retain their strength at much higher temperatures. Such tools have been developed and friction stir welding of many steel and titanium alloys has been demonstrated [e.g.,2-10].

Similarly, much of the research on determining the microstructural evolution within friction stir welds has been focused on aluminum alloys. It is the purpose of this study to extend some of the more recent research on the evolution and development of grain structure and crystallographic texture in aluminum alloys [11-19] to the understanding of the comparable processes within a near-alpha titanium alloy, 5-1-1-1. This alloy was developed to exhibit a high toughness, good weldability, and good stress-corrosion cracking resistance, and is primarily considered for marine applications that require a superior toughness and corrosion resistance [20].

Experimental

The ½" (12.7 mm) thick 5-1-1-1 titanium material was provided by Concurrent Technologies Corporation. The weld examined in this study is a bead-on-plate (no seam) weld that was prepared at the Edison Welding institute with a tungsten-based alloy tool. The tool geometry, shown schematically in Figure 1, is strongly tapered and contains no threads, flats, or other features. Because of the poor thermal conductivity of titanium, the shoulder is drastically reduced from the conventional tool geometry in order to introduce an even distribution of heat during the welding process. Welding was performed at 140 rpm advancing at 2 inches per minute (51 mm/min).

The tool was extracted from the plate immediately upon completion of the weld and the weld end was quenched with cold water to preserve the microstructure surrounding the welding tool. Both transverse cross sections of the deposited weld and a plan view cross section through the plate mid-thickness at the weld end were prepared for detailed examination by optical microscopy, scanning electron microscopy (SEM), and electron backscattered diffraction (EBSD) analysis. Final polishing of the surface was accomplished with a solution of 20% hydrogen peroxide (30%) and 80% colloidal silica solution. Samples were etched with Krolls to remove the surface layer and to reveal the underlying microstructure. An automated microhardness tester was used to determine the microhardness across the transverse cross section with a 1 kg load and with a spacing of 0.5 mm.

Figure 1. Tool design for friction stir welding the ½" 5-1-1-1 titanium plate.

Results and Discussion

The transverse cross section of the friction stir weld (Figure 2) reveals that the base plate microstructure consists of very large grains, particularly towards the center of the plate where some grains on the order of 8 mm in diameter are observed. This coarse base plate microstructure serves to clearly demarcate the boundary between the base plate and the weld-affected regions, and is likely to reveal microstructural evolution occurring in the outer regions of the weld more clearly than a fine-grained microstructure would. As this figure shows, there is no apparent transition between the base plate and the fine-grained region at the center of the weld. No distinction can be made between the various friction stir welded regions typically observed, such as the heat affected zone (HAZ), thermo-mechanically affected zone (TMAZ), and weld nugget.

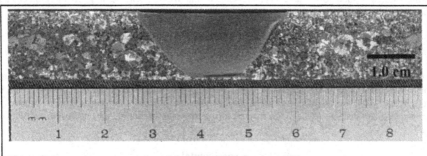

Figure 2. Transverse cross section of the 5111 Ti friction stir weld.

To help distinguish between the various portions of this friction stir weld, microhardness measurements were performed across the transverse cross section of the weld. The map of these microhardness measurements is shown in Figure 3. This microhardness map reveals that the coarse-grained regions of the transverse cross section vary between approximately 260 and 360 HVN without any systematic variations associated with the weld. In addition, the deposited weld at the center is also uniform in microhardness (300 to 320 HVN), with no evidence of different weld regions. Thus, at this level of resolution, both the image of the weld and the microhardness

296

variations across the weld reveal only a single, homogeneous weld affected region surrounded by the apparently unaffected base plate.

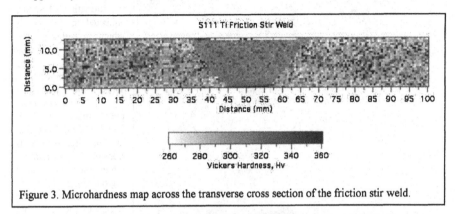

Figure 3. Microhardness map across the transverse cross section of the friction stir weld.

The plan view cross section was examined to determine how the material ahead of the tool is influenced by the heat and deformation fields surrounding the tool, how that material evolves as it is swept around the tool, and finally what additional microstructural changes occur as it is deposited in the wake of the tool and cooled to ambient temperature. An optical macrograph of the plan view cross section from the plate mid-thickness is shown in Figure 4. This image reveals that although the transition from base plate to deposited weld is fairly abrupt downstream from the weld, there are features preserved around the weld tool that reflect the shear deformation field introduced by the rotating tool.

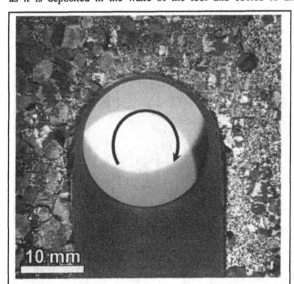

Figure 4. Image of the plan view cross section of the end of the weld showing the mid-plane microstructure and the tool rotation direction.

Closer examination of the microstructure development occurring ahead of the tool (Figure 5) reveals the transition between the base plate and the grain-refined region adjacent to the tool. The basketweave α microstructure (and the grain boundaries) observed in the base plate far from the tool becomes deflected in the tool rotation direction by the deformation of welding. As the heat generated by the welding process increases towards the tool, the β transus temperature will be achieved. At this temperature, nucleation of β commences within the basketweave microstructure, which is comprised of laths

297

Figure 5. Montage of optical micrographs from directly ahead of the tool.

of the low-temperature α phase. This results in the formation of small, equiaxed grains of β which subsequently transform to α as the weld is cooled to ambient temperature. Thus, the deformed basketweave microstructure occurs in regions of the weld that did not reach the β transus temperature while the fine-grained region surrounding the tool corresponds to the region of the weld that exceeded the β transus temperature during welding.

As the material is swept around the retreating side of the tool, further microstructural features develop at the boundary between the deformed basketweave microstructure and the grain-refined regions near the tool – see Figure 6. The curved bands that develop at this location appear to have a relatively regular spacing, although that spacing changes with position around the tool and does not necessarily correspond to the advance per revolution of the welding process. At this level of resolution, the basketweave microstructure appears to extend between the bands, while the bands themselves are composed of smaller, more equiaxed grains.

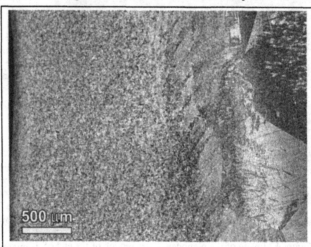

Figure 6. Montage of optical micrographs from the retreating side of the tool.

The curved bands were examined in greater detail by EBSD. A map of the crystal orientation across this region is shown in Figure 7. This figure reveals that the α phase comprising the basketweave microstructure has a single crystal normal direction in the upper right of the figure, as evidenced by the similar grey level of those crystallites. Approaching the tool (towards the lower left), a new orientation of the α crystallites develops in ever increasing amounts up to the line of arrows. This new orientation

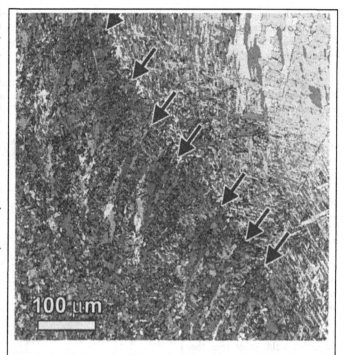

Figure 7. Electron backscattered diffraction map of the crystal orientations in the region of the curved bands (arrowed).

develops within the region that exhibits deformation of the basketweave microstructure in response to the rotational shear of welding. This new orientation therefore appears to correspond to a rotation of the basketweave components in response to the welding-induced shear. The curved bands that develop closer to the tool, indicated by arrows in Figure 7, have a coarser grain structure than the surrounding microstructure. Between these bands, a finer acicular microstructure develops that is similar to the deformed basketweave microstructure, but

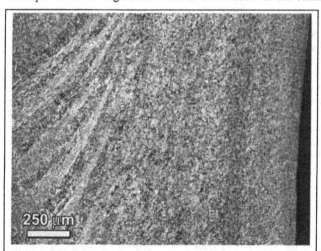

Figure 8. Montage of optical micrographs from the advancing side of the tool.

exhibits a different crystal orientation. Further analysis of the crystal orientations and evolutions is ongoing. A fine, equiaxed microstructure exists closer to the tool (lower left corner of Figure 7) that arises from the nucleation of fine β grains during welding and the subsequent transformation of that microstructure to α upon cooling.

Similar bands are also present on the advancing side of the weld, and are shown in Figure 8. The bands on the advancing side are larger than those on the retreating side, but otherwise exhibit the same features as those discussed above for the retreating side bands.

Behind the tool, in the deposited weld, the microstructure consists of refined, equiaxed grains (Figure 9). The grains in the region adjacent to the tool are the most refined in size. These grains rapidly transition to a coarser, more stable grain size within 200 μm of the tool. The grain size gradually coarsens further over a length of more than a millimeter, similar to the observations of Su et al in an aluminum alloy [21]. No banding is evident in the deposited weld, other than the coarse bands on the advancing side of the weld that can be seen in Figure 4.

Figure 9. Optical micrograph of the refined grain microstructure deposited immediately behind the tool.

Summary

The microstructures present within a friction stir weld of 5-1-1-1 titanium were examined. The transverse cross section exhibited a single weld region with refined grains, but did not reveal differences between the typical weld-affected regions observed in aluminum alloys. Examination of the mid-plane microstructure at the weld end revealed some interesting features. The basketweave α microstructure within the prior β grains became deformed within a narrow band around the weld. The microstructure adjacent to the tool, as well as within the deposited weld, consists of fine, equiaxed grains of α. There is often a transition region between these two regions that contains curved bands that appear to be separated by regions of the basketweave microstructure. Further analysis of these regions is required to determine the evolutionary path of the microstructures during friction stir welding of this alloy.

Acknowledgements

The authors would like to acknowledge Kevin Klug from Concurrent Technologies Corporation and Ernie Czyryca from the Naval Surface Warfare Center at Carderock for supplying the plate used in this study. In addition, we would like to thank Leroy Levenberry for his assistance with preparing the samples and Dave Rowenhorst for his assistance with the EBSD. Finally, we would like to thank the Office of Naval Research for funding this research.

References

1. W.M. Thomas, E.D. Nicholas, J.C. Needham, M.G. Murch, P. Templesmith, and C.J. Dawes, "Friction Stir Butt Welding", Int. Patent App. PCT/GB92/02203 and GB Patent App. 9125978.8, Dec. 1991. U.S. Patent No. 5,460,317, Oct. 1995.
2. W.M. Thomas, P.L. Threadgill, E.D. Nicholas, *Sci Tech Weld Join*, **4**, 365-372 (1999).
3. A.P. Reynolds, W. Tang, M. Posada, and J. DeLoach, *Sci Tech Weld Join*, **8**, 455-460 (2003).
4. P.J. Konkol, J.A. Mathers, R. Johnson, and J.R. Pickens, *J Ship Production*, **19**, 159-164 (2003).
5. M. Posada, J. DeLoach, A. P. Reynolds, R. W. Fonda, and J. P. Halpin, *Proc 4th Int Symp on FSW*, TWI Ltd., S10A-P3 (2003).
6. A.J. Ramirez and M.C. Juhas, *Mat Sci Forum*, **426-4**, 2999-3004 (2003).
7. T.W. Nelson, J.-Q. Su, and R.J. Steel, *Proc 14th Int Offshore and Polar Eng Conf*, 50-54 (2004).
8. A.P. Reynolds, E. Hood, and W. Tang, *Scripta Metall*, **52**, 491-494 (2005).
9. Y.S. Sato, T.W. Nelson, and C.J. Sterling, *Acta Metall*, **53**, 637-645 (2005).
10. W.B. Lee, C.Y. Lee, W.S. Chang, Y.M. Yeon, S.B. Jung, *Mat Sci Lett*, **59**, 3315-3318 (2005).
11. K.V. Jata and S.L. Semiatin, *Scripta Mat*, **43**, 743-749 (2000).
12. J.F. Bingert and R.W. Fonda, *Proc 4th Intl Symp on FSW*, TWI, Cambridge (2003).
13. J.-Q. Su, T.W. Nelson, R. Mishra, and M. Mahoney, *Acta Mater*, **51**, 713-729 (2003).
14. R.W. Fonda, J.F. Bingert, and K.J. Colligan, *Scripta Mat*, **51**, 243-248 (2004).
15. J.A. Schneider and A.C. Nunes, Jr, *Met Mat Trans B*, **35B**, 777-783 (2004).
16. R.W. Fonda, J.F. Bingert, and K.J. Colligan, *Proc 5th Intl Symp on FSW*, TWI, Cambridge (2004).
17. P.B. Prangnell and C.P. Heason, *Acta Mat*, **53**, 3179-3192 (2005).
18. R.W. Fonda and J.F. Bingert, *Met Mat Trans A*, in press.
19. R.W. Fonda, J.A. Wert, A.P. Reynolds, and W.Tang, *Sci Tech Weld Join*, in press.
20. Titanium Metal Corporation: TIMETAL 5111, Report # TMC-0170, Titanium Metal Corporation, Denver, Co, (2000).
21. J.-Q. Su, T.W. Nelson, C.J. Sterling, *Phil Mag*, **86**, 1-24 (2006).

EFFECTS OF FRICTION STIR WELDING ON THE COEFFICIENT OF THERMAL EXPANSION OF INVAR 36

Bharat K. Jasthi[1], Stanley M. Howard[1], Casey D. Allen[2], William J. Arbegast[2]

[1]Department of Materials and Metallurgical Engineering
[2]Advanced Materials Processing and Joining Laboratory
South Dakota School of Mines and Technology
501 E. St. Joseph St., Rapid City, SD 57701 USA

Keywords: Friction stir welding, Invar 36, Coefficient of thermal expansion

Abstract

The purpose of this investigation was to determine if the change in coefficient of thermal expansion observed in the fusion welding of Invar 36 could be avoided by friction stir welding (FSW). Invar 36 steel plates (0.5x3.75x24 inch) were friction stir welded using both polycrystalline cubic boron nitride and W-25% Re pin tools. Coefficient of thermal expansion is critical to Invar applications such as aircraft composite tooling and high precession measuring devices. Since FSW requires no filler material, problems associated with the filler metal and liquid state were avoided. This paper describes the process parameter development, the effect of process parameters, and pin tool materials on friction stir welded Invar's CTE.

Introduction

Advances in the aircraft designs have led to the use of composite materials for commercial and military aircrafts [1]. These new designs increased the demand for low coefficient of thermal expansion (CTE) materials such as Invar 36 for composite tooling. Invar 36 is an iron- 36% nickel alloy with a face centered cubic structure. It has excellent toughness, good formability, and very low CTE. Invar also finds applications in bimetal joints, cryogenic piping, measuring devices, and compensating pendulums. Fusion welding of Invar 36 has solidification and reheating cracking issues when matching filler metal was used. However, this can be addressed by changing the filler metal composition leading to CTE mismatch in the weld regions [1]. Witherell [2] reported that Invar 36 fusion welds are porous and sensitive to hot cracking. However, their work also reported that a crack-free weld can be obtained by additions of some deoxidants to the filler metal. Gottleib and Shira [3] reported that fillerless fusion welds may have unpredictable ductility and are to be avoided for structures where reliability is needed. In this investigation, an attempt was made to weld Invar 36 using friction stir welding (FSW) with polycrystalline cubic boron nitride (PCBN) and W-25%Re pin tools. Since FSW is a solid state joining process and requires no filler material, problems associated with the filler metal and liquid state were avoided. Also the effect of pin tools on the CTE of Invar 36 welds was evaluated.

Experimental

Invar 36 plates of 0.5 X 3.75 X 24 inch (12.7 X 95.2 X 609.6 mm) were friction stir welded with the selected processing parameters. PCBN and W-Re pin tools of 0.25 inch (6.35 mm) tip length were used for welding these materials. A series of 3 inch (76.2 mm) long bead-on-plate welds were made to develop the process parameters. A typical FSW setup for Invar 36 plates is shown in Figure 1. All of the welds were made at a position controlled plunge depth of 0.005 inch (0.0127 mm). The process parameters used for parameter development are shown in Table I. Specimens were polished and etched for microscopic examination. The specimens were evaluated for microstructure, weld quality, and defects.

Figure 1. Friction stir welding setup for Invar 36.

Linear coefficient of thermal expansion of the material was determined using a thermomechanical analyzer. ASTM 831-03 standard test method procedure was followed. Linear coefficient of thermal expansion was measured in both transverse and longitudinal directions of the weld. Samples of 0.31 X 0.17 X 0.15 inch (8.0 X 4.5 X 4.0 mm) were extracted from both parent metal and the nugget regions. In this analysis a plot is drawn between temperature and change in length of the sample. CTE measurements were from 25 °C to 500 °C at the heating rate of 5 °C/min. CTE experiments were conducted on welds made with both PCBN and W-25%Re pin tools to determine the effect of pin tool material. A double-sided butt joint was made with the optimized processing parameters (600 rpm and 5 IPM) determined by a series of preliminary welds. Tensile strength of the butt welded joint made with the W-25%Re pin tool was evaluated and is compared with the strength of the parent material. X-ray radiography was performed on the welds made with both W-25%Re and PCBN tools to determine if pin tool remnants appeared in the welds. SEM-EDX analysis was also performed to identify the morphology of the remnants in the weld nugget region.

Table I. Welding processing parameters used for parameter development

Pin Tool	Weld Speed (IPM, inches/minute)	Spindle Speed (RPM)
PCBN 0.25 –in (6.35 mm)	3, 4, 5	500
PCBN 0.25 –in (6.35 mm)	3, 4, 5	600
W-25%Re 0.25 -in (6.35 mm)	3, 4, 5	600

Results

Invar 36 was plasticized for all the welds made with PCBN and W-25%Re pin tools. One PCBN 0.25 inch (6.35 mm) pin tool was broken in the process of developing the process parameters. The W-25%Re pin tool showed noticeable wear during the parameter development. Some sticking of the Invar 36 to W-25%Re pin tool was observed. No sticking has been observed on the PCBN pin tool. Figure 2 and 3 shows the CTE results in transverse and longitudinal directions made with W-Re and PCBN pin tools respectively. Figure 4 shows the CTE of Invar 36 welds with W-Re and PCBN pin tools compared with the parent metal. The X-ray radiographic analysis of the friction stir welds made with PCBN and W-25%Re pin tools are shown in Figure 5. Figure 6 shows X-ray radiographic pictures of welds made with PCBN pin tool. A defect-free weld was achieved at 500 RPM and 3 IPM. Where as a worm hole defect was observed at 500 RPM and 5 IPM. Tensile testing of Invar 36 butt welds made with W-25%Re was performed and the results are tabulated in Table II. Figure 7 shows the crack initiation site in the butt welded Invar 36 specimen during tensile testing. The SEM back scattered image of weld nugget showing the W/Re pin tool remnants in the weld nugget is shown in Figure 8.

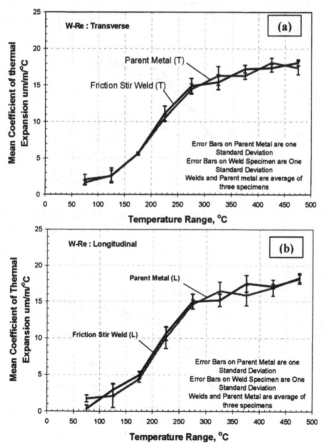

Figure 2. Comparison of mean CTE for Invar 36 parent metal and welds made with W-Re pin tool; (a) in transverse direction, (b) in longitudinal direction.

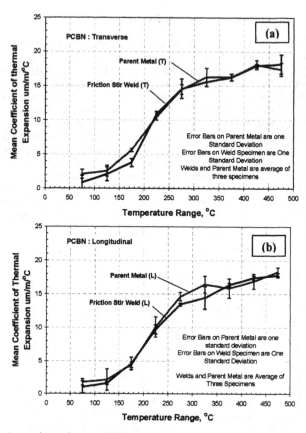

Figure 3. Comparison of mean CTE for Invar 36 parent metal and welds made with PCBN pin tool; (a) in transverse direction, (b) in longitudinal direction.

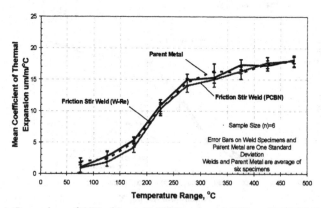

Figure 4. Comparison of CTE of Invar 36 FSW made with W-Re and PCBN pin tools with parent metal.

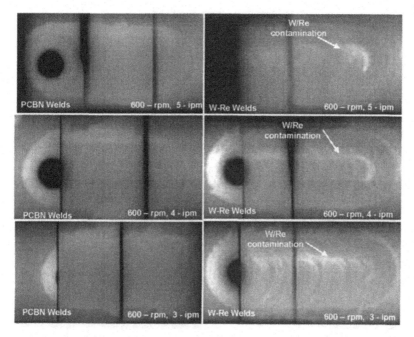

Figure 5. X-ray radiographic analyses of welds made with PCBN and W-Re pin tools

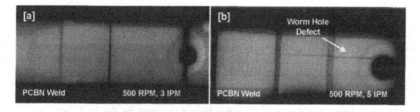

Figure 6. X-ray radiographic analysis of welds made with PCBN pin tools
[a] Good weld with no defects [b] Weld with worm hole defect.

Table II. Comparison of mechanical properties of weld and parent metal for Invar 36

Invar 36	Parent	Welds made with W-25% Re pin tool	Comparison with Parent Metal
UTS (psi)	66,800 ± 950	65700 ± 530	98 %
% Elongation (in/in)	52 ± 0.9	30 ± 0.9	58 %

Figure 7. Invar 36 butt welded joint showing crack initiation during tensile testing

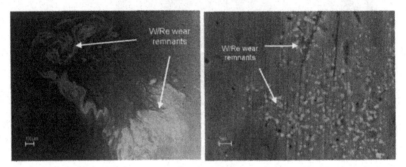

Figure 8. SEM back scattered image of Invar 36 nugget region showing W-Re wear remnants.

Discussion

Invar 36 plates were plasticized when stirred with PCBN and W-Re pin tools. TMA analysis showed no significant difference in CTE between the weld nugget and the parent material in both longitudinal and transverse directions. Analysis also showed that pin tool material (PCBN and W-Re) has no effect on the CTE of Invar 36 plates. X-ray radiographic analysis located W-Re pin tool remnants in the friction stir welded Invar 36 plates. Wear in the W-Re pin tools was greatest during the initial plunge. Little wear was observed elsewhere in the weld. Preheating the Invar 36 plates before welding could possibly reduce the wear for the W-Re pin tools. X-ray radiography showed no wear debris in the welds made with the PCBN pin tool. This suggests that defect-free and contamination-free welds can be produced with a PCBN pin tool. The transverse tensile testing of welded specimens revealed joint efficiencies (ratio of weld strength to parent material strength) of about 98% for UTS and 58% for elongation.

Conclusions

Invar 36 was successfully welded with both PCBN and W-25%Re pin tools. The pin tool material has no effect on the CTE of Invar 36. CTE for friction stir welds is comparable with the parent metal. No significant difference was observed in CTE in transverse and longitudinal directions. Weld process parameters exhibited no effect on the CTE of Invar 36 friction stir welds. PCBN pin tools did not show any wear and performed better than W-Re pin tools. High strength joints with no change in CTE can be produced in Invar 36 plates using friction stir welding.

Acknowledgements

This research was sponsored by the Army Research Laboratory (ARL) and was accomplished under Cooperative Agreement Number DAAD19-02-2-0011. The views and conclusions contained in this document are those of the authors and should not be interpreted as representing the official policies, either expressed or implied, of the Army Research Laboratory or the U.S. Government. The U.S. Government is authorized to reproduce and distribute reprints for government purposes notwithstanding any copyright notation hereon. The authors also acknowledge the support from Ellsworth Air Force Base for performing the X-ray radiography.

References

1. Otte et al., "Welding Low Thermal Expansion Alloys for Aircraft Composite Tooling". Welding Journal 75(7), (1996), 51-55
2. Witherell C. E., "Welding nickel-iron alloys of the Invar type". Welding Journal 43(4), (1964) 161-2 to 169-s.
3. Gottlieb, T., and Shira, C.S., "Fabrication of iron nickel alloys for cryogenic piping service" Welding Journal 44(3), (1965), 116-s to 123-s.

STUDY OF PLUNGE MOTION DURING FRICTION STIR SPOT WELDING – TEMPERATURE AND FLOW PATTERN

Harsha Badarinarayan[1], Frank Hunt[1], Kazutaka Okamoto[1], Shigeki Hirasawa[2]

[1]Hitachi America Ltd.;
34500 Grand River Avenue, Farmington Hills, MI 48335, USA

[2]Kobe University;
Department of Mechanical Engineering, 1-1 Rokkodai, Nada, Kobe, Hyogo 657-8501, Japan

Keywords: plunge study, spot FSW, particle method

Abstract

This paper is the first step towards understand the complex phenomena of tool plunge during friction stir spot welding (FSSW). Both experimental and numerical approaches were undertaken. For a given welding condition and material, welds were made with varying tool penetration depth (from shallow to deep). Surface temperature for the workpiece, tool and anvil were measured. Energy consumption for each weld was also measured by monitoring the machine power. Actual weld depth and stir zone size were quantified from the weld cross-sections. Furthermore, an existing numerical code was used to simulate the temperature profile and material mixing phenomena during FSSW. Simulation results for temperature and flow had good correlation with experimental data.

Introduction

Friction Stir Spot Welding is an evolving technique that has been rapidly gaining momentum since the beginning of this decade and has already found its place in commercial applications in the automotive industry [1]. Although the principle of FSSW is based on that of linear FSW, the process is much more complex in the sense that the actual welding time itself is very short but the process dynamics involved is still the same – tool plunge, material mixing (during dwell time) and tool retract. Linear FSW can be considered as a steady state process whereas FSSW is a transient process.

The key parameters important for FSSW are tool geometry, tool rotational speed, plunge depth and hold (dwell) time. Each of these parameters have an influence on the weld in terms of heat input, material mixing and weld cycle time, all of which are the key to achieving a sound weld in terms of strength and morphology. A previous study [2] carried out to evaluate the effect of some of these parameters on the static strength of friction stir spot welded coupons concluded that the variation in plunge depth had a significant effect on the weld strength which leads us to the fact that plunge depth is an important parameter for FSSW. The effect of key geometrical parameters of the tool such as pin length and plunge depth on weld quality is still unknown.

There are a number of fundamental aspects for FSSW that remain to be studied in detail; study of the effect of tool plunging into the workpiece is one such aspect where there is little available open literature. The plunging of the tool into the workpiece essentially decides the geometry of the welded joint. Actual plunge depth, formation of hook and 1st sheet thinning are key geometric features that will eventually dictate the quality of the welded joint. The scope of the

311

plunge motion study is to understand the chain of events that occur during the tool plunging period since this is a dominant stage of FSSW. Unfortunately, these events occur within a very short duration of time (usually 1-2 seconds) and are also very difficult to understand by just observing the workpiece surface, hence the ultimate goal is to develop a numerical technique that can incorporate all the physics (temperature and flow) that occurs during this transient phase.

There have been numerous attempts to mathematically model the friction stir welding process. Most of the work has been done for linear FSW wherein the simulation is run under steady state conditions. Langerman et al. [3], and Colegrove et al. [4] analyzed the temperature distribution and the flow of metals by using a computational fluid dynamic (CFD) code for non-Newtonian fluids. Khandkar et al. [5], Schmidt et al. [6], and McCune et al. [7] analyzed the temperature distribution and the plastic deformation by using stress analysis code and an elastic-plastic model. Hirasawa et al. [8-10] analyzed the temperature distribution during FSW process by using the finite element method, and also analyzed the plastic flow by using the particle method. Tomimura et al. [11] measured the dynamic friction coefficients between the rotating tool and the work pieces. Nevertheless, numerical simulation of friction stir spot welding remains a challenge primarily because of its highly transient nature. Approaches for the computational modeling of the FSW process are still under development and a great deal of work is underway particularly the application of explicit finite element codes for a verifiable simulation [12]. Muci-Küchler et al. [13] reported results on a simplified isothermal three-dimensional Finite Element Method (FEM) model of the initial plunge phase of the FSSW process. The model, based on a solid mechanics approach, was developed using the commercial software ABAQUS/Explicit. However, there is no numerical code that can simulate both the temperature profile and the flow pattern during the transient period of the plunge motion.

Hence, the scope of this paper was to use an existing FEM code [8-10] to simulate this complex motion. Experimental data would be used as an input to fine tune the code. Since the plunge period is such a complex phenomenon, getting the required experimental data was a challenge. Due to the nature of process, it was necessary to divide this period into several smaller segments and observe each one individually to understand the whole process completely. Hence an 'in-situ' observation of an 'ex-situ' process is undertaken for this study, in order to obtain relevant experimental data. Subsequent to that, an existing FEM code was used to predict the temperature profile and the material flow pattern.

From the experimental standpoint, welds were made with varying plunge depth (from shallow to deep). The surface temperature of the workpiece, tool and anvil were measured. This surface temperature was used as a reference value to validate the simulation results and secondly, in the future, this data can be used to develop a thermal heat balance model of the entire system considering the input and output energies. The weld force was measured which served as an input to simulation. Cross-section of all the welds were taken in order to visual the weld morphology and quantify the stir zone size as a function of plunge depth. The simulation model was then 'calibrated' with appropriate boundary conditions to predict the temperature at the measured locations. Finally, a comparison between the experimental and simulation results for the temperature and material flow were made.

Experimental Setup

Figure 1 shows the machine setup for the experimental runs. The welds were performed on Hitachi's FSSW machine called "Swing-Stir" [14] with rated spindle speed of 3000 RPM and that is designed for aluminum welding with up to 3mm of tool plunge depth. A typical FSSW tool made of H13 was utilized in this study. The key dimensions of this tool were: shoulder diameter of 8mm, pin diameter of 3mm and pin length of 1.41mm. In order to simplify the modeling, the welds were made on only 1 sheet and hence called as 'Spot-on-Plate' type weld as seen in Fig. 2. The workpiece was mounted on the work table and securely clamped. 'Spot-on-Plate' welds were performed on the center of the sheet of A6022-T4 with a size of 150^L x 40^W x 2.0^T mm. The tool rotation speed was 2500 RPM, while the plunge speed was 250 mm/min. Once the tool was plunged to a desired depth, it was immediately retracted (absence of hold time). The initial tool plunge depth was 0.2 mm and was incremented in steps of 0.1 mm, till the total plunge depth was 1.7mm. Hence, there were a total of 16 welds made. There was a time interval of 15 minutes between each weld to eliminate the effect of variation of initial tool temperature.

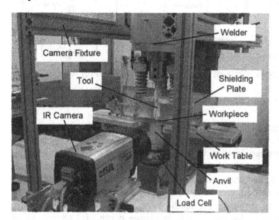

Figure 1. Welding Setup

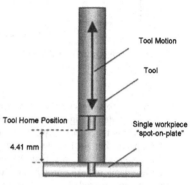

Figure 2. Spot-on-Plate Technique

A load cell, with a capacity of up to 2500 lbs, was mounted directly below the anvil (as seen in Fig. 1) to measure the plunge force during welding. A data acquisition unit was connected to the welder that could read the load cell output as well as the power consumed by the welder. The power consumption of 2 motors were tracked – one was the spindle motor and the other one was the Z motor which facilitated the up and down motion of the tool.

An IR camera (FLIR ThermoVision A series) was mounted directly onto the welder to measure the surface temperature of the tool, specimen and anvil. In order to eliminate the reflected temperature from the field of view, all the objects (tool, specimen and anvil) were coated by a commercial grade foot powder. The emissivity value set on the camera was 0.95, this value was decided based on experiments done in-house that were not part of this experiment. A 'shielding plate' coated with the foot powder was also used in the background so that there would be no instance of any stray reflected temperature from the far field view. Virtual thermocouples were 'placed' on the workpiece, tool and anvil. There were 7 thermocouples on the workpiece (S1 through S7), 4 on the tool (T1 through T4), and 6 on the anvil (A1 through A6) as illustrated in Fig. 3. Measuring the temperature of the tool especially at locations T1 and T2 was challenging because the material extruded by the pin occupies the circumferential region around the shoulder after a given plunge depth value causing the tool shoulder surface to be obstructed at the measured points. Therefore, it was decided to measure the tool surface temperature once the tool retracted back to its 'home position', which is 4.41mm away from the top surface of the workpiece. A high tool retract speed of 1500 mm/min (25 mm/sec) was chosen to minimize the delay of the tool returning to its 'home position' after the weld was done. At the deepest plunge position (1.68 mm), the tool had to travel a distance of 6.09 (1.68+4.41) mm to reach the 'home position' after welding that took about 0.25 seconds. Although, there is some error in

313

temperature measured by the IR camera due to this short delay, for the purpose of this study it is assumed that the temperature drop is negligible.

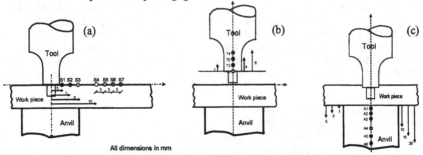

Figure 3. Virtual thermocouple locations on (a) Workpiece (b) Tool and (c) Anvil

Experimental Results

Figure 4 shows macro-structure images of the workpiece at different plunge depth value (from shallow to deep). Fig 4 (a) is for a plunge depth of 0.2 mm, this is just an indent on the surface. Fig. 4(b) is for a relatively shallow plunge depth of 0.9 mm, here only the pin plunges into the workpiece. Fig. 4(c) shows the instant when the shoulder is just in contact with the workpiece. It may seem contradictory that the shoulder comes in contact with the workpiece at an actual plunge depth of 1.18 mm when the pin length is actually 1.41 mm, however, this occurs because the material extruded upward by the pin at this plunge depth comes in contact with the shoulder surface. Fig. 4(d) shows the cross-section for the maximum plunge depth of 1.68mm. Sheet thinning is clearly visible in this image, and is about 0.3mm.

Fig 4(b) shows a sharp profile (illustrated by a circle) at the top of the keyhole region, this is because only the pin is in contact with the workpiece when the tool is plunged into the workpiece, and then as the pin is retracted up it removes some amount of material along with it. However, in Fig. 4(c) and (d), the top of the keyhole region has a smooth profile (illustrated by dotted circle). This occurs because the tool has a concave shoulder which comes in contact with the workpiece during welding.

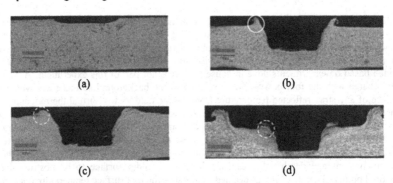

Figure 4. Cross-sections at different plunge depths: (a) 0.2 mm, initial plunge depth; (b) 0.9 mm, shallow plunge depth; (c) 1.18 mm, here the shoulder is just in contact with the work piece and (d) 1.68 mm, maximum plunge depth position

314

Figure 5 shows the growth of the stir zone area as a function of plunge depth. The stir zone region was distinguished from the base metal by the grain size. The macro-structure images for each plunge depth were imported into CAD software and the stir zone region was traced manually. Once a closed loop sketch was drawn, the software automatically calculated the area enclosed within that sketch which signified the stir zone area. The inset in Fig. 7 shows the macro cross-section for the actual plunge depth of 1.5 mm. Here, the stir zone is shown in dotted lines and the corresponding area is about 5.5 mm².

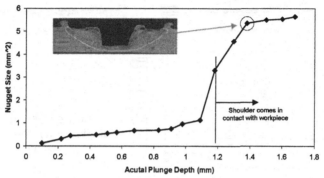

Figure 5. Variation of stir zone size with plunge depth

Table 1 shows the variation in the actual plunge depth from the set plunge depth. The set plunge depth is the value that is input into the weld controller. The actual plunge depth is what is observed after the weld. This was measured using a dial caliper gauge. It is very difficult to exactly achieve the set plunge depth. One of the factors for this deviation is the compliance of the C-frame structure of the welder. However, it is evident that we can track desired plunge depth very closely. Figure 6 illustrates the surface temperature profile on the workpiece and the tool. Figure 6(a) shows the variation in the workpiece surface temperature as a function of distance from the weld center. As expected, the temperature increases with increase in plunge depth. A peak surface temperature of 220^0C was measured at thermocouple location S1 at an actual plunge depth of 1.68 mm. Figure 6(b) shows the variation in the tool surface temperature measured from the shoulder

Table 1. Variation in actual plunge depth from set plunge depth

Set Plunge Depth (mm)	Actual Plunge Depth (mm)
0.20	0.22
0.30	0.28
0.40	0.44
0.50	0.51
0.60	0.57
0.70	0.68
0.80	0.83
0.90	0.91
1.00	0.98
1.10	1.09
1.20	1.18
1.30	1.30
1.40	1.38
1.50	1.50
1.60	1.60
1.70	1.68

region upwards. A peak surface temperature of 180^0C was measured at thermocouple location T1 at an actual plunge depth of 1.68 mm. The anvil temperature profile is shown in Fig. 6(c). A peak surface temperature of 53^0C was measured at thermocouple location A1 at an actual plunge depth of 1.68 mm. A common observation in all these graphs is a sudden jump in the temperature magnitude seen at the actual plunge depth value of 1.18 mm. This is because at this plunge depth, the shoulder comes in contact with the material that is squeezed out causing a rise in the heat generated.

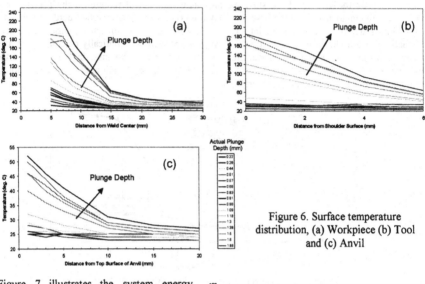

Figure 6. Surface temperature distribution, (a) Workpiece (b) Tool and (c) Anvil

Figure 7 illustrates the system energy consumption (in Joules) as a function of weld depth. The energy values shown here is the energy required for making the weld only, i.e. the energy consumed by the Z motor and spindle motor from the time it comes in contact with the workpiece till it leaves the workpiece. Idle energy i.e. the energy consumed by the Z motor and spindle motor during tool approach and retract is not included in the graph. Similar to the trend observed for temperature, there is a sudden increase in the energy consumption once the shoulder comes in contact with the workpiece. The energy

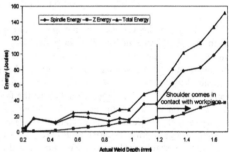

Figure 7. Machine energy consumption v/s plunge depth

consumed by the spindle is more prominent as the plunge depth increases. The spindle energy consumption almost linearly increases with increase in plunge depth. This can be explained by the fact that as the tool plunges progressively into the workpiece, it consumes more power in order to overcome the frictional losses and keep the tool rotating at a constant speed.

Numerical Approach

The approach here is to model the temperature distribution of the work piece during FSSW and also to model the material flow pattern using the particle method [8-10]. The model setup is shown in Fig. 8. A rotating tool is plunged into a metal plate of 2 mm thickness on an anvil. The calculation region is 14 mm in diameter and the plastic flow region is assumed to be a circular of 12 mm in diameter. The diameter of the pin of the rotating tool is 3 mm, however the pin length was 1.25 mm as opposed to the experimental pin length of 1.4 mm, this was because the pitch of the particles (in the particle flow algorithm) was set at 0.25mm mainly due to the computational hardware limitation. Diameter of the shoulder of the rotating tool is 8 mm. The

316

shoulder is assumed to be flat. There are two different techniques used for simulation – FEM and particle flow. The temperature of the tool was calculated using FEM whereas the material flow was calculated using the particle flow algorithm.

The plastic deformation is modeled by a flow of fluid with temperature dependent viscosity in the plastic flow region with a free surface. The temperature distribution and the plastic flow are analyzed by the Moving-Particle Semi-implicit Method [15]. Material is assumed to be numerous particles, and heat transfer and movement of these particles are calculated. The continuity, momentum and the energy equation are shown in equations (1)-(3).

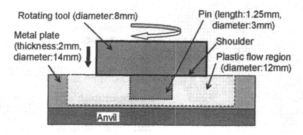

Figure 8. Calculation model

$$\frac{D\rho}{Dt} = 0 \tag{1}$$

$$\frac{Du}{Dt} = -\frac{1}{\rho}\nabla P + \nu\nabla^2 u \tag{2}$$

$$\frac{DT}{Dt} = \frac{\lambda}{c\rho}\nabla^2 T + \frac{q}{c\rho} \tag{3}$$

where, u is velocity vector, T is temperature, P is pressure, ρ is density, ν is dynamic viscosity, λ is thermal conductivity, c is specific heat, t is time, and q is heat generation rate per unit volume. The calculation scheme of the velocity is as follows; movement of the particles is calculated by summing viscous force due to the surrounding particles at each time step. Then, the position of the particles is corrected to keep the density of particles constant. These calculations are repeated, with time-step of 0.0001 seconds, until a final time of 0.7 seconds (which corresponds to a final plunge depth of 1.68mm). Figure 9 shows the initial condition of particles. The metal plate and tool are modeled by 30000 particles with a pitch of 0.25 mm. The particles that make up the model have different boundary conditions as illustrated in the figure. The rotating tool is moving down at 2.4 mm/s with rotation speed of 2500 RPM. The outside temperature of the metal plates at diameter of 14 mm is fixed to be 25°C and other thermal boundary conditions are adiabatic. Heat conduction loss through the rotating tool is neglected. Also, since the temperature rise in the anvil above the ambient temperature is relatively low, the numerical results for the temperature rise in the anvil are not discussed in this report. The metal plate is aluminum alloy A6061 with thermal conductivity 180 W/mK, density 2700 kg/m³ and specific heat of 896 J/kgK. Dynamic viscosity of metal plates during plastic flow is assumed to be a function of the local temperature as shown in Fig. 10, which is a simplified model of the work reported by Langerman et al. [3]. Figure 11 shows the plunge force of rotating tool, which was obtained from experimental data. Heat is assumed to be generated at a contact surface between the tool and the plate. Local contact surface heat generating rate q_0 is calculated by equation (4).

317

$$q_0 = 2\pi\mu pNr \qquad\qquad\qquad (4)$$

where, μ is the dynamic friction coefficient ($\mu = 0.6$ reported in [11]), p is contact pressure (Pa) which is obtained from Fig. 11, N is rotating speed (rps), and r is radius (m). Slip at contact surface between the metal plates and the rotating tool is neglected during plastic flow.

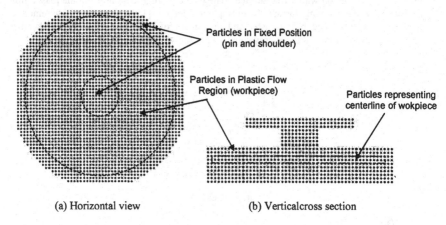

(a) Horizontal view　　　　　　　　(b) Verticalcross section

Figure 9. Initial condition of particles

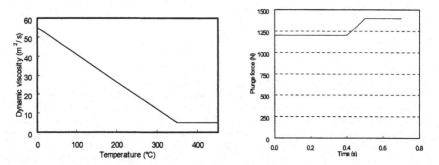

Figure 10. Dynamic viscosity of metal during plastic deformation

Figure 11. Plunge force of rotating tool

Numerical Results

Figure 12 shows the simulation result of the particle movement and temperature distribution at plunge depths of 0.2, 0.98, 1.18 and 1.68 mm. The tool temperature is computed by particle method. Frictional heat is generated near the shoulder pin of the rotating tool. The simulation also illustrates the movement of the particles in the center of the workpiece; this is useful in realizing the hook formation when welding 2 sheets. Material near the outside of the rotating tool is extruded out due to the plunging of the tool into the workpiece. The calculation results are similar to experimental results. Furthermore, Fig. 13 is a comparison of the weld cross section obtained by experiment and simulation.

Fig. 14 and Fig. 15 illustrate the temperature profile at different thermocouple locations of the workpiece and the tool respectively. The tool temperature profile predicted by simulation correlates well with the experimental results for all the 3 thermocouple locations shown, however for the workpiece temperature, the simulation results are a little skewed from the experimental results especially for thermocouple locations away from the weld center. The main reason for this could be the coarseness of the mesh density which was due to the computer hardware limitation.

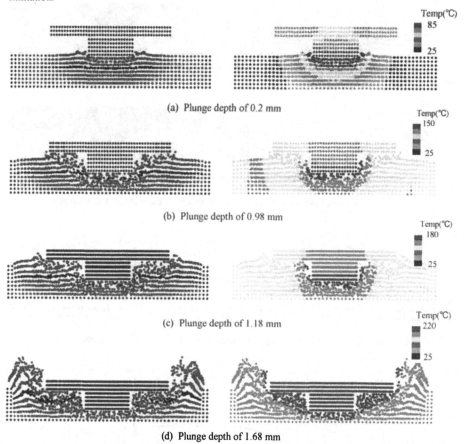

(a) Plunge depth of 0.2 mm

(b) Plunge depth of 0.98 mm

(c) Plunge depth of 1.18 mm

(d) Plunge depth of 1.68 mm

Figure 12 Position of particles and temperature distribution at different plunge depths

(a) Simulation result at plunge depth 0.2mm   (b) Experiment result at plunge depth of 0.2 mm

(c) Simulation result at plunge depth 0.98 mm   (d) Experiment result at plunge depth of 0.98 mm

(e) Simulation result at plunge depth 1.18mm   (f) Experiment result at plunge depth of 1.18 mm

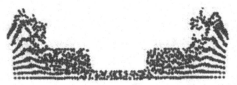 

(g) Simulation result at plunge depth 1.68mm   (h) Experiment result at plunge depth of 1.68 mm

Figure 13. Comparison of weld cross section (flow pattern) – experiment and simulation results

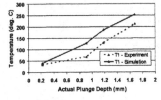 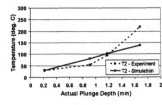 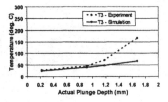

Figure 14. Comparison of workpiece temperatures at various thermocouple locations

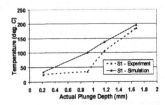 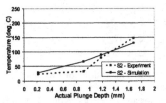 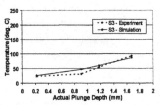

Figure 15. Comparison of tool temperatures at various thermocouple locations

Conclusion and Future Direction

An experimental and numerical approach was undertaken in an attempt to understand the complex phenomena that occurs during the plunging period of the tool in FSSW. 'Spot-on-plate' welds were made on A6022-T4 with 2 mm thickness with varying plunge depth.

1. At the deepest plunge depth position of 1.68 mm, the peak surface temperature of the tool at 5 mm from the weld center was recorded to be 220^0C using an IR camera. The tool shoulder temperature at the same depth was recorded as 180^0C.
2. The geometry of the nugget is significantly influenced by the initial interaction of the shoulder with the workpiece; however the nugget size plateaus out with further increase in plunge depth.
3. The energy consumed by the spindle motor was sensitive to the plunge depth once the shoulder was in contact with the workpiece surface, and increased almost linearly with increase in plunge depth. This implies the fact that if welds are to be made with different plunge depth, then a suggested control strategy to ensure that the desired plunge depth is achieved is to monitor the spindle energy.
4. A unique particle flow simulation technique was implemented to numerically simulate the tool plunge period. Preliminary simulation results showed good correlation with the experiments.

The simulation modeled developed can be used as a visual aid to observe the material flow during the plunge period. Geometric features such as formation of hook and thinning of first sheet can be visualized with this simulation technique. In the next planned experimental runs, welds will be made on dissimilar alloys (e.g. A6xxx on A5xxx). Since these alloys etch out differently, the cross-sectional images will clearly indicate the flow pattern, hook formation and top sheet thinning. This can be a good 'calibration' technique for the simulation, wherein the boundary conditions can be tweaked to match the experimental results. Once the model reaches a level of confidence, it can be used as a predictive tool to try out different material and/or thickness combination to predict these geometric features and temperature distribution. Furthermore, from the thermal profile obtained, it is possible to determine the joining efficiency of the system by considering motor energy as input and the heat absorbed by the workpiece as output. This approach will be reported in detail in a future publication.

References

1. R.Sakano *et al.*: "Development of spot FSW robot system for automobile body members", Proc. of 3rd Int. symp. on FSW, Kobe, Japan, 2001.
2. Hunt, F., Badarinarayan, H., Okamoto, K., "Design of Experiments for Friction Stir Stitch Welding of Aluminum Alloy 6022-T4", 2006 SAE World Congress, Paper No. 2006-01-0970, Michigan, USA, 2006.
3. Langerman, M., and Kvalvik, E., 2003, "Modelling Plasticised Aluminium Flow and Temperature Fields During Friction Stir Welding," Proc. 6th ASME-JSME Thermal Engineering Joint Conference, TED-AJ03-133.
4. Colegrove, P. A., and Shercliff, H. R., 2004, "Modelling the Friction Stir Welding of Aerospace Alloys," Proc. 5th International FSW Symposium.
5. Khandkar, M. Z. H. and Khan, J. A., 2003, "Predicting Residual Thermal Stresses in Friction Stir Welding," Proc. 2003 ASME International Mechanical Engineering Congress and Exposition, IMECE2003-55048.
6. Schmidt, H., and Hattel, J., 2004, "Modelling Thermomechanical Condition at the Tool/Matrix Interface in Friction Stir Welding," Proc. 5th International FSW Symposium.
7. McCune, R. W., Ou, H. Armstrong, G., and Price, M., 2004, "Modelling Friction Stir Welding with the Finite Element Method, A comparative Study," Proc. 5th International FSW Symposium.
8. Hirasawa, S. Haneda, M., Hirano, S., and Tomimura, T., 2004, "Analysis of Flow and Temperature Distribution during Friction Stir Welding," Proc. 2004 ASME Heat Transfer/Fluids Engineering Summer Conference HT-FED2004-56303.

9. Hirasawa, S. Okamoto, K., Hirano, S., and Tomimura, T., 2005, "Combined Analysis of Plastic Deformation Flow and Temperature Distribution during Friction Stir Welding," Proc. 2005 ASME International Mechanical Engineering Congress and Exposition, IMECE2005-79328.

10. Hirasawa, S. Okamoto, K., Hirano, S., and Tomimura, T., 2006, "Analysis of Temperature Distribution and Plastic Flow during Spot Friction Stir Welding," Proc. 13th International Heat Transfer Conference, MNF-03.

11. Tomimura, T., Okita, K., and Hirasawa, S. 2006, "Experimental Evaluation of Apparent Dynamic Coefficient of Friction between Rotating Rod and Metal Plate" Proc. Materials Science & Technology 2006.

12. Awang, M., Mucino, V.H. Feng, Z., David, S.A., 2006, "Thermo-Mechanical Modeling of Friction Stir Spot Welding (FSSW)". SAE Technical Paper no. 2006-01-1392, Society of Automotive Engineers, Warrendale, PA.

13. Muci-Küchler, K.H.; Kakarla, S.S.T.; Arbegast, W.J. and Allen, C.D. "Numerical Simulation of the Friction Stir Spot Welding Process" SAE Paper 2005-01-1260.

14. K. Okamoto, et. al. "Development of Friction Stir Welding Technique and Machine for Aluminum Sheet Metal Assembly –Friction Stir Welding of Aluminum for Automotive Applications (2)". 2005-01-1254.

15. Koshizuka, S., and Oka, Y, 1996, "Moving-Particle Semi-implicit Method for Fragmentation of Incompressible Fluid," Nuclear science and engineering, 123, 421-434.

SPOT FRICTION WELDING OF THIN AZ31 MAGNESIUM ALLOY

Tsung-Yu Pan[1], Michael L. Santella[2], P. K. Mallick[3], Alan Frederick[2]

[1]Manufacturing & Processes Research and Advanced Engineering
Ford Research and Advanced Engineering, Ford Motor Company
Dearborn, MI 48124, U.S.A.
[2]Metals and Ceramic Division, Oak Ridge National Laboratory
Oak Ridge, TN 37831-6096, U.S.A.
[3]Mechanical Engineering Department, The University of Michigan-Dearborn
Dearborn, MI 48128-1491, U.S.A.

Keywords: Spot friction welding, magnesium AZ31, lap shear strength, friction stir welding, resistance spot welding

Abstract

Spot friction welding (SFW) is a novel variant of the linear friction stir welding process with the potential to create strong joints between similar, as well as dissimilar sheet metals. It is particularly suitable for soft, low melting point metals such as aluminum, magnesium, and their alloys where resistance spot welding can cause defects such as voids, trapped gas and micro-cracks due to the intense heat requirement for joint formation. Until now, spot friction welding has focused primarily on aluminum alloys. This paper presents a feasibility study on spot friction welding of AZ31, a wrought magnesium alloy available in sheet form. Lap joints of 1.58-mm-thick magnesium alloy AZ31B-O sheet were produced by spot friction welding. The spot welds were made in 2 sec with a 15-mm-diameter pin tool rotating at 500-2,000 rpm. The tool was inserted into 2-sheet stack-ups to depths of either 2.4 or 2.8 mm relative to the top sheet surface. Tensile-shear testing showed that joint strengths up to 4.75 kN could be obtained. The removal of surface oxides from the sheets prior to welding increased lap shear strengths by about 50% at the 2.4-mm insertion depth and it promoted failure by nugget pull-out rather than by interface separation.

Introduction

Magnesium alloys are receiving a great deal of attention in the automotive industry due to the weight saving potential they offer over both steel and aluminum alloys. Magnesium has the lowest density among the structural metals considered for automotive applications. Its density is less than one fourth of that of steel and about two-third of that of aluminum alloys. It also has a high strength to density ratio and a modulus higher than common engineering plastics and glass-fiber reinforced polymer composites. These advantages have spawned interest in magnesium alloys for applications ranging from instrument panels to engine blocks.

Most current magnesium parts are produced from cast magnesium alloys, such as AZ91 and AM60, however, wrought magnesium alloys such as AZ31, AZ61, AZ80 and ZK60 are beginning to draw attention for body and chassis parts [1]. Among these wrought alloys, AZ31 is the most common magnesium alloy available in sheet form. They are produced by direct-chill casting, followed by hot rolling, or by a twin-roll continuous casting process. The formability of AZ31 and other wrought magnesium alloys is relatively poor compared to that of steel or aluminum alloys, and therefore, elevated forming temperature is required to produce deep drawn or stretch formed parts. Recently, some research effort has been directed toward warm stamping, warm hydroforming and super-plastic forming of automotive parts using AZ31 [2].

There have been relatively few studies on the joining compared to forming of wrought magnesium alloys. Much of the joining studies so far involved cast magnesium alloys [3, 4]. Fusion welding, such as gas metal arc welding, gas tungsten arc welding, resistance welding, laser welding, and e-beam welding, can be used with wrought magnesium alloys, but can cause the formation of a brittle inter-metallic phase ($Mg_{17}Al_{12}$) in the welds [1] causing reduced weld strength. Adhesive joining and bolting can be used but require additional processing steps. Adhesive joining requires cleaning and surface preparation, while bolting requires drilling. These additional steps increase the cost of assembly. Furthermore, magnesium alloys in bolted joints are susceptible to creep causing eventual loss in joint strength. Clinching and self-piercing riveting have also been attempted with magnesium alloys. These processes require more formability than magnesium alloys typically possess so that the resulting attempts have been less than successful. Linear friction stir welding (FSW), a solid-phase joining process developed in 1991, has become one of the more promising joining methods for aluminum alloys in lightweight vehicles [5]. It also shows a great deal of potential for joining magnesium alloys. In this process, a rotating tool, consisting of a pin-shaped probe attached at the bottom of a larger diameter shoulder, is traversed under pressure along the interface between sheets being joined. The frictional heat generated by the rotating tool produces a thermally softened zone at the interface and the adjoining area. This thermally softened material, forced to flow in a rotary motion around the tool pin, coalesces and as it cools upon tool retraction, a butt weld is formed.

Friction stir welding offers a significant quality advantage over the conventional fusion welding processes for both aluminum and magnesium alloys. Fusion welding requires parent metal to melt and solidify causing potential defects such as voids, trapped gas and solidification cracking. Friction stir welding does not require melting of the parent metal and, thus, eliminates the formation of these defects. As a result, the strength of friction stir welded joints in aluminum alloys is found to be close to or identical to that of the base alloy. Additionally, since no melting occurs in friction stir welding, the energy input for FSW is relatively low, the heat-affected zone (HAZ) is relatively small and residual stresses associated with the weld are negligible. Lower residual stresses result in less distortion for friction stir welded joints over fusion welded joints.

Spot Friction Welding

Spot friction welding (SFW) is based on the same principle as the linear friction stir welding process [6, 7]. They both use a cylindrical tool with a shoulder and a pin, centered at the lower face of the shoulder, to be plunged into the sheets to be joined. However, unlike linear friction stir welding, the tool in spot friction welding does not move laterally. Instead, the tool is retracted from the sheets after the tool reaches a predetermined load or plunge depth. The process is schematically illustrated in Figure 1. The heat and deformation, caused by friction and normal force from the rotating tool, create a stirred zone (Figure 2), forming a spot weld at the interface between the sheets upon cooling to room temperature.

Resistance spot welding, though the most common assembly process in the automotive industry, may not to be suitable for magnesium alloys. High welding currents are required for resistance spot welding magnesium alloys due to the relatively high thermal conductivity and electric conductivity of magnesium alloys [10]. Furthermore, frequent cleaning of the electrode is required to prevent copper contamination of the magnesium surfaces since magnesium is highly susceptible to galvanic corrosion when in contact with copper. These drawbacks of resistance spot welding of magnesium alloys have shifted attention to other alternatives including, most notably, spot friction welding. A few technical articles exploring spot friction welding of

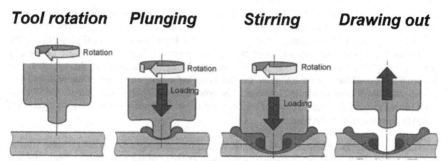

Tool rotation   Plunging   Stirring   Drawing out

Figure 1: A schematic illustration of spot friction welding (SFW) process. (Courtesy of Kawasaki Heavy Industries.)

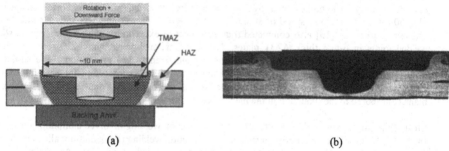

(a)   (b)

Figure 2: (a) A schematic illustration of the Thermo-Mechanically Affected Zone (TMAZ) and Heat Affected Zone (HAZ) in a spot friction welding joint. (b) A typical cross-section micrograph of an aluminum spot friction welding joint. [8, 9]

magnesium alloys are now appearing in reference literature. This is a preliminary study to examine the feasibility of spot friction welding AZ31, a wrought magnesium alloy available in sheet form.

<u>Prior Research on Friction Stir Welding of Magnesium Alloys</u>

A brief review of prior studies regarding linear and spot friction welding of magnesium alloys is provided in this section.

General comparison between the mechanical properties of friction-stir welded joints versus those of parent (base) metals was inconsistent from historical data. This may be due to the variability of production methods for magnesium materials including die-cast shapes, extrusions, or wrought sheets and the variability of referenced friction stir welds. Iwasaki et al. [11] developed a prototype magnesium swing arm for motorcycles. Both gas tungsten arc welding (TIG) and friction stir welding were used for making butt joints of AZ31 and AZ61 extruded specimens. TIG welds retained over 90% of the tensile strength and yield strength of the parent (base) material. The friction stir welds retained only about 89% of the strength of parent materials [11,12]. One hundred percent (100%) of the friction stir welded joints passed X-ray radiographic inspection while 89% of TIG welded samples passed inspection. Joint efficiencies for the fatigue strength were around 60% for both joining methods.

Okamoto et al. [13] found that the hardness and mechanical properties of linear friction stir welded AZ31 was lower than the base material. This was attributed to the fact that the base metal of AZ31 sheet was work hardened. The reduction in hardness and mechanical properties in the joint was due to the annealing effect of the friction-generated heat. In the case of die-cast magnesium alloys, such as AZ91D, AM50A and AM60B, the hardness in the stir zone was higher than that of the base metal since the stir zone had undergone grain refinement and was free of porosity or inter-dendritic □ phases. The proof stress of friction stir welded AM50A and AM60B was almost the same as that of the base metal, but the tensile strength and elongation of AZ91D were lower than the base metal.

In contrast, France and Freeman [14] reported the tensile and fatigue properties of linear friction stir welded joints in AZ91 were essentially the same as those of the base metal. The same was not true for the friction stir welded joints welds in AM50.

Nagasawa et al. [15] studied the strength of linear friction welded 6-mm thick AZ31 plates, and found that the mechanical strength of the weld was comparable to the base material but with only half the ductility. The highest measured temperature from their study, at a tool rotational speed of 1750 rpm and a traverse speed of 88 mm/min, was 460°C, showed this was truly a solid state process. Park et al. [16] also conducted transverse tensile experiments with linear friction stir welded joints in 6-mm thick AZ31 plates. The tool rotational speed was 1220 rpm and the traversing speed was 90 mm/min. A much lower yield strength and elongation and a slightly lower ultimate tensile strength of the welded joint was reported compared to the base material. Using a micro-texture analysis, it was concluded that a heterogeneously distributed (0002) basal plane contributed to the reduction in the mechanical properties.

Most of the studies reported in the literature on spot friction welding involved aluminum alloys [8, 9, 17-20]. One recently published study on spot friction welding of magnesium alloys by Su et al. [21] who conducted experiments with 1.5-mm thixo-molded AM60 alloy. A correlation between the energy input (mainly from the torque to rotate the tool) during the process and the tensile shear strength was reported. The displacement-controlled process included the control parameters of rotational speed (2000 and 2500 rpm), plunge rates (1 mm/sec and 2.5 mm/sec) and the final penetration depth (2.3 mm and 2.8 mm). It was shown that higher tensile shear strength was related to higher energy input with limited data. The highest lap shear strength reported was 3 kN.

Experimental

Magnesium alloy AZ31B-H24 sheets, 1.58 mm in thickness, were obtained from Mg Elektron N.A. The nominal composition was 3.0 wt.% Al, 1.0 wt.% Zn, 0.2 wt.% Mn, and balance Mg. The sheets were heated to O temper at 412°C for 10 minutes before spot friction welding experiments. Sheets were cut to 100-mm long x 30-mm wide coupons. The spot friction welded specimens were prepared in lap-shear configuration with 30-mm overlap. Two different sample surface conditions were prepared. One was as received with a dark-gray oxide layer on the surface. The second group was cleaned and wire-brushed to remove most of the surface oxides. The SFW process was carried out using an MTS Systems friction stir welding system at Oak Ridge National Laboratory. The SFW tool was a fixed pin probe-type tool, made of hardened H13 tool steel, with a 15-mm diameter at the tool shoulder. The 15-mm diameter SFW tool was chosen because it was close to the recommended electrode diameter (15.9 mm) for resistance spot welding of 1.6 mm thick AZ31B [22,23]. The process used a displacement-control algorithm with insertion depth of either 2.4 mm or 2.8 mm and four different rotational speeds, specifically, 500, 1,000, 1,500 and 2,000 rpm. The process time was fixed at 2 sec. for each joint, which was considered to be typical of a production operation. Tensile tests of lap shear coupons were conducted on an Instron mechanical testing machine with a crosshead speed of 12.5 mm/min. Three replicates were produced for each joining condition.

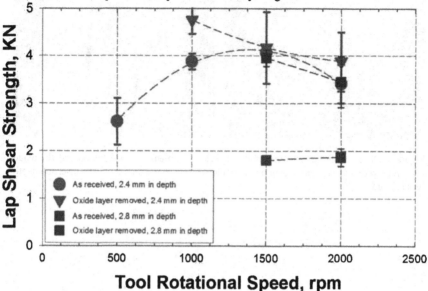

Tool Rotational Speed, rpm

Figure 3: Lap shear strength vs. tool rotational speed in spot friction welding of AZ31 at two different insertion depths (2.4 and 2.8 mm) and two different surface conditions (as received and oxide layer removed).

Results

Lap Shear Strengths

Maximum tensile loads, identified as lap shear strengths, of spot friction welded AZ31 versus tool rotational speed are presented in Figure 3 with standard deviation associated with each averaged data point from 3 replicates. In general, the specimens with surface oxide cleaned had higher strength than the as-received specimens. The highest lap shear strength achieved was 4.75 KN with oxide-removed specimens at 2.4-mm insertion depth and 1,000 rpm. Surface oxide cleaned specimens showed failure occurring as circumferential cracking around the weld button, hereas for the as-received specimens, bond failure occurred between the two sheets at the spot weld location (interfacial), as shown in Figure 4.

The lap shear strength was also dependent on the rotational speed and the insertion depth. Increasing the rotational speed for the surface oxide cleaned specimens from 1,000 rpm to 2,000 rpm reduced the lap shear strength from 4.75 KN to 3.9 KN. The low rotational speed of 500 rpm produced the weakest spot friction welded joints. The as-received specimens produced with a 2.4 mm insertion depth and 500 rpm rotational speed had an average lap shear strength of 2.6 kN and was much lower than the those made at 1,000 (3.9kN), 1,500 (4.1 kN) and 2,000 (3.4 kN) rpm. It indicated that the slower rotational speed could not generate enough frictional heat and/or stirring to produce satisfactory joints.

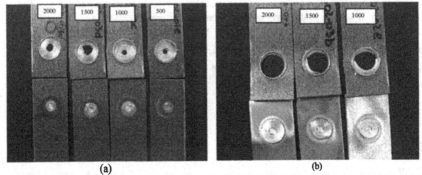

(a)   (b)

Figure 4: Photographs of tension tested lap-shear specimens (a) as-received and (b) oxide layer removed. The number shown on each figure refers to the tool rpm. The insertion depth was 2.4 mm.

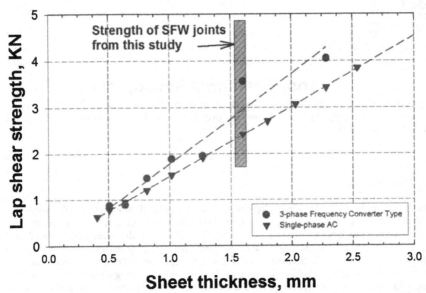

Figure 5: Comparison of lap shear strengths of spot friction welded and resistance spot welded AZ31. (The data of resistance spot welded joint strength are from [22, 24].)

SFW joints produced from surface oxide cleaned specimens had higher lap shear strength than those made of as-received specimens. This was especially evident on the specimens with 2.8-mm insertion depth. The differences were approximately 1.5 kN (at 2,000 rpm) to 2 kN (at 1,500 rpm); however, the difference was less for the specimens with 2.4-mm insertion depth.

Comparison with Resistance Spot Welding

The most widely published spot joining method for wrought magnesium is resistance spot welding. The study was first pioneered by Klain et al. in 1953 [24] and later adopted as an industrial standard with recommended equipment, electrodes and process parameters [22,23].

The Resistance Welder Manufacturers' Association (RWMA) [22] lists the minimum average tensile shear strength requirements for various magnesium alloys and thicknesses. The data were generated using two different RSW equipment set-ups including a three-phase frequency converter type machine and a single-phase alternating current machine and is shown in Figure 5. The minimum average tensile shear strength, using RSW for magnesium specimens with a thickness of 1.58 mm, was between 2.4 kN (single-phase A/C machine) to 3.6 kN (3-phase machine with post heat). All data for SFW joint strength from this study are overlaid and presented in Figure 5. It is shown, without optimization in this study, joint strengths as high as 4.75 kN are possible using spot friction welding process.

The lap-shear strength, t-peel strength and fatigue performance of SFW welds of aluminum are as good or better than those of RSW, self-piercing rivets (SPR), or clinching [6, 25]. In a recent cost analysis, SFW showed favorable cost compared to that of RSW and SPR. SFW does not require a large power source, cooling water or a compressed air supply as does RSW. The electrical consumption of SFW is less than 1/20 th of RSW [26, 27]. Other merits of SFW include long tool life, high productivity, high reliability and better working environment. [26]

Conclusions

Lap joints of 1.58-mm-thick magnesium alloy AZ31B-O sheet were produced by spot friction welding. Tensile-shear testing showed that joint strengths up to 4.75 kN were obtained. The spot friction weld strengths agree favorably with minimum spot weld strengths specified by the Resistance Welder Manufacturers' Association for similar Mg alloy sheet using the recommended 15.9-mm-diameter resistance spot welding. The removal of surface oxides from the sheets prior to welding increased lap shear strengths about 50% at the 2.4-mm insertion depth and it promoted failure by nugget pull-out rather than by interface separation.

Acknowledgement

Research sponsored by the U.S. Department of Energy, Assistant Secretary for Energy Efficiency and Renewable Energy, Office of FreedomCAR and Vehicle Technologies, as part of the High Strength Weight Reduction Materials Program (VT0502020/VT0602010, CEVT023), under contract DE-AC05-00OR22725 with UT-Battelle, LLC.

The assistance of Peter Friedman and Craig Miller of Ford Research and Advanced Engineering is greatly appreciated for providing magnesium sheet materials and performing heat treatment.

References

1. A. Luo, "Wrought Magnesium Alloys and Manufacturing Processes for Automotive Applications", SAE Technical Paper, 2005-01-0734, Society of Automotive Engineers, Warrendale, PA, 2005.

2. E. Doege and K. Dröder, "Sheet Metal Forming of Magnesium Wrought Alloys-Formability and Process Technology", J. Materials Processing Technology , Vol. 115, pp. 14-19, 2001.

3. L. K. France and R. Freeman, "Welding and Joining of Magnesium", SAE Technical Paper No. 2001-01-3443, 2001, Society of Automotive Engineers, Warrendale, PA, 2001.

4. C. Friedrich, "Reliable Light Weight Fastening of Magnesium Components in Automotive Applications", SAE Technical Paper 2004-01-0136, Society of Automotive Engineers, Warrendale, PA, 2004.

5. S. Kallee and D. Nicholas, "Application of Friction Stir Welding to Lightweight Vehicles", SAE Technical Paper 982362, Society of Automotive Engineers, Warrendale, PA, 1998.

6. R. Sakano, K. Murakami, K. Yamashita, T. Hyoe, M. Fujimoto, M. Inuzuka, Y. Nagao and H. Kashiki, "Development of Spot FSW Robot System for Automobile Body Members", Proc. 3rd International Symposium on Friction Stir Welding, Kobe, Japan, 2001.

7. T. Iwashita, "Method and apparatus for joining", US Patent 6601751 B2, August, 5, 2003.

8. D. Mitlin, T. Pan, M.L. Santella, Z. Feng, "The Effect of Spot Friction Welding (SFW) on the Strength and the Microstructure of Aluminum 6111-T4 Lap Joints", TMS2005, 134th TMS Annual Meeting & Exhibition, Feb 13-17, 2005, San Francisco, CA.

9. T. Pan, A. Joaquin, D.E. Wilkosz, L. Reatherford, J. M. Nicholson, Z. Feng and M. L. Santella, "Spot Friction Welding for Sheet Aluminum Joining", Proc. 5th International Symposium on Friction Stir Welding, Metz, France, 2004.

10. M. M. Avedesian and H. Baker (eds.), "Magnesium and Magnesium Alloys", ASM Specialty Handbook, ASM International, pp. 106-118, 1999.

11. H. Iwasaki, A. Mizuta, T. Hasegawa and H. Yoshitake, "Development of a Magnesium Swing Arm for Motorcycles", SAE Technical Paper 2004-32-0048, Society of Automotive Engineers, Warrendale, PA, 2004.

12. M. Tsujikawa, H. Somekawa, K. Higashi, H. Iwasaki, T. Hasegawa, and A. Mizuta, "Fatigue of Welded Magnesium Alloy Joints", Materials Transactions, Vol. 45, No. 2, pp. 419-422, 2004.

13. K. Okamoto, F. Hunt and S. Hirano, "Friction Stir Welding of Magnesium for Automotive Applications", SAE Technical Paper 2005-01-0730, Society of Automotive Engineers, Warrendale, PA, 2005.

14. L.K. France, and R. Freeman, "Welding and joining of magnesium", SAE Technical Paper 2001-01-3443, Society of Automotive Engineers, Warrendale, PA, 2001.

15. T. Nagasawa, M. Otsuka, T. Yokota and T. Ueki, "Structure and Mechanical Properties of Friction Stir Weld Joints of Magnesium Alloy AZ31", in "Magnesium Technology 2000", TMS, Warrendale, PA, pp. 383-387, 2000.

16. S.H. C. Park, Y.S. Sato and H. Kokawa, "Texture Effects on Tensile Properties in Friction Stir Weld of a Magnesium Alloy AZ31", Proceedings of 4th International Friction Stir Welding Symposium, Park City, Utah, May 14-16, 2003.

17. P.-C. Lin, J. Pan and T. Pan, "Fracture and Fatigue Mechanisms of Spot Friction Welds in Lap-Shear Specimens of Aluminum 6111 -T4 Sheets", SAE Technical Paper No. 2005-01-1247, Society of Automotive Engineers, Warrendale, PA, 2005.

18. S. G. Arul, T. Pan, P.-C. Lin, Z. Feng and M. L. Santella, "Microstructure and Failure Mechanisms of Spot Friction Welds in Lap-Shear Specimens of Aluminum 5754 Sheets", SAE Technical Paper No. 2005-01-1256, Society of Automotive Engineers, Warrendale, PA, 2005.

19. C. D. Allen and W. J. Abregast, "Evaluation of Friction Spot Welds in Aluminum Alloys", SAE Technical Paper No. 2005-01-1252, Society of Automotive Engineers, Warrendale, PA, 2005.

20. T. Pan, W. Zhu, W.J. Schwartz, "Spot Friction Welding - A New Joining Method for Aluminum Sheets", Proceedings of the 2005 International Automotive Body Congress. Vol. 2, pp. 95-99, September 20-21, 2005, Ann Arbor, Michigan.

21. P. Su, A. Gerlich and T.H. North, "Friction Stir Spot Welding of Aluminum and Magnesium Alloy Sheets", SAE Technical Paper 2005-01-1255, Society of Automotive Engineers, Warrendale, PA, 2005.

22. "Resistance Welding Manual", Revised 4th edition, Resistance Welder Manufacturers' Association, 2003.

23. "Metals and Their Weldability", in "Welding Handbook", 7th edition, volume 4, pp. 418-424, American Welding Society, Miami, FL, 1982.

24. P. Klain, D. L. Knight and J.P. Thorne, "Spot Welding of Magnesium with Three-Phase Low Frequency Equipment", The Welding Journal, Vol. 32, No. 1, 1953, pp. 7-18.

25. M. Fujimoto, M. Inuzuka, M. Nihio, Y. Nakashima, "Development of Friction Spot Joining (Report 2) - Mechanical Properties of Friction Spot Joints", Preprints of the National Meeting of Japan Welding Society, No.74, pp.6-7, 2004.

26. "Mazda Develops World's First Aluminum Joining Technology Using Friction Heat", Mazda Media Release, February 27, 2003.

27. R. Hancock, "Friction Welding of Aluminum Cuts Energy Cost by 99%," Welding Journal, vol. 83, p. 40, 2004.

THE EFFECT OF SURFACE TREATMENTS ON THE FAYING SURFACE OF FRICTION STIR SPOT WELDS

Tweedy, B.T., Widener, C.A., and Burford, D.A.

Advanced Joining Laboratory
National Institute for Aviation Research - Wichita State University
1845 Fairmount St.; Wichita, KS 67260, USA

Keywords: friction stir, spot welding, FSSW, Alodine, anodized, Alclad, 2024-T3

Abstract

Friction stir spot welding (FSSW) has shown great potential for reducing manufacturing steps, simplifying structural design, and lowering overall costs in automotive and aerospace applications. A variant of the friction stir welding process, FSSW involves, in its simplest form, plunging and retracting a weld tool into the materials to be joined at single location or at a series of locations along a joint line. Due to rapid processing times and increased reliability, the process is an attractive alternative to existing joining techniques such as riveting and resistance spot welding. The purpose of this project was to investigate the effect of engineered materials, such as chemical conversion coatings, aluminum cladding, anodizing, etc., placed on the faying surface to prevent or reduce corrosion of FSSW spot welds and the adjoining structure. This paper presents the results of this study on the use of interfacial/faying surface materials and the effectiveness of processing parameters developed for a new pin tool design and swept spot pattern. Effectiveness of the process was measured by the degree to which the interfacial/faying surface material is dispersed without producing detrimental effects on the mechanical integrity of the weld.

Introduction

Friction stir welding has seen rapid growth in research and industrial applications since being patented by TWI in 1991 [1]. It has been shown to have many advantages over conventional welding processes, including increased strength, fewer defects, lower residual stress and lower distortion. Friction stir spot welding (FSSW) is a discontinuous form of friction stir welding that is used to penetrate through a faying surface for creating joints between two overlapped sheets, similar to riveting and resistance spot welding. Joints produced by either FSW or FSSW can be tailored to be stronger than traditional riveted or resistance spot welded joints and are potentially less labor intensive and expensive. Applications for FSSW have been shown in the automotive industry [2,3,4] in the 5xxx and 6xxx aluminum alloys with exceptional reproducibility and low operating costs with as much as a 99% energy savings compared to resistance spot welding [5]. FSSW also lends itself as a manufacturing assist technology for fixturing or clamping of components and in situations where the part is highly contoured and traditional joining methods are not feasible or practical.

A method for increasing the strength of FSSW spot welds is to increase the shear area of the metallurgical bond in the plane of the faying surface by passing the tool through a swept tool path. To increase strength, the pin tool is programmed to follow a prescribed weld track [6] that creates a larger stirred zone. The swept spot utilized in this research program, referred to as an

Octaspot™, is shown in Figure 1. As the faying surface shear area is increased in this manner, a limit is reached in which shear failure during testing is forced to occur through the thickness of the weaker sheet. When failure occurs through the sheet thickness, spot strength can continue to be increased by increasing the periphery of the spot. In addition to changing the size and shape of the swept spot, the size of tool probe and shoulder can be varied to control the ratio of the faying surface shear area to the through-thickness shear area.

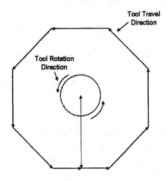

Figure 1. Octaspot™ swept spot weld travel path (0.160 inches across)

Whether using FSW or FSSW to manufacture lap joints, a solution for crevice corrosion must be addressed for production implementation in aerospace applications. One approach identified to prevent or reduce the corrosion of FSSW spot welded joints is to place sealants, aluminum cladding, epoxy primers, etc. at the faying surface. Concerns generally arise as to whether the presence of such materials will lead to the contamination of the weld joint and thereby weaken it.

Experimental Setup and Procedure

The primary control variables in FSSW for a given tool geometry are the pin rotation speed, plunge depth and plunge rate. When using a swept volume FSSW, additional variables are used to control the process, specifically the spot radius, the travel speed of the tool path, and the relative direction of travel and tool rotation (which determine the orientation of the advancing and the retreating side relative to the travel path). These three additional parameters have a significant affect on the mechanical properties of the joint. Preliminary testing for this specific study has shown that placing the advancing side to the outside of the path yielded joints with a higher tensile strength in unguided lap shear. This affect was also observed and studied in friction stir welding of lap joints by Cederqvist and Reynolds [7].

For this investigation, 2024-T3 aluminum, 0.040" thick (1 mm), was chosen as the upper and lower sheet in the lap welds. The welds were tested in unguided lap shear to measure the load carrying capability of individual spots. In addition to Alclad, anodize, and Alodine, several coating materials were screened for evaluation, namely an epoxy primer and two thermoplastics. However, because of difficulties in forming consolidated joints these coatings were removed from the test plan.

The four faying surface treatments selected to investigate their effect on FSSW spot weld mechanical strength were:

1. Alclad
2. Alodine
3. Phophoric Acid Anodized (PAA)
4. Hard Anodized

FSSW of bare sheet was selected as a baseline for comparison. A central star composite design of experiments (DOE), provided in Table 1, was generated using Statgraphics™, a statistical analysis software package. The same DOE was used for each faying surface treatment. The parameter development window used was selected to cover a wide variation in each parameter, and even though it was considered rather coarse, the selected process window revealed the potential for optimizing FSSW for each treatment further. The pin rotation speed, travel speed around the travel path (Figure 1), and shoulder plunge depth was varied while the plunge rate, spot size, tool lead angle and relative direction of travel were held constant. The travel and rotation direction was always counterclockwise and the tool had a lead angle of 1 degree. A single bare sheet was placed on top of a treated sheet to test each treatment, as illustrated in Figure 2. All samples were naturally aged at least 96 hours prior to testing.

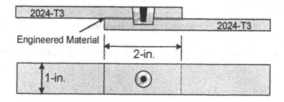

Figure 2. FSSW test setup

Table 1: DOE used for investigation of faying surface materials on FSSW

Run No.	Rotation Speed (RPM)	Travel Speed (IPM)	Plunge Depth (in)
1	1830	10.0	0.006
2	2000	5.0	0.003
3	2500	10.0	0.009
4	2250	18.4	0.006
5	2250	10.0	0.011
6	2250	10.0	0.006
7	2000	15.0	0.009
8	2670	10.0	0.006
9	2250	10.0	0.006
10	2250	10.0	0.006
11	2250	10.0	0.006
12	2500	5.0	0.003
13	2250	1.6	0.006
14	2000	15.0	0.003
15	2250	10.0	0.006
16	2250	10.0	0.006
17	2250	10.0	0.001
18	2250	10.0	0.006
19	2500	5.0	0.009
20	2250	10.0	0.006
21	2500	15.0	0.003
22	2250	10.0	0.006
23	2000	5.0	0.009

Initially a simple tapered pin with three beveled flats was chosen for this study, however, with coatings at the interface, the pin failed to produce adequate mixing resulting in a detrimental affect on the mechanical properties. The pin tool was modified by adding two flutes spaced 180°

apart. This resulted in a flute overlapping a flat and one spaced between to flats, which may promote chaotic mixing that possibly aids in dispersing the faying surface treatments. The pin had a diameter of 0.140-inch and a pin length of 0.055-inch. The shoulder has a diameter of 0.400-inch. A picture of the pin tool is shown in Figure 3.

Figure 3. FSSW pin tool

Results and Discussion

The DOE runs that produced the greatest strength for each faying surface treatment are shown in Table 2.

Table 2. DOE runs producing greatest unguided lap shear tensile strengths for faying surface treatments [8]

Treatment	Run #	Tensile Strength (lbs)	Shoulder plunge depth (inches)	Spindle Speed (rpm)	Tool Travel Speed (ipm)
Bare	14	1298 (5.8 kN)	-0.003 (0.07 mm)	2000	15 (6.4mm/s)
Alclad	23	1241 (5.5 kN)	-0.009 (0.23 mm)	2000	5 (2.1 mm/s)
Alodine	21	1266 (5.6 kN)	-0.003 (0.07 mm)	25 00	15 (6.4mm/s)
PAA	14	1204 (5.4 kN)	-0.003 (0.07 mm)	2000	15 (6.4mm/s)
Hard Anodize	17	862 (3.8 kN)	-0.001 (0.025mm)	2250	10 (4.2mm/s)

Bare (Baseline Configuration)

For bare material, DOE run number 14 (-0.003-inch shoulder plunge depth, 2000 rpm tool rotation speed, and 15 ipm travel speed) produced the maximum tensile strength. This indicates that in the absence of a faying surface coating, the maximum tensile strength is obtained under the "coldest" welding conditions. This result may point to increasing the plunge rate to further lessen the heat input to the weld and, potentially, further increasing the strength of the spot weld. A macroscopic view of the cross section in bare material for the swept spot with the highest strength condition is shown in Figure 4.

Figure 4. FSSW in bare 2024-T3

Alodine and Phosphoric Acid Anodized (PAA)

The unguided lap shear tensile strengths for Alodine and PAA were found to be comparable to the baseline bare 2024-T3 maximum strength coupon. For PAA, the DOE run which produced the maximum tensile strength was the same as for the bare coupons, run number 14 (-0.003-inch shoulder plunge depth, 2000 rpm tool rotation speed, and 15 ipm travel speed). For Alodine, DOE run number 21 (-0.003 inch shoulder plunge depth, 2500 rpm tool rotation speed, and 15 ipm travel speed) produced the maximum tensile strength. This is essentially the same set of parameters that produced the highest strength in bare 2024-T3, but with a moderately higher spindle speed. This suggests that a slightly higher heat input may be necessary to induce sufficient material flow to disperse the interface material. A cross section of the joints produced with PAA and Alodine at the interface are shown in Figures 5 and 6, respectively.

Figure 5. FSSW in 2024-T3 with PAA

Figure 6. FSSW in 2024-T3 with Alodine

Alclad

For Alclad, DOE run number 23 (-0.009-inch shoulder plunge depth, 2000 rpm tool rotation speed, and 5 ipm travel speed) produced the maximum tensile strength, which is comparable to the tensile results from the bare, PAA, and Alodine maximum strength unguided lap shear coupons. The joints produced with Alclad at the interface showed some remnant cladding in the nugget of the FSSW, as shown in Figure 7. The joints produced with PAA and Alodine at the interface appear to have a more complete mixing than the Alclad joints. It appears that the greater plunge depth of 0.009 inches (versus 0.003 for bare, PAA, and Alodine) may have been needed to sufficiently reduce the amount of retained Alclad in the spot in order to achieve nearly equivalent spot strengths to that of bare material.

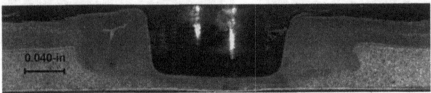
Figure 7. FSSW in 2024-T3 Alclad

Hard Anodized

For hard anodized, DOE run number 17 (-0.00-inch shoulder plunge depth, 2250 rpm tool rotation speed, and 10 ipm travel speed) produced the maximum tensile strength, which was considerably less than the tensile strength than achieved with the bare material and other treatments. This could be due to two effects. One, these parameters resulted in the highest heat input to the weld zone causing the most overaging in the heat-affected zone. Two, the hard anodized layer interferes with vertical heat conduction between the two sheets, partially insulating the bottom sheet, and negatively influencing the flow dynamics. The macroscopic cross section gives a slight indication of a shift in the flow dynamics of the spot weld. In Figure 8, the hard anodized coating appears to prevent mixing at the interface enough where the nugget actually forms below the interface on the right side.

Figure 8. FSSW in 2024-T3 with Hard Anodized

Summary

FSSW with Alclad, Alodine and PAA coatings at the interface of the joint produced unguided lap shear tensile strengths comparable to that of the FSSW produced in bare material over nearly the entire DOE, as shown in Figure 9. This suggests that the pin tool used was effective in dispersing the faying surface materials. Conversely, the joints produced with the hard anodized at the interface showed a noticeable decrease in the tensile strengths. It should be noted, however, that the lowest overall tensile strength of the FSSW DOE, produced with a hard anodized coating, was still higher than MIL-HDBK-5H design data for riveting and resistance spot welding. Consequently, the DOE produced results in all conditions exceeding that of MIL-HDBK-5H table values for equivalent strengths when using rivets, as shown in Figure 10. The results of the DOE also show an improvement over resistance spot welding data from MIL-HDBK-5H table 8.2.2.3.2(c) of 276 pounds per spot.

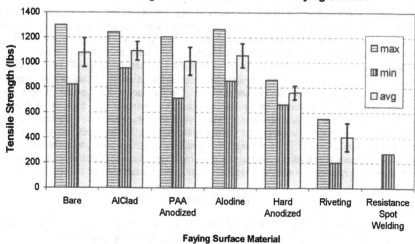

Figure 9. Static tensile strengths of FSSW through coatings

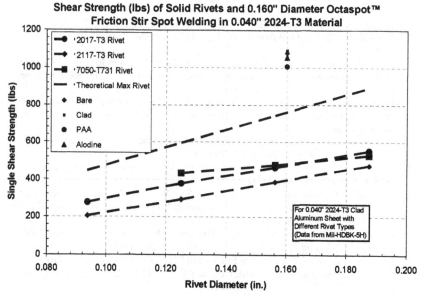

Figure 10. Static tensile strengths of FSSW through coatings compared to riveting (data from MIL-HDBK-5H)

Conclusion

Preliminary results of faying surface treatments to help prevent or reduce corrosion of friction stir spot welds, including aluminum cladding, conversion coatings, anodizing, etc. can be used without adversely reducing static mechanical tensile strengths in 2024-T3 lap joints. When Alclad and Alodine are applied to the faying surface, the FSSW joints tend to prefer "hotter" processing conditions than bare material, evidenced by a higher tool rotation speed needed to produce nearly equivalent unguided lap shear tensile strengths. Further, Alclad requires a greater shoulder plunge depth to break up the clad material. The presence of PAA on the faying surface does not necessitate a parameter change as compared bare material. At the other end of the spectrum, the presence of hard anodize on the faying surface requires the "hottest" welding parameters. Hard anodize also appears to inhibit material flow, further producing weaker joints compared to bare 2024-T3. All treatments produced unguided lap shear strengths that exceed handbook minimums for riveting and resistance spot welding. Fatigue, exfoliation corrosion and stress corrosion cracking were not investigated in this study, but are planned for future work.

Acknowledgements

This work was funded by grants from the state of Kansas – NIS program, and the Federal Aviation Administration (FAA). Special thanks to the following students in the National Institute for Aviation Research's Advanced Joining Laboratory for their work on this project: Jeremy Brown, Josh Merry, and Paul Nething.

References

1. Thomas, W.M.: Friction Stir Butt Welding, International Patent App. No. PCT/GB92/02203 and GB Patent App. No. 9125978.8, Dec. 1991, U.S. Patent No. 5,460,317.
2. Pan, T. Y. et al. "Spot Friction Stir Welding for Sheet Aluminum Joining," *Proceedings of the 5th International Symposium on Friction Stir Welding*, Metz, France, September 2004.
3. Hinrichs, J. F. et al. "Friction Stir Welding for the 21st Century Automotive Industry," *Proceedings of the 5th International Symposium on Friction Stir Welding*, Metz, France, September 2004.
4. Gerlich, A. et al. "Friction Stir Spot Welding of Aluminum and Magnesium Alloys," *Materials Science Forum*, 29, (2005), 290-294.
5. Iwashuta, T. "Spot Friction Welding to Achieve Light-Weight Automobile-Body," *Welding in the World, 48, (2004), 71-77*
6. Addison, A. C., Robelou, A. J., "Friction Stir Spot Welding: Principal Parameters and their effects," *Proceedings of the 5th International Symposium on Friction Stir Welding*, Metz, France, September 2004.
7. Cederqvist, L., Reynolds, A., "Factors Affecting the Properties of Friction Stir Welded Lap Joints," *Welding Research Supplement*, 80 (12), (2001), 281-287.
8. Tweedy, B.M. et al. "The Effect of Engineered Materials on the Faying Surface of Friction Stir Spot Lap Welds," SAE General Aviation Technology Conference, Wichita, KS, August 30, 2006, Brown, J. presenter.

FRICTION STIR SPOT WELDING OF 6016 ALUMINUM ALLOY

R. S. Mishra[1], T. A. Freeney[1], S. Webb[1], Y. L. Chen[2], X. Q. Gayden[2], D. R. Herling[3] and G. J. Grant[3]

[1]Center for Friction Stir Processing, Materials Science and Engineering, University of Missouri, Rolla, MO 65409, USA
[2] GM R&D Center, Warren, MI 48090, USA
[3]Pacific Northwest National Laboratory, Richland, WA 99356, USA

Keywords: Friction Stir Spot Welding, 6016 aluminum, tool design

Abstract

Friction stir spot welding (FSSW) of 6016 aluminum alloy was evaluated with conventional pin tool and new off-center feature tools. The off-center feature tool provides significant control over the joint area. The tool rotation rate was varied between 1000 and 2500 rpm. Maximum lap-shear strength was observed in the tool rotation range of 1200-1500 rpm. The results are interpreted in the context of material flow in the joint and influence of thermal input on microstructural changes. The off-center feature tool concept opens up new possibilities for plunge-type friction stir spot welding.

Introduction

There is a strong interest in the automotive industry for using aluminum alloy sheet for vehicle application, particularly the body, where spot welding is the principal joining method. Currently resistance spot welding (RSW) is the primary method for spot welding aluminum sheets. However, RSW have excessive porosity and surface indentation that adversely impact their strength and fatigue performance [1]. Also, RSW of aluminum alloys require higher electric power and severe electrode tip wear problems have been encountered [2,3]. The high heat input of RSW can also be detrimental to alloys which gain strength from heat treatment. Hence, mechanical fastening processes such as self-piercing riveting and clinching are frequently used instead of RSW [4]. Currently FSSW has been investigated as a tool for spot welding thin sheet aluminum alloys. Recently a feasibility study was conducted by Briskham et al. [5] showing an approximate 90 percent reduction in energy costs when compared to RSW. Allen et al. [6] have done a study on refill FSSW of 7075-T6 and 2024-T3 reporting strengths that near values required for the direct replacement of RSW and rivet technologies. Refill FSSW was patented by GKSS [7] and is a relatively new technique that has the potential to provide an alternative to other spot welding/joining techniques in aerospace and automotive applications.

Similar to friction stir welding (FSW) [8], FSSW is a solid-state technology. Figure 1, shows a schematic of plunge type FSSW. During plunge type FSSW, the rotating pin tool is inserted into the workpiece to a selected depth and is then retracted leaving a keyhole. Frictional heat is generated at the tool/workpiece interface and inter-mixing of the plasticized material at the workpiece interface leads to the development of a weld. The process is very fast and the weld cycle can be finished within a few seconds. The plunging action of the tool gives rise to a vertical force along the tool axis that is a function of the processing parameters. The properties at the weld interface can be optimized by varying the tool rotation rate and tool design. A study of FSSW of AA5052 using a conventional centered pin tool showed that maximum strength was obtained at lower tool rotation rates when the frictional condition at the tool workpiece interface was in the sticking regime [9]. This condition produces very high process loads and torques which may be undesirable in application of FSSW on car bodies where the load support may come solely from the frame. In this paper, the effect of tool design on process loads and mechanical properties of FSSW is reported. A comparison of a conventional centered pin tool to a newly designed off-center featured tool is made. The concepts behind the off-center featured tool are briefly described in the next section.

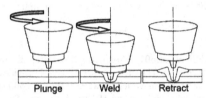

Fig 1. Schematic of plunge type FSSW process with counterclockwise rotation.

Concept of Off-center Feature Tool

The off-center feature tool concept has been developed to address the following key aspects:
  (a) lap shear strength of the FSSW depends on the stir zone around the pin,
  (b) the centered pin dictates the overall size of the stirred zone,
  (c) independent control of stir zone and process temperature is desirable,
  (d) the mixing of faying surface oxide film is inadequate in most cases with the conventional pin tool, and
  (e) reduction of process load and torque without lowering the joint strength is desirable.
There have been previous attempts to increase the stirred region in linear FSW. For example, the A-Skew[TM] tool increases the stirred zone [10]. Also, Herling et al. [11] reported a tool with three pins arranged in a circular path.

A key concept with off-center feature tool is that the outermost feature controls the overall size of the stirred zone. The size, shape, number and arrangement of these features can be changed to control the overall temperature, material flow, process load and torque. In this study, only one tool is shown in Fig. 2, along with a conventional tool for comparison. Tool A, which is a conventional pin centered tool, has a concave shoulder diameter of 10 mm, and a 1.77 mm long conical pin with the root diameter of 4.5 mm and tip diameter of 3 mm. Tool B is the newly designed offset feature tool with a concave 10 mm shoulder diameter and 3 offset features of 1.2 mm height (Fig 2). It is important to note that with the off-center feature tool, the spindle at the end of the weld cycle has to be stopped while in contact with the workpiece and then retracted.

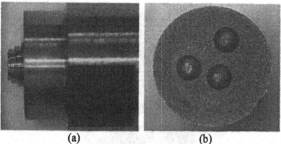
(a)   (b)

Fig 2. Macro images of (a) tool A - conventional centered pin tool and (b) tool B - offset feature pin tool. Shoulder diameter is 10 mm for both tools.

Experimental Approach

Heat treatable aluminum 6016 of 1mm thickness was selected for study. AA6016 is a low Cu, Mg-Si alloy that has gained popularity as skin material for body panels due to good dent resistance and relatively high formability [12]. Coupons were sheared from the parent sheets and spot welded in configuration shown in Fig 3. Although the paint bake cycle in automotive applications is of consideration, this was not covered in the scope of this study. In AA6111 the paint bake cycle was found to insignificantly affect the strength of the post treatment weld [13]. The FSSW machine used in this study (Fig 4) has a linear actuator which is capable of applying an axial load of up to 22.2 kN, while the spindle itself can achieve rotational speeds up to 3000 rpm. The FSSW machine also gives the capability of varying the plunge rate up to 25 mm/s and the dwell time to a maximum of 490 milliseconds (ms). In order to determine the effects of conventional tool versus offset feature tool, two tools were used in this study.

In order to determine the effects of tool rotation rate and plunge rate on the mechanical strength of the joint, welds were made at various combinations of tool rotation rates and plunge rates. The plunge depth and dwell time were held constant with the dwell time being 490 ms. During the weld, axial force, torque and time were data logged. The target depth was determined by increasing depth until the target depth was sufficient to fill the shoulder completely.

After the welds were completed, fracture loads for various conditions were determined using an MTS testing machine. Spacer plates were coupled with the coupons to ensure correct alignment of the sample during shear testing. The welded coupons were pulled at a rate of 0.02 mm/min and the load to fracture was recorded. In addition to mechanical testing, one weld in each condition was also mounted through the cross section and etched using a modified Keller's reagent to determine nugget morphology.

343

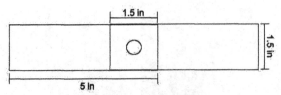

Fig 3. Schematic of lap shear specimens.

Fig 4. Friction stir spot welder at the University of Missouri-Rolla, Center for Friction Stir Processing.

Results and Discussion

Welding process forces and strength of lap shear specimens for tool A are shown in Fig 5. Heat input into the weld is a function of tool rotation rate and time in contact with the material, controlled by plunge rate. As tool rotation rate increases for a given plunge rate, temperature of the weld zone increases and the flow stress of the material decreases causing a drop in torque and plunge force. For the conventional tool the lap-shear strength increased to a maximum and then decreased when the tool rotation rate increased from 1000 to 2500 rpm. Welding done with tool B (Fig 6) shows an expected decrease in process forces but an increase in weld strength. In the strongest condition, the lap shear weld failure occurred in the HAZ of the welded joint. For the conventional tool, maximum average failure load of 2.65 kN was obtained at tool rotation rate of 1500 rpm and plunge rate of 0.5 mm/s. The corresponding process torque was ~ 17 N-m. In comparison, the maximum average failure load of 2.75 kN was obtained at tool rotation rate of 2500 rpm and plunge rate of 0.5 mm/s for the new off-center feature tool. The corresponding process torque was ~12.5 N-m. It is noteworthy that the joint strength kept increasing in the rotation range used in this study for the new tool. The material flow around the off-center features is more complicated than the conventional centered pin. It appears that higher temperature allow better material flow and joint strength with the new tool.

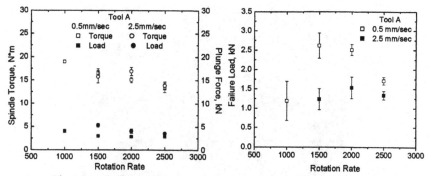

Fig 5. Process data and lap shear strength of conventional center pin tool.

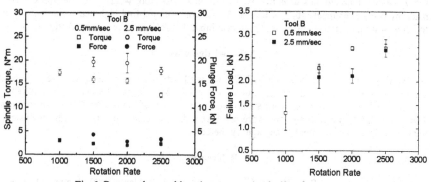

Fig 6. Process data and lap shear strength of offset feature tool.

Selected cross-sections of welds made at low and high heat input are shown in Fig 7. It was observed from lap shear testing that strength decreased as the heat input for tool A increased. By examining the macrograph of the weld nugget it can be seen that the conventional tool at high heat input produced a localized region of plastic flow causing a decrease in joint area. The increase in strength at higher heat input when examining tool B can be attributed to the increase in bond area caused by the increased mixing and breakage of the oxide film of the faying surface.

Conclusions

Initial results for a new off-center featured tool design provides better lap shear strength when compared to the conventional centered pin tool at higher rotation rates. The use of high rotation rates provides lower process loads and torques.

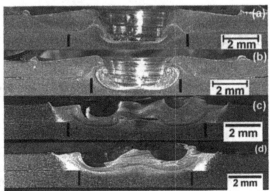

Fig 7. Optical macrographs of conventional tool and offset tool welds on 1 mm thick sheets made by (a) conventional tool low heat input, (b) conventional tool high heat input, (c) offset tool low heat input, and (d) offset tool high heat input. Black lines indicate stirred region.

Acknowledgments

This work was performed under the NSF-IUCRC for Friction Stir Processing and the additional support of NSF, Boeing and Friction Stir Link for the UMR site is acknowledged. This report was prepared as an account of work sponsored by an agency of the United States Government. The views and opinions of authors expressed herein do not necessarily state or reflect those of the United States Government or any agency thereof.

References

1. Gean, A., Westgate, S.A., Kucza, J.C. and Ehrstrom, J.C., March 1999 "Static and fatigue behavior of spot welded 5182-O aluminum alloy sheet," *Welding Journal*, vol. 78, pp. 80.
2. Sakano, R., Murakami, K., Yamashita, K., Hyoe, T., Fujimoto, M., Inuzuka, M., Nagao, Y. and Kshiki, H., 2001, "Development of spot FSW robot system for automobile body members." *Proc. 3rd International Symposium on Friction Stir Welding*, Kobe; Japan, pp. 27.
3. Zhou, Y., Fukumoto, S., Peng, J., Ji, C. T., Brown, L., 2004 "Experimental simulation of surface pitting of degraded electrodes in resistance spot welding of aluminum alloys." *Mat. Sci. and Tech.*, vol. 20, pp. 1226.
4. S. Westgate, 1996 "High speed sheet joining – by mechanical fastening", TWI Bulletin.
5. Briskham, P. et al. "Comparison of SPR, RSW, and SFJ for Aluminum Automotive Sheet". SAE 2006-01-0774. *Welding, Joining, Fastening & Friction Stir Welding. SP-2034*
6. Allen, C. Arbegast, W. "Evaluation of Friction Spot Welds in Aluminum Alloys" SAE 2005-01-1252.*Journal of Materials and Manufacturing*,2006
7. Shilling, C. and Santos, J. dos. "Method and Device for Joining at Least Two Adjoining Work Pieces by Friction Welding", *US Patent Application* 2002/0179 682.
8. W.M. Thomas et al., "Friction Stir Welding", G.B. Patent application no. 9125978.8; US patent no. 5460317, Oct. 1995.

9. Freeney, T.A., Sharma, S.R., Mishra, R.S. "Effect of Welding Parameters on Properties of 5052 Al Friction Stir Spot Welds". SAE 2006-01-0969. *Welding, Joining, Fastening & Friction Stir Welding. SP-2034.*

10. W.M. Thomas, K.I. Johnson, and C.S. Wiesner, Adv. Eng. Mater. 5 (2003) 485.

11. D.R. Herling, G.J. Grant, R.W. Davies, and Moe Khaleel, "Superplasticity in Aluminum and Magnesium Alloys Prepared by Friction Stir Processing", presented at the 2001 TMS Fall Conference Friction Stir Welding and Processing, 2001, Indianapolis, IN.

12. Hirth, S.M. et al. "Effects of Si on the aging behavior and Formability of Aluminum Alloys based on AA6016" *Mater. sci. eng. A Struct. mater. prop. microstruct. proces.* ICSMA-12, 27 September, 2000, Asilomar, CA, USA

13. Blundell, N. et al. "The Influence of Paint Bake Cycles on the Mechanical Properties of SFJ Aluminum Alloys". SAE 2006-01-0968. *Welding, Joining, Fastening & Friction Stir Welding. SP-2034*

Friction Stir Welding and Processing IV
Edited by Rajiv S. Mishra, Murray W. Mahoney, Thomas Lienert, Kumar V. Jata
TMS (The Minerals, Metals & Materials Society), 2007

PRELIMINARY STUDY OF MATERIAL FLOW IN FRICTION STIR SPOT WELDING USING COPPER AS MARKER MATERIAL

Sindhura Kalagara[1], Karim H. Muci-Küchler[1], William J. Arbegast[2]

[1]Mechanical Engineering Department, Computational Mechanics Laboratory (CML),
South Dakota School of Mines and Technology;
501 East Saint Joseph Street; Rapid City, SD 57701-3995, USA
[2]Advanced Materials Processing and Joining Laboratory (AMP),
South Dakota School of Mines and Technology;
501 East Saint Joseph Street; Rapid City, SD 57701-3995, USA

Keywords: Friction Stir Spot Welding, Flow Visualization

Abstract

Friction Stir Spot Welding (FSSW) is a solid state joining technology that has the potential to substitute other joining methods in certain automotive and aerospace applications. One FSSW approach currently being explored is the refill method. Although initial research efforts have been performed to visualize the material flow in refill FSSW, additional work is needed to determine the circumferential motion of the material in the vicinity of the tool. In this paper, a preliminary experimental study aimed at visualizing the material flow in the refill FSSW of an aluminum alloy is presented. Copper strips are placed at a certain depth from the plate surface and experiments corresponding to the plunge phase are performed. Metallographic samples are prepared and examined to identify the final location of the marker material. Based on the results, inferences are made regarding the path of motion of the plate material during the process.

Introduction

Friction Stir Spot Welding is a new and evolving solid state joining technology that is capable to find applications in the aerospace and automotive industries. Since the process is fairly complex, research efforts aimed at understanding the physics of the process are being performed by different researchers. Once the technology reaches a higher level of maturity, it may be a possible substitute for conventional joining methods such as rivets and Resistance Spot Welding (RSW). In FSSW, the rotating tool plunges into the metal sheets to be joined and the stirring motion produces heat due to friction and plastic deformation. Existing FSSW technologies can be classified into two categories namely Plunge FSSW and Refill FSSW. The basic difference between the two lies in the tool used to join the metal sheets.

In plunge FSSW, which was patented by Iwashita et al. [1], the tool is a single piece comprising of pin and shoulder that is similar to the one used in Friction Stir Welding (FSW). The process consists of three stages: Plunge, Stir and Retract. During the plunge phase, the rotating tool plunges into the plates to be welded until the shoulder reaches the top surface of the plate. In the stir phase, the stirring action of the pin and shoulder causes the material to deform plastically. Finally, the tool retracts leaving a spot weld during the retract stage. The cycle time to perform a weld with the plunge method is relatively low [2] which allowed its successful application in the automotive industry. Mazda reported the first application of this process on one of its mass production vehicle where rear doors were welded using the plunge FSSW process [3]. One disadvantage of the plunge FSSW process is that the pin leaves a key hole in the middle of the joint during the retract stage which, in some cases, can act as a stress raiser when the welded part is put into service.

Refill FSSW, which is the other approach currently being used, is a hole-filling process and was first proposed by the German company GKSS [4]. The tool used in the refill method consists of three basic parts: pin, shoulder and clamp. The pin and shoulder move with independent translational speeds but with the same angular velocity. In comparison with the plunge FSSW process, the cycle time to perform the weld is more with the refill method due to the additional motions of the tool components. Once, the cycle time is reduced, this method has the potential to find applications in the automotive industry replacing RSW and in the aerospace industry as a replacement for rivets. The method presented in this paper is a modified version of the original refill process developed by GKSS. To overcome some of the difficulties encountered in welding thick plates with the original process, a new approach is being explored at the Advanced Material Processing and Joining Center (AMP) of South Dakota School of Mines and Technology (SDSM&T). In what follows, a brief description of the later is presented. It must be pointed out that sequence of tool motions different than the ones described in this paper are possible. Initially, the two plates to be spot welded are placed above each other and are clamped so that the deformations that result during the spot welding process can be restricted. The rest of the process can be divided into four stages that are described in the following paragraph.

During the first stage, the pin and shoulder rotate with the same angular velocity ω and move towards the plate with the same plunge speed as shown in Fig. 1. During the second stage, the pin plunges into the plate until the specified plunge depth displacing the material beneath the pin which fills up the reservoir between the shoulder, clamp and the plate. Some amount of flash formed by material displaced by the pin escapes through the gap between shoulder and the clamp during this stage. In the third stage, the pin retracts back while the shoulder keeps moving down towards the plate pushing the material in the reservoir back into the plate. The translational speed of the pin (V_p) is greater than that of the shoulder (V_s) in the second and third stages of the process. Finally, once the pin and shoulder have simultaneously reached the surface of the plate, the pin along with the shoulder and clamp are retracted back in the last stage.

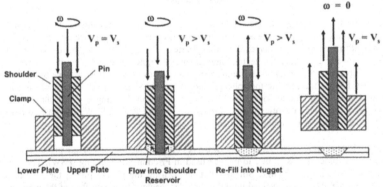

Stage 1 Figure 1. Stages of the modified Refill FSSW process Stage 4

As the material flow around the pin plays an important role in determining the quality of the weld, understanding how the material moves during different stages of the modified re-fill FSSW process is essential to improve the process parameters. Substantial efforts were done in the past in order to understand the flow visualization in FSW. A brief overview of the most relevant information found in the literature is presented in the following paragraph.

One of the earliest investigations into the material flow in FSW of aluminum alloys was carried out by Colligan [5], who used steel balls embedded into slots at different positions relative to the joint line as tracer particles. Although the general flow of the material could be visualized using this approach, the flow pattern was changed due to the presence of the steel balls. Reynolds [6] used a dissimilar aluminum alloy as marker material to show how the process parameters could influence the material flow. Dickerson et al. [7] used a marker technique in which a copper strip was placed in the original joint-line prior to welding and were successful in tracing the flow during FSW with minimal amount of copper. London et al. [8] conducted experiments using Al-SiC and Al-W composite markers to highlight flow patterns in FSW. For FSSW, few flow visualization experiments are reported in the literature. Muci-Küchler et al. [9] performed marker studies using aluminum foil as marker material to trace the vertical flow of the plate material. Also, Muci-Küchler et al. [10] used rods of Al 1100 as marker to try to visualize the circumferential motion during the plunge phase of the modified refill FSSW process. In FSW, the flow patterns were successfully classified into different flow zones by Arbegast [11]. However, apparently, there is no such classification in the FSSW process. In this paper, an experiment conducted using copper as marker material to study the circumferential material flow in the vicinity of the pin during the plunge phase of the modified FSSW process is presented.

Experimental Approach

The experiments reported in this paper were performed using a MTS Friction Stir Welding machine located at AMP and in all cases the same FSSW tool and process parameters were employed. The tool used was such that the pin and shoulder could have different translational velocities but rotated with the same angular speed. The clamp did not rotate and was placed in contact with the surface of the top plate before the plunging phase started to help keep the plates together firmly during the process. The key dimensions of the tool employed to perform the test were as follows. The diameter of the pin was 0.1875 inches , the outside diameter of the shoulder was 0.376 inches , and the outside diameter of the clamp was 1.25 inches . It must be pointed out that there is only a very small clearance between the pin and shoulder and between the shoulder and clamp.

To visualize the material flow during the plunge tests, it is necessary to select an appropriate marker material. The marker material should be easily traceable when treated with an appropriate etching agent. Also, it should not produce significant change in the flow patterns during the process. For this to be achieved, ideally, the plate and marker materials should have the same material properties, but, practically, this is impossible. The marker material chosen for the experiments presented in this paper was copper and the plate material was Al 7075-T6. Copper has lower yield strength and higher thermal conductivity than Al 7075. However, from the literature search (see Dickerson et al. [5]), copper was readily distinguishable from aluminum alloys when etched with Keller's reagent. Also, Dickerson et al. concluded that copper will not have a significant effect on the material flow as long as the amount of copper used as the marker material is very minimal. The details of the experimental set up are discussed in the following paragraphs.

The thickness of the top plate was 0.08 inches and that of the bottom plate was 0.125 inches. A 0.025-inches thick and 0.09375-inches wide copper strip was chosen for the marker material. Slots were made in the bottom plate using a milling machine in order to insert the copper strips into them as shown in Fig. 2.

351

Figure 2. Copper strips inserted in the slots made in the bottom plate

Initially, the two plates were clamped together firmly as shown in Fig. 3 and spot welds were made with velocity control by plunging the pin at the center of the grove. All the experiments performed were full plunge tests. During the experiments, the distance between the bottom surface of the pin and the bottom surface of the shoulder was kept fixed at 0.20 inches. The pin was plunged at a rate of 2 inches/min to a depth of 0.15 inches from the surface of the Al 7075-T6 plate. For all the tests, the spindle rotation rate was kept constant at 1600 rpm. Forge force and torque data for the pin and shoulder were recorded.

FSSW Tool

Plates to be Welded

Figure 3. Experimental set up for the spot welds

As shown in Fig. 4, a considerable amount of flash was formed in the space between the pin and the clamp after the plunge tests were performed. A line was engraved at the surface of the plate before the pin was plunged to identify the original location of the marker. After each test, a square section of the plate of about twice the size of the outside diameter of the flash was cut, and prepared for metallurgical examination.

Figure 4. Flash formed during the plunge tests

For the metallographic studies, the sample was mounted on bakelite and transverse metallographic sectioning was done successively in two perpendicular directions as shown in Fig. 5. The sample was etched with Keller's reagent and observed under both a stereomicroscope and an inverted metallurgical microscope. This procedure was repeated for six sections schematically shown in Figs. 5 and 6. Micrographs and macrographs were taken for each section. In what follows, the results corresponding to those sections are presented.

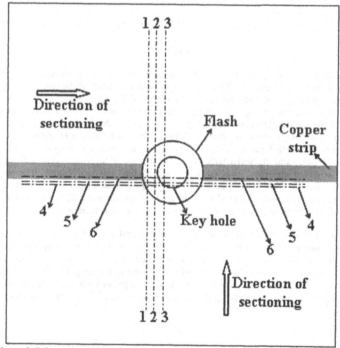

Figure 5. Schematic representation showing different sections considered during the metallographic study

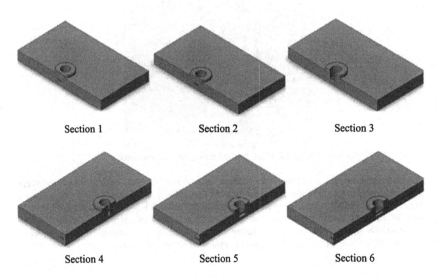

Section 1   Section 2   Section 3

Section 4   Section 5   Section 6

Figure 6. Different sections considered during the metallographic study

Results

Based on the metallographic study of the six sections, the observations presented in the following paragraphs were made. Before proceeding with the details of the results, it is important to provide information about some common features in the figures corresponding to the different sections. As can be seen in Fig. 8, the copper particles are black in color in all the macrographs taken with the stereomicroscope and white in color in all the micrographs taken using the inverted microscope. A relatively large gap between the two plates can be observed which was due to a low clamping force on the plates during the spot welding process. In all cases, the direction of rotation was clockwise with respect to the plane of the images shown. From Fig. 10, it can be observed that the copper particles have reached the flash region. This was to be expected because the material tries to flow vertically upwards as the pin tool is plunged into the plate. The same pattern was observed in all the six sections, but was highlighted only in Fig. 10. In what follows, the results corresponding to each section are discussed in detail.

In the case of Section 1, it was observed from Fig. 7 that most of the copper strip was still present as a single piece. However, the cross-sectional area decreased by approximately 27% indicating that some of the copper material was displaced. In addition, copper particles were seen in the flash region and the micrograph revealed that some of the copper material tried to move upwards in the direction of rotation of the pin. It must be pointed out that in this initial section, the work pieces were not completely joined based on the extent of the gap between two plates.

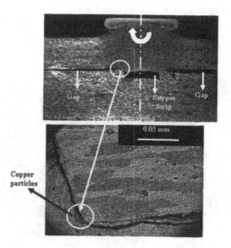

Figure 7. Macrograph and Micrograph corresponding to Section 1

From the results corresponding to Section 2, it can be seen that most of the copper strip still remained as a single piece but the cross sectional area of the strip was decreased and the extent of joined region was well defined when compared to the previous section. The strip formed an asymmetric shape which might be due to the rotation of pin tool. Based on the results for Sections 1 and 2, it was evident that the material in the region corresponding to those cuts did not experience substantial stirring. Also, as expected, it was observed that the copper strip moved down from its initial position. Initially the position of marker material was such that the top surface of the copper strip was coincident with the line of separation between the two plates.

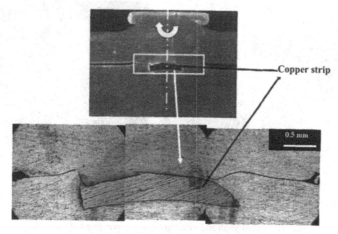

Figure 8. Macrograph and Micrograph corresponding to Section 2

355

The results corresponding to Section 3 are presented in Fig. 9. In this section, the material of the copper strip experienced a significant motion upwards, as in a backward extrusion, indicating that the material in the vicinity of the pin experienced significant deformation. From the figure, it can be observed that the strip was initially pushed down by the pin and later it tried to move upwards towards the flash.

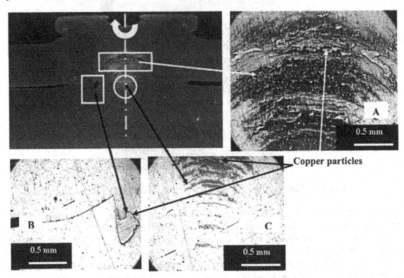

Figure 9. Macrograph and Micrograph corresponding to Section 3.

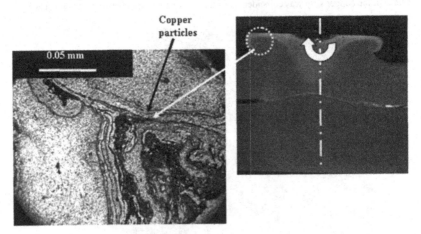

Figure 10. Macrograph and Micrograph corresponding to Section 4

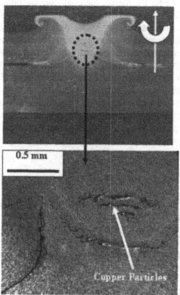

Figure 11. Macrograph and Micrograph corresponding to Section 5

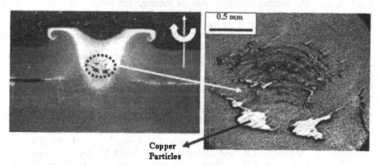

Figure 12. Macrograph and Micrograph corresponding to Section 6

From Figs. 10, 11 and 12, a similar flow pattern was observed where, the material far from the pin tool tried to move upwards towards the flash region and the material in the vicinity of pin tried to move upwards and was simultaneously subjected to stirring motion. Also, the intermixing of material was random in nature which made it difficult to predict the circumferential motion in the vicinity of pin. It is important to note that copper particles in Sections 4 and 5 are a consequence of stirring since those sections do not intersect the original location of the copper strip.

Conclusions

A preliminary effort was made to study the material flow in the vicinity of the pin during the refill FSSW process. Based on the results corresponding to the initial sections, it was observed

357

that copper was a good choice for the marker material from the point of view that it was easy to identify in the metallographic samples. Although the number of sections considered was relatively few, it was possible to draw some basic conclusions. Only the material that is very close to the pin experiences large deformations. The evidence of copper particles in the flash region indicated that a vertical flow of material similar to a backward extrusion was one important component of the overall motion of the plate material. The presence of copper particles in sections parallel and away from the copper strip indicated that stirring of the material in the vicinity of the pin took place. Given the limited number of sections considered, it was not possible to assess the extent of the stirring motion. A more detailed study is needed to understand the flow patterns during all the four stages of the process and to classify the material flow zones corresponding to the process. At the present time, the authors are still working on the results corresponding to additional sections towards the centerline and the results will be reported in the future.

Acknowledgements

This research was sponsored by the Army Research Laboratory and was accomplished under Cooperative Agreement Number DAAD19-02-2-0011. The views and conclusions contained in this document are those of the authors and should not be interpreted as representing the official policies, either expressed or implied, of the Army Research Laboratory or the U.S. Government. The U.S. Government is authorized to reproduce and distribute reprints for government purposes notwithstanding any copyright notation hereon.

References

1. Iwashita, T. et al. "Method and Apparatus for Joining," US Patent 6601751 B2, August 5, 2003.
2. Pan, T.; Joaquin, A.; Wilkosz, D.E.; Reatherford, L.; Nicholson, J.M.; Feng, Z. and Santella, M.L. "Spot Friction Welding for Sheet Aluminum Joining," Proc. 5th International Symposium on Friction Stir Welding, Metz, France, 2004.
3. Mazda News Release, "Mazda Develops World's First Aluminum Joining Technology Using Friction Heat," February 27, 2003:
   http://www.mazda.com/publicity/release/2005/200506/050602.html
4. Schilling, C. and dos Santos, J. "Method and Device for Joining at Least Two Adjoining Work Pieces by Friction Welding," US Patent Application 2002/0179 682.
5. Colligan, K. "Material Flow Behavior During Friction Stir Welding of Aluminum," Welding Research Supplement, pp. 229-237, July 1999.
6. Reynolds, A.P. "Visualization of Material Flow in Autogenous Friction Stir Welds," Science and Technology of Welding and Joining, Vol. 5, No. 2, pp. 120-124, 2000.
7. Dickerson, T.; Shercliff, H.R. and Schmidt, H. "A Weld Marker Technique for Flow Visualization in Friction Stir Welding," 4th International Symposium on Friction Stir Welding, Park City, Utah, USA, May 14-16, 2003.
8. London, B.; Mahoney, M.; Bingel, W.; Calabrese, M.; Bossi, R.H. and Waldron, D. "Material Flow in Friction Stir Welding Monitored with Al-Sic and Al-W Composite Markers," in Friction Stir Welding and Processing II (Eds. Jata, K.V.; Mahoney, M.W.; Mishra, R.S. and Lienert, T.J.), pp. 3-12, TMS (The Minerals, Metals and Materials Society), 2003.
9. Muci-Küchler, K.H.; Kakarla, S.S.T.; Arbegast, W.J. and Allen, C.D. "Numerical Simulation of the Friction Stir Spot Welding Process," SAE Paper 2005-01-1260.
10. Muci-Küchler, K.H.; Itapu, S.K.; Arbegast, W.J. and Koch, K.J. "Visualization of Material Flow in Friction Stir Spot Welding Process," SAE Paper 2005-01-3323. Transaction Journal of Aerospace (Eds. Alkidas, A.C. and Yerkes, K.L.), Volume 114, Section 1, pp 1062-1072, February 2006.
11. Arbegast, W.J. "Modeling Friction Stir Joining as a Metal Working Process," in Hot Deformation of Aluminum Alloys III (Ed. Jin, Z.). The Minerals, Metals, and Materials Society, 2003.

SCREENING FOR PROCESS VARIABLE SENSISTIVITY IN REFILL FRICTION SPOT WELDING OF 6061 ALUMINUM SHEET

Clark Oberembt[1], Casey Allen[1], William Arbegast[1], Anil Patnaik[2]

[1]Advanced Materials Processing Center, South Dakota School of Mines & Technology;
501 E St Joseph St; Rapid City, SD, 57701, USA
[2]Department of Civil Engineering, South Dakota School of Mines & Technology;
501 E St Joseph St; Rapid City, SD, 57701, USA

Keywords: Spot Welding, Aluminum, 6061, Screening, Optimization, Process, Parameters

Abstract

The screening of process variable sensitivity to strength and metallurgical quality of Refill Friction Spot Welds (RFSW) has been completed for 6061-T6 Aluminum in 0.080" sheet. These results will be used to optimize process parameters and tool size for several thicknesses of 6061-T6 sheet. The screening portion is a fully designed experiment of type 2^{k-1} half-fraction factorial. There are seven variables which have been investigated in the screening experiment, yielding 65 combinations including 1 center point. Ultimate tensile strength was chosen as the response variable, additionally; the effects of processing parameters on void size, effective shear area, surface indention, and sheet thinning have also been investigated. The results of the screening experiment have shown the sensitivity and trend of each variable on strength and metallurgical quality. The most sensitive variable to strength was found to be plunge depth. Pin retract rate and dwell were also observed to have significant effects on strength. Void size and effective shear area were most affected by plunge depth. Surface indent is most effected by pin retract rate. Upper sheet thinning is most effected by plunge depth; however, pin plunge rate and dwell were also shown to have significant effects. Interactions between process variables which had significant effects were discovered for strength and sheet thinning. The variables to optimize are: Plunge Depth, Pin Retract Rate, Dwell, and possibly Pin Plunge Rate.

Introduction

The objectives of this experiment are to determine process variable sensitivity to weld quality and to identify their trends for process optimization of RFSW. Presented herein are the screening results for 0.080" Al 6061-T6, using a 0.232" flatted pin tool with a 0.405" shoulder. This experiment was setup, managed, and analyzed utilizing design of experiments (DOE) methods. A 2^{k-1} Half-Fraction Factorial design was developed looking at 7 process variables (Plunge RPM, Retract RPM, Pin Plunge Rate, Pin Retract Rate, Dwell, Plunge Depth, & Final Forge Load) and 1 main response (Tensile Strength). Additionally, the effects of process variables on four metallurgical response variables (Void Area, Effective Shear Diameter, Surface Indent, & Sheet Thinning) have been investigated. Statistical analysis methods were used to investigate the relations between the process and response variables, and the interactions between the process variables. With this data an optimization DOE will be developed.

359

Process

All welds were made using an MTS ISTIR 10 Friction Stir Welding System equipped with a custom RSFW Head. The ISTIR provides for independent control of a forge axis and a pin axis. This system allows for the precise control of all process variables while providing response feedback. The RFSW process can be adjusted to meet specific task requirements, but there are two core stages which must occur: plunge and retract. This process was first developed and patented by GKSS (GmbH) [1]. This process can be added to producing more complex but task specific variants. The process investigated here ranges from two to a total of 4 stages. The additional stages of dwell and static final forging were added to observe their effects. Figure 1 shows a process diagram as step-by-step breakdown of stages. The pin and shoulder move simultaneously and proportionally. As the pin plunges, the shoulder retracts to create a cavity which should be equal to or slightly larger then the volume displaced by the pin. This material is then trapped under the shoulder and inside the clamping ring around the pin. Since the pin and shoulder are rotating together this material is constantly being worked and kept at temperature until the pin retract is initiated. At this point the displaced material is extruded back into the hole created by the retracting pin [2].

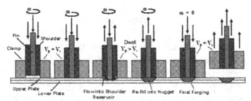

Figure 1: RFSW Process Diagram

Apparatus

To refill a friction spot weld there are three key tooling components: pin, shoulder, and clamping ring [2]. These pieces can be seen in Figure 2. The shoulder used was cylindrical with a diameter of 0.405". The pin used was a flatted tool with a major diameter of 0.232" and a flat width of 0.1575". The clamping ring had a 0.410" inner diameter. The clamping ring is a stationary annulus which is held in position under constant force, 500 lb for this experiment. The shoulder is free to rotate and travel within the clamping ring, these two components work together to contain the plasticized material.

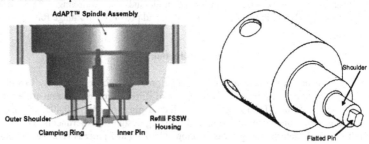

Figure 2: Custom Tooling Assembly with Pin & Shoulder

Experimental

Seven process variables were chosen to screen; Table 1 shows these variables with there corresponding high and low values. Two distinguishable Plunge Depths were chosen. This

360

process separates the Pin Plunge and Retract stages, and sometimes divides them with a Dwell period, which maintains Plunge process parameters. During the Plunge and Retract stages, the tool rpm and the rates of relative pin-shoulder movement are varied according to Table I. The ratio of relative motion for the pin to shoulder movement was held constant. The volume displaced by the pin area over the plunge depth was used to calculate the final shoulder position, ensuring a constant volume exchange. The Final Forge Load refers to a stamping operation which was performed on half of the welds. The spindle was lifted from the part, brought to a full stop, pressed down with a specified force, and then removed. The 2^{k-1} experimental design with these seven variables, and one center point, produced 65 combinations [3]. Three welds were made at each combination, 2 for unguided lap shear and 1 for metallurgy.

Table I: Process Variables

Parameter	Values (∸)	Units
RPM Plunge	1500 1800	rpm
RPM Retract	1500 1800	rpm
Pin Plunge Rate	2 6	in/min
Pin Retract Rate	2 6	in/min
Dwell	0 1	sec
Plunge Depth	0.088 0.12	in
Final Forge Load	0 2000	lbf

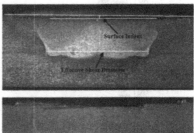

Figure 3: Metallurgical Responses (3x)

The main response variable was Tensile Strength, determined through unguided lap shear tests. Two samples from each weld set were broken and their strengths (lb/spot) averaged for analysis. The strengths will be compared with values for resistance spot welds for similar material and thickness. Failure modes were recorded for each sample.

One sample from each weld set was cut, polished, and etched with Tucker's Etch for macro evaluation. Four measurements were made for each sample: central void area, effective shear diameter, surface indent, and upper or lower sheet thinning, shown in Figure 3 [2]. These measurements were treated as secondary response variables in the statistical analysis. Some samples also had corner voids, which were not always continuous; the presence of these voids was recorded. The central void area represents void size. The effective shear diameter was used to calculate the effective shear area. In many cases a void would be directly in the shear path. The diameter of the void was measured on the interface line and the corresponding area was subtracted resulting in an annular shear area.

Results and Discussion

Software was primarily used to perform the statistical analysis. This experiment was fully designed in Statgraphics, and it was used to evaluate the effects and interactions of the process variables on all response variables. Along with these responses; processing times and failure modes have also been observed.

Tensile Strength
There are seven effects for Tensile Strength which were found to be significantly different from zero at a 95.0% confidence level. Three of them are the process variables: Plunge Depth, Pin

361

Retract Rate, and Dwell. Plunge Depth is by far the most significant variable for strength. The other four effects are interactions between process variables: Pin Retract Rate–Plunge Depth (DF), Plunge Rate–Dwell (CE), Plunge Rate–Plunge Depth (CF), RPM Plunge–Retract Rate (AD). These effects can be seen on the Pareto Chart in Figure 4.

Standardized Pareto Chart for Tensile Strength

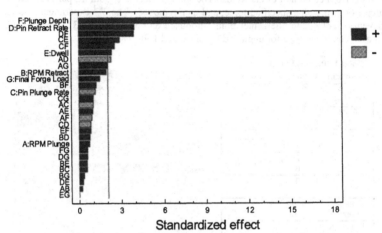

Figure 4: Pareto Chart with all variables & interactions

The minimum and maximum strengths observed were 721 lb/spot and 1260 lb/spot, respectively. It was shown that the shallow plunge depth (0.088") produced an average tensile strength of 837 lb/spot with a minimum of 721 lb/spot. The deeper plunge depth (0.120") had an average tensile strength of 1105 lb/spot, with a minimum of 901 lb/spot; these strengths are within SAE AMS-W-6858 strength requirements for resistance spot welds of this type [4].

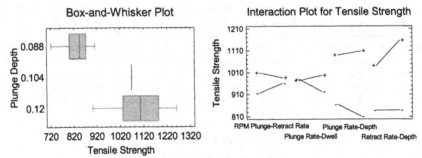

Figure 5: Plots for Strength

Figure 5 shows a Box-and-Whisker Plot of Tensile Strength and Plunge Depth, and Interaction Plots of significant interactions. The first variable in each interaction pair would have values on the axis with left being a low value and right being a high value. The second variable in the pair corresponds to the '+' (high) and '-' (low) lines.

Plunge Depth and Pin Retract Rate should be further investigated, and the trend is deeper plunge and faster retract rates produce stronger welds. There is evidence that at least some dwell will help to increase strength, however the difference in the average tensile strength of welds with no dwell and welds with 1 sec dwell is only 31 lb/spot. This effect is doubled (64 lb/spot) when looking at the interaction of plunge rate–dwell, but still not nearly as significant as plunge depth alone. The faster the plunge rate the more desirable it is to have a dwell, but with a slow plunge rate a dwell makes almost no difference on strength. In terms of process time, the samples with a high plunge rate and a dwell still averaged almost 1 second quicker then the slow plunge rate with zero dwell samples. From the interaction plots it was also noticed that with a high retract rate a slower plunge rpm might help to maximize strength, but the difference is small. At shallow plunge depths the plunge rate interaction is negligible.

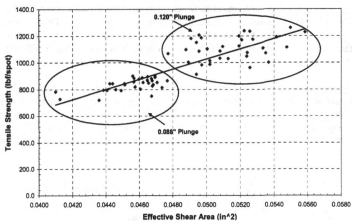

Effective Shear Area vs Tensile Strength

Figure 6: Effective Shear Area vs. Tensile Strength

In the deep (0.120") plunge samples the central void was mostly, and in many cases entirely, below the shear line, but for the shallow (0.080") plunge samples the void was directly on the shear line. Depending on the application, voids may be acceptable, but their location can effect both strength and failure mode. The presence of a void on the shear line decreases the shear area and consequently strength. Figure 6 shows Effective Shear Area vs. Tensile Strength and the trend is larger shear areas yield greater strength. The increased scatter in the deep plunge samples can be attributed to variable interactions which are sensitive to high plunge depths and possibly the effects of process variables on weld material properties.

Metallurgical Response Variables
Central Void Area - Void area was found to be most significantly effected by Plunge Depth, although Pin Retract Rate has some correlation. The shallow plunge depth produced an average Void area of 0.0002 in² while the deeper plunge depth averaged 0.0007 in², Figure 7. It was observed that a slower pin retract rate may help to minimize void size, but the uncertainty is too high to be statistically confident in this effect. It appears that shallower plunge depths will produce smaller voids. Figure 7 shows a comparison of the main effects for Central Void Area; the slope of each variable corresponds to the strength of its effect.

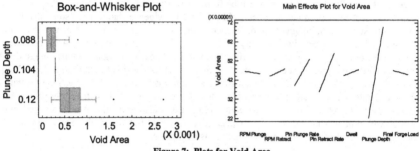

Figure 7: Plots for Void Area

High plunge depths displace more material which must be extruded back into the hole as the pin retracts. This material must also be extruded further into the work piece. These increases in the extrusion process are responsible for the increase in void size at high plunge depths. It is possible that alterations to the retract stage which promote material extrusion will help to better consolidate the welds.

Effective Shear Diameter – Shear diameter was measured on the faying surface across the joint, but was found to be inaccurate when calculating area for samples with an annular shear area (voids on the interface). Additional measurements were taken for these samples to account for their geometry. The calculated annular shear areas were substituted into the analysis and more consistent results were obtained. It was found that shear area was most effected by Plunge Depth, shown in Figure 8.

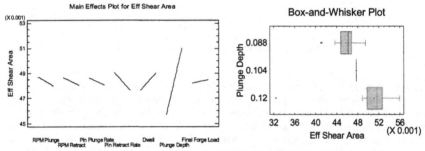

Figure 8: Plots for Effective Shear Area

The shallow plunge depth yielded an average shear area of 0.046 in², and the deep plunge depth samples averaged 0.051 in². Deeper plunge depths produce larger shear areas. It should be noted that Pin Retract Rate was only slightly outside of the 95% confidence level, and that its effect was opposite of that for strength. Shear area and strength should be directly correlated and the effects should be consistent for both; here they are not. Error in measurements and differences in the nugget material properties or failure modes may help to explain this.

Surface Indent - Surface indent is a measured average across the weld, and was found to be most significantly effected by Pin Retract Rate. The lower Pin Retract Rate (2 in/min) resulted in an average indent of 0.0033", while the higher rate (6 in/min) gave an average indent of 0.0050", Figure 9. The minimum Surface indent observed was 0.0" (not measurable); however, this sample had the largest void. The largest indent observed was 0.0089". The trend observed is that lower retract rates will produce less surface indent. Figure 9 shows the main effects and their trends for Surface Indent.

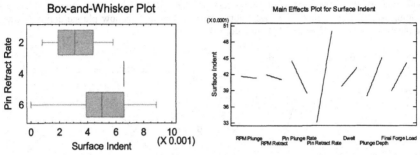

Figure 9: Plots for Surface Indent

There is always some material loss during RFSW which can be observed as surface indent, but the surface conditions are significantly improved when compared to non-refill Friction Spot Welding [5,6]. The reason Pin Retract Rate had such a strong effect on indent is likely linked to a tuning issue. From the plot in Figure 10 it can be seen that the pin is not at its commanded position at the end of the weld and is still sticking out 0.005" past the shoulder. For the majority of the samples with a high Pin Retract Rate some pin extension was observed. These retract rates are higher then typical values used on this machine, so RFSW application specific tuning is needed to correct this. The effect of Pin Retract Rate on surface indent cannot be certain for this experiment.

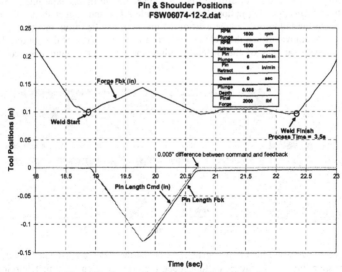

Figure 10: Process Graph of Pin & Shoulder Positions

Sheet Thinning – Sheet thinning was measured from the interface to furthest point of the defect. There was only one sample with lower sheet thinning (0.0058"). It was a shallow plunge sample. No statistical data could be drawn from this one sample. There were 13 samples with

365

measurable upper sheet thinning, 12 were deep plunge samples and the other was the center point sample. Plunge Depth, Pin Plunge Rate, and Dwell all had significant effects on this response, Figure 11. The interactions, also seen in Figure 11, of Plunge Rate–Depth and Dwell-Depth had significant effects which agree with the process variable effects. The average measurable sheet thinning was 0.0043". From the evaluation it was evident that shallow plunge depths, slower plunge rates, and a dwell will all help to reduce sheet thinning.

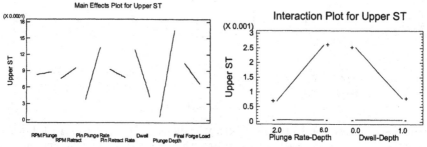

Figure 11: Plots for Upper Sheet Thinning

At high plunge depths more material is flowing past the faying surface which can produce upper sheet thinning. High plunge rates move this material quickly and under "colder" conditions deforming the interface. However, if a dwell period is introduced more heat is input and the interface material will stir more completely reducing sheet thinning. The RFSW method offers significantly less sheet thinning than non-refill FSW methods [5,6].

Failure Modes - There were four different failure modes present in this experiment: partial plug pullout in both sheets (1), plug pullout (2), nugget pullout (3), and nugget shear (4). Here, plug pullout refers to the failure mode where all of the material is pulled from one sheet, and nugget pullout refers to removal of just the weld nugget while leaving the rest of the bottom sheet intact. Figure 12 shows examples of the different failure modes observed and their corresponding macrographs.

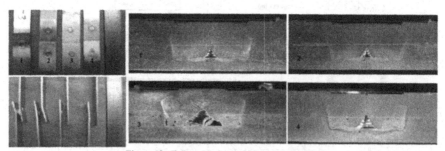

Figure 12: Failure Modes and Macrographs (3x)

Over 95% of all samples failed by nugget shear (4) and the rest occurred equally sparsely. The macro corresponding to the nugget pullout mode (3) also represents the metallurgically worst weld but not the weakest. This shows why optimization based on more than just strength is necessary. Calculations suggest that more samples will be likely to fail by plug pullout in the 0.060" sheet and even more in the 0.040" sheet, with this tool size. Sheet thinning will play an important role in determining failure modes for thinner sheets. Pin tool size and sheet thickness will be included in future optimization efforts.

Process Time - Process time is defined as the time the tool is in contact with the processed material. For all samples in this experiment the process times ranged from 1.5 to 9.8 seconds. Process time is not a response variable in this experiment but it will be of interest for industrial applications. Trends to minimize process time while maintaining acceptable quality will be more thoroughly investigated in future efforts.

Conclusions

In this experiment the Pin Plunge RFSW process was divided into stages and the screening of process variables was performed. A 2^{k-1} Factorial DOE was utilized to setup and manage the experiment and analysis. An analysis of the effects of the process variables on strength and metallurgical qualities was completed. From the statistical analysis the following conclusions for the optimization of tensile strength can be drawn:

1. The Deep Plunge samples' average and minimum strengths are within SAE AMS-W-6858 strength requirements for resistance spot welds of this type
2. Deep Plunge Depths produce stronger welds
3. Fast Retract Rates produce stronger welds
4. Some Dwell time produces stronger welds
5. Final Forge stage may be removed

The trends observed show that for the optimization of metallurgical quality the following conclusions can be drawn:

1. Shallow Plunge Depths decrease Void Area, slow Retract Rates may help
2. Deep Plunge Depths increase Shear Area, slow Retract Rates and Dwell may help
3. Application specific tuning is needed to minimize Surface Indent at high Retract Rates, conclusions for process variable effects cannot be drawn for the Surface Indent
4. Shallow Plunge Depths, Slow Plunge Rates, & Dwell minimize Upper Sheet Thinning

The optimization must be focused on more than one response. Some of the trends to optimize metallurgical quality are opposite of the trends to optimize strength. Optimization of the RFSW process will require compromise to ensure all around quality. Parameters between the high and low values of this experiment will be included in optimization.

Acknowledgments

This work was performed under the multi-university National Science Foundation - Industry / University Co-operative Research (I/UCRC) Center for Friction Stir Processing (CFSP). The financial support at the SDSMT CFSP site provided by Army Research Laboratory, Boeing, Pacific Northwest National Laboratory, MTS Systems, BAE Systems, and Sikorsky is gratefully acknowledged. The views and conclusions contained in this paper are those of the authors and should not be interpreted as representing the official policies, either expressed or implied, of CFSP, the center members or the National Science Foundation.

References

1.   C. Shilling and J. dos Santos, "Method and Device for Joining at Least Two Adjoining Work Pieces by Friction Welding", US Patent Application 2002/0179 682

367

2.   C.D. Allen, and W.J. Arbegast, "Evaluation of Friction Spot Welds in Aluminum Alloys", 2005 SAE World Congress, Detroit, MI, USA

3.   Douglas C. Montgomery, *Design and Analysis of Experiments* (New York, NY: John Wiley & Sons, Inc 1997) 171-216, 304-317

4.   SAE-AMS-W-6858, "Welding: Resistance, Spot and Seam" , April 2000

5.   A.C. Addison, and A.J. Robelou, "Friction Stir Spot Welding: Principal parameters and their effects", Proceeding of the 5th International Symposium on Friction Stir Welding, Metz, France, 2004

6.   T. Pan, et al, "Spot Friction Welding for Sheet Aluminum Joining", Proceeding of the 5th International Symposium on Friction Stir Welding, Metz, France, 2004

NUMERICAL SIMULATION OF A REFILL FRICTION STIR SPOT WELDING PROCESS

Sindhura Kalagara and Karim H. Muci-Küchler

Mechanical Engineering Department, Computational Mechanics Laboratory (CML),
South Dakota School of Mines and Technology;
501 East Saint Joseph Street; Rapid City, SD 57701-3995, USA

Keywords: Friction Stir Spot Welding, Fully Coupled Thermo-Mechanical Model

Abstract

Friction Stir Spot Welding (FSSW) is a solid state joining technology that has the potential to find applications in the automotive and aerospace industries. One approach currently used is the refill method. Having effective and reliable numerical models capable to represent this process can help to optimize process parameters and explore new tool designs. In this paper, a fully coupled thermo-mechanical finite element model of the initial part of the plunge phase of a refill FSSW process is presented. The commercial software ABAQUS/Explicit is employed to obtain the temperature distribution, deformations and stresses induced in the plates being joined. An Arbitrary Lagrangian-Eulerian formulation together with an adaptive meshing strategy is used for the numerical simulation. In addition, a contact algorithm with a modified Coulomb friction law is employed to take into account the interaction between the tool and the plate material. A maximum shear stress value is defined to control the stick/slip behavior between the parts that are in contact. The results presented show that the value of the shear stress used to define the stick/slip behavior has an important effect in the predicted temperature and strain fields. Also, the numerical results indicate that the plastic deformation of the plate material is the primary energy source for the temperature increase experienced by the plate material during the process.

Introduction

Friction Stir Spot Welding is a new and evolving solid state joining technology that is capable to find applications in the aerospace and automotive industries. Since the process is fairly complex, research efforts aimed at understanding the physics of the process are being performed by different researchers. Once the technology reaches a higher level of maturity, it may be a possible substitute for conventional joining methods such as rivets and Resistance Spot Welding (RSW). In FSSW, the rotating tool plunges into the metal sheets to be joined and the stirring motion produces heat due to friction and plastic deformation. Existing FSSW technologies can be classified into two categories namely Plunge FSSW and Refill FSSW. The basic difference between the two lies in the tool used to join the metal sheets.

In plunge FSSW, which was patented by Iwashita et al. [1], the tool is a single piece comprising of pin and shoulder that is similar to the one used in Friction Stir Welding (FSW). The process consists of three stages: Plunge, Stir and Retract. During the plunge phase, the rotating tool plunges into the plates to be welded until the shoulder reaches the top surface of the plate. In the stir phase, the stirring action of the pin and shoulder causes the material to deform plastically. Finally, the tool retracts leaving a spot weld during the retract stage. The cycle time to perform a weld with the plunge method is relatively low [2] which allowed its successful application in the automotive industry. Mazda reported the first application of this process on one of its mass production vehicle where rear doors were welded using the plunge FSSW process [3]. One disadvantage of the plunge FSSW process is that the pin leaves a key hole in the middle of the joint during the retract stage which, in some cases, can act as a stress raiser when the welded part is put into service.

Refill FSSW, which is the other approach currently being used, is a hole-filling process and was first proposed by the German company GKSS [4]. The tool used in the refill method consists of three basic parts: pin, shoulder and clamp. The pin and shoulder move with independent translational speeds but with the same angular velocity. In comparison with the plunge FSSW process, the cycle time to perform the weld is more with the refill method due to the additional motions of the tool components. Once, the cycle time is reduced, this method has the potential to find applications in the automotive industry replacing RSW and in the aerospace industry as a replacement for rivets. The method presented in this paper is a modified version of the original refill process developed by GKSS. To overcome some of the difficulties encountered in welding thick plates with the original process, a new approach is being explored at the Advanced Material Processing and Joining Center (AMP) of South Dakota School of Mines and Technology (SDSM&T). In what follows, a brief description of the later is presented. It must be pointed out that sequence of tool motions different than the ones described in this paper are possible. Initially, the two plates to be spot welded are placed above each other and are clamped so that the deformations that result during the spot welding process can be restricted. The rest of the process can be divided into four stages that are described in the following paragraph.

During the first stage, the pin and shoulder rotate with the same angular velocity ω and move towards the plate with the same plunge speed as shown in Fig. 1. During the second stage, the pin plunges into the plate until the specified plunge depth displacing the material beneath the pin which fills up the reservoir between the shoulder, clamp and the plate. Some amount of flash formed by material displaced by the pin escapes through the gap between shoulder and the clamp during this stage. In the third stage, the pin retracts back while the shoulder keeps moving down towards the plate pushing the material in the reservoir back into the plate. The translational speed of the pin (V_p) is greater than that of the shoulder (V_s) in the second and third stages of the process. Finally, once the pin and shoulder have simultaneously reached the surface of the plate, the pin along with the shoulder and clamp are retracted back in the last stage.

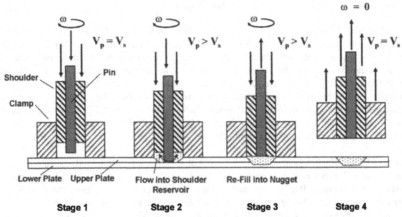

Figure 1. Stages of the modified refill FSSW process

370

Modeling Approaches of Friction Stir Spot Welding Process

To optimize the modified refill FSSW process, it is important to understand the physics of the process. For this purpose various numerical approaches and carefully designed experiments are being used by different researchers. Different numerical models varying in their level of complexity, type of analysis performed and the output provided are reported in the literature for plunge FSSW process. However, numerical modeling of the modified refill FSSW process is still under development indicating that more sophisticated numerical models capable to represent the actual process are required. In what follows, a brief overview of different approaches reported in the literature for the numerical simulation of FSSW process is presented.

Thermal, mechanical and thermo-mechanical computer simulations of the FSSW process have been performed. For example, Gerlich et al. [5] employed a finite volume approach based on Computational Fluid Dynamics (CFD) to predict the temperature distribution in the plunge FSSW process. The amount of heat being input into the model was calculated by considering the viscosity and shear rate terms at each control volume. Heat generation resulting from the viscous dissipation was used to predict the temperatures in the plasticized material beneath the pin. Awang et al. [6] developed a thermo-mechanical model for plunge FSSW to predict the deformations, stresses and temperature distribution immediately after the operation was completed. A finite element approach using an explicit formulation in the commercial software ABAQUS/Explicit was employed which allowed obtaining the coupled thermo-elasto-plastic response. Buck et al. [7] developed a simple Couette flow fluid mechanics based model assuming the material as a Bingham plastic and tried to predict the velocity and temperature distribution for a range of non-dimensionalized heat flux values. Kakarla et al. [8] presented an isothermal model of the first part of the plunge phase of the modified refill FSSW process using the commercial finite element software ABAQUS/Explicit [9]. Also, Itapu and Muci-Küchler [10] developed a three dimensional isothermal solid-mechanics based model of the complete plunge phase of modified refill FSSW process and tried to predict the material flow by incorporating virtual tracers in the simulation.

Another numerical approach reported in the literature for FSW that could potentially be extended to FSSW is using Smooth Particle Hydrodynamics (SPH) which is a Lagrangian particle method that can simulate large material deformations. Tartakovsky et al. [11] simulated FSW using SPH technique and were successful in studying the effects of process parameters on the temperature field and the associated mixing mechanisms.

FSSW involves a coupled thermo-mechanical phenomena that requires sophisticated models to accurately represent the actual process. For example, in the case of plunge FSSW, contact between the two plates needs to be defined in addition to the interaction between the tool and work piece. Similarly, the refill method involves interactions between different tool components and the plates to be joined. It should be pointed out that interaction between the two plates is not needed in modeling the refill method if the clamp exerts enough force over the plates to keep them in very close contact during the process. Therefore, due to its complex nature, various idealizations were made in the numerical models of FSSW reported in the literature. In this paper, an initial attempt was made to develop a fully coupled thermo mechanical model for the plunge phase of the modified refill FSSW.

Description of the FEM Model

The initial plunge phase of the modified refill FSSW process is simulated by using the commercial finite element software ABAQUS/Explicit [9]. The shoulder is not included in the

371

simulation as it does not come in contact with the plate during the plunge phase. Therefore, as shown in Fig. 2, only the pin, plate and clamp are included in the model. The dimensions of the pin and clamp are chosen based on the tool currently being used at the AMP laboratory. The two plates to be spot-welded are modeled as a single deformable solid plate. Since in the refill method only the plate material very close to the pin experiences large plastic deformations, if dimensions of the plate included in the model are several times the diameter of pin it is expected that the numerical results should represent the physics of the process with reasonable accuracy. For the simulation, the pin and clamp are treated as analytical rigid bodies as they do not undergo significant deformations during the process.

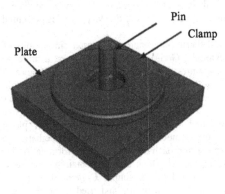

Figure 2. Parts included in the FEM model

The plate material chosen for the simulations is Al 7075-T6. The mechanical behavior of the material is represented using an elastic-perfectly plastic constitutive relation. The material properties included in the simulation comprises of both mechanical properties (i.e. Young's modulus, Poisson's ratio, yield strength and coefficient of linear expansion) and thermal properties (thermal conductivity, specific heat and density) which vary with temperature. The temperature-dependent material properties included in the model correspond to the range from room temperature to 700 K and were taken from graphs provided in MIL-HDBK-5H [12]. The temperature-dependent values provided in that reference were available only up to 550 K for some of the properties such as Young's modulus, thermal conductivity, etc. In those cases, data points were extracted from the graphs and curve fitting methods were used to obtain an estimate for these material properties up to 700 K. Graphs showing the variation with temperature of various material properties are presented in Figs. 3, 4 and 5. The properties are given at equal temperature intervals in order to facilitate the software to quickly interpolate the input data during the analysis, thereby reducing the computational time. The density and Poisson's ratio are assumed to be constant in the range of temperatures under consideration.

The interaction between the parts in the model was defined using the surface to surface contact algorithm provided by ABAQUS/Explicit [9] that employs the modified Coulomb friction law. The coefficient of friction between the contacting surfaces of the pin and plate was taken as 0.75 and a maximum shear stress value was defined to control the stick/slip behavior of the material around the pin. The important effect of the later on predicting the temperature field in the plate material is discussed in the results section. As the material in the vicinity of the pin undergoes large deformations, the elements around the pin will be excessively distorted. An adaptive meshing technique available in ABAQUS/Explicit [9] was used to minimize the distortion of the elements during the analysis. Also, to increase the computational efficiency, the mass-scaling option was employed. Mass-scaling artificially increases the density of the material to increase the stable time increment used by the solver. However, the value chosen for the mass-scaling should be appropriate so that the accuracy of the results is not affected.

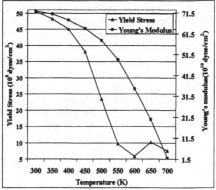

Figure 3. Plots showing the variation of Young's modulus and yield stress with temperature

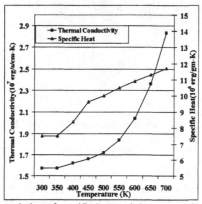

Figure 4. Plots showing variation of specific heat and thermal conductivity with temperature

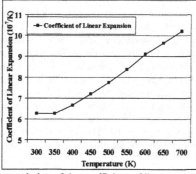

Figure 5. Plots showing variation of the coefficient of linear expansion with temperature

The loads and boundary conditions are applied to the pin and clamp at specified nodes known as reference nodes. The clamp is constrained in all the directions. The motion of the pin is constrained in all the directions except for the translation and rotation about the vertical axis. The process parameters are chosen as those used to perform the spot welds at the AMP laboratory. The plunge speed of the pin is taken as 0.085 cm/sec and the angular velocity is considered as 1600 rpm. The plunge depth of the spot weld is taken as 0.38 cm. The initial temperature of the pin and clamp correspond to room temperature. As can be seen in Fig. 6, the bottom surface of the plate is constrained in the vertical direction and the lateral surfaces are constrained in the direction of their normals. As an initial condition, the plate is considered to be at room temperature. A convective heat transfer boundary condition is applied to the top surface of the plate. The required convective heat transfer coefficient needs to be determined based on experimental results. However, for the model, the value for still air is considered in the simulations. A total of 173,000 hexahedral linear elements were used for the discretization of the plate.

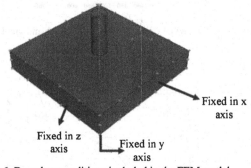

Figure 6. Boundary conditions included in the FEM model

Numerical Results

At the present time, work is being done to solve problems that have been encountered during the first simulations that have been performed and to obtain results for the complete plunge phase of the FSSW process. In what follows, only results corresponding to the initial part of the plunge phase are presented.

For comparison purposes, first the results corresponding to a model with temperature dependent material properties are compared against the ones provided by a model with constant material properties corresponding to room temperature. All the other parameters are kept same for both models including the maximum shear stress for the stick/slip condition which was taken as half the yield stress at room temperature (2.1×10^8 dyne/cm^2). Figure 7 presents a comparison of the von Mises stress contours obtained with the model with temperature dependent material properties and the model with material properties defined at room temperature at a plunge depth of 0.015 inches. Figures 8 and 9 present a similar comparison for the equivalent plastic strains and the nodal temperatures, respectively. From those figures, it can be observed that the results obtained in both simulations are very similar. This was due to the fact that a low temperature increase was predicted by both models for the plunge depth under consideration.

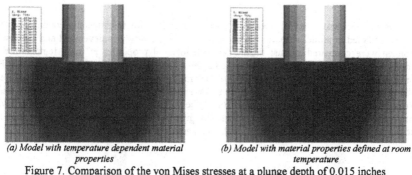

(a) Model with temperature dependent material properties   *(b) Model with material properties defined at room temperature*

Figure 7. Comparison of the von Mises stresses at a plunge depth of 0.015 inches

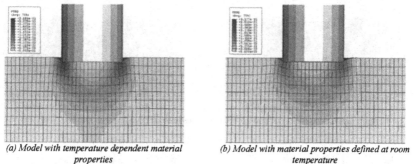

(a) Model with temperature dependent material properties   *(b) Model with material properties defined at room temperature*

Figure 8. Comparison of the equivalent plastic strains at a plunge depth of 0.015 inches

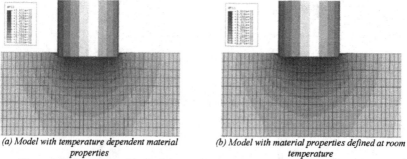

(a) Model with temperature dependent material properties   *(b) Model with material properties defined at room temperature*

Figure 9. Comparison of the nodal temperatures at a plunge depth of 0.015 inches

From Figure 9 it can be seen that the maximum temperature increase predicted for a plunge depth of 0.015 inches was about 60 K. Obviously, this does not correspond to the expected value based on the experimental evidence. However, it provides important information regarding changes needed in the modeling approach since the incorrect results are most likely due to the value of the maximum shear stress used to control the stick/slip behavior of the material around the pin. The von Mises stress distribution presented in Figure 7 shows that the material in the vicinity of pin reached the yield stress which is considerably higher than the considered maximum shear stress value for the stick/slip condition. Thus, the plate material around the pin predominantly slipped causing the heat due to plastic deformation to be very small.

In order to check the effect of maximum shear stress value for the stick/slip condition on the results predicted, a model with very high value for that quantity was developed and the results obtained were compared against the model with material properties at room temperature. It should be pointed out that all the other parameters are the same for both the models. For convenience, in what follows the model with material properties defined at room temperature is referred to as the model with low shear stress value and the other as the model with high shear stress value. Figure 10 presents a comparison of the von Mises stress distribution obtained with both models at a plunge depth of 0.015 inches. Figures 11 and 12 present a similar comparison for the equivalent plastic strains and the nodal temperatures, respectively.

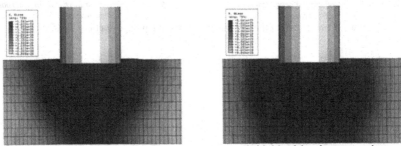

(a) Model with high shear stress value   (b) Model with low shear stress value
Figure 10. Comparison of the von Mises stresses at a plunge depth of 0.015 inches

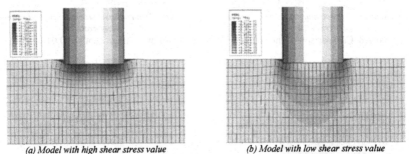

(a) Model with high shear stress value   (b) Model with low shear stress value
Figure 11. Comparison of the equivalent plastic strains at a plunge depth of 0.015 inches

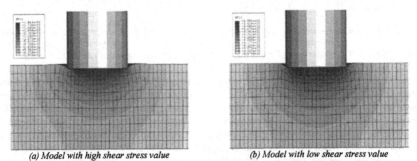

(a) Model with high shear stress value   (b) Model with low shear stress value
Figure 12. Comparison of the nodal temperatures at a plunge depth of 0.015 inches

Regarding Fig. 12, the temperature increase in the vicinity of the pin predicted by the model with high shear stress value was about 700 K, which was far more than the one predicted by model with low shear stress value for the same plunge depth. Also, the equivalent plastic strains predicted by the high shear stress model are greater than the ones corresponding to the low shear stress model. Obviously, the temperature results predicted by the model with high shear stress are too high since they are above the melting point of the aluminum alloy considered. This is most likely due to the continuous sticking of the material to the pin resulting in large plastic deformations, which is evident from the comparison of equivalent plastic strains in Figure 9. Therefore, for future simulations, it is necessary to determine an adequate value for the maximum shear stress that controls the stick/slip condition so that the temperatures predicted by the simulation are close to the expected values.

Conclusions

An initial attempt to simulate the first part of the plunge phase of the modified refill FSSW process was presented. The preliminary results obtained showed that the modeling the stick/slip behavior of the plate material in the vicinity of the pin has a strong influence in the temperature distribution predicted at the spot weld. When a limiting shear stress value is used to control the stick/slip condition, a low value for that quantity results in a very low temperature increase in the simulations whereas a very high one results in predicted temperatures that are above the melting point of the aluminum alloy under consideration. Thus, an adequate value needs to be found by running the simulation with different limiting shear stresses until the temperature increase is similar to the one corresponding to experimental data.

The results of the simulations also suggest that the primary mechanism responsible for the temperature increase during the process is the plastic deformation of the plate material that takes place when it sticks to the pin and is stirred. The contributions corresponding to the heat generated by friction, which occurs when the plate material slips around the pin, and the one due to plastic deformation cause by the plunge alone (i.e., without the material sticking to the tool) are not enough to account for the temperature increase at the spot weld.

Acknowledgements

This research was sponsored by the Army Research Laboratory and was accomplished under Cooperative Agreement Number DAAD19-02-2-0011. The views and conclusions contained in this document are those of the authors and should not be interpreted as representing the official policies, either expressed or implied, of the Army Research Laboratory or the U.S. Government. The U.S. Government is authorized to reproduce and distribute reprints for government purposes notwithstanding any copyright notation hereon.

References

1. Iwashita T. et al. "Method and Apparatus for Joining", US Patent 6601751 B2, August 5, 2003.
2. Pan, T., Joaquin, A., Wilkosz, D. E., Reatherford, L.,Nicholson, J. M., Feng, Z., and Santella, M. L., 2004,"Spot Friction Welding for Sheet Aluminum Joining," Proc. 5th International Symposium on Friction Stir Welding, Metz, France.
3. "Mazda Develops World's First Aluminum Joining Technology Using Friction Heat," Mazda News Release, February 27, 2003:
   http://www.mazda.com/publicity/release/2005/200506/050602.html
4. Schilling, C. and dos Santos, J. "Method and Device for Joining at Least Two Adjoining Work Pieces by Friction Welding." US Patent Application 2002/0179 682.

377

5. A. Gerlich, P. Su, G.J. Bendzsak, and T.H. North, "Numerical Modeling of FSW Spot Welding: Preliminary Results" in Friction Stir Welding and Processing III (Eds. K.V. Jata, M.W. Mahoney, R.S. Mishra, and T.J. Lienert), The Minerals, Metals and Materials Society, 2005.

6. M. Awang, V.H. Mucino, Z. Feng, and S.A. David, "Thermo-Mechanical Modeling of Friction Stir Spot Welding Process; Use of an Explicit Adaptive Meshing Scheme." SAE Paper 2005-01-1251.

7. G.A. Buck and M. Langerman, "Non-Dimensional Characterization of the Friction Stir/Spot Welding Process Using a Simple Couette Flow Model Part I: Constant Property Bingham Plastic Solution." AIP Conference Proceedings, pp. 1283-1288, Issue 1, Volume 712, June 10, 2004.

8. S.S.T. Kakarla, K.H. Muci-Küchler, W.J. Arbegast, and C.D. Allen, "Three-Dimensional Finite Element Model of the Friction Stir Spot Welding Process." Friction Stir Welding and Processing III (Eds. K.V. Jata, M.W. Mahoney, R.S. Mishra, and T.J. Lienert), pp. 213-220. The Minerals, Metals and Materials Society, 2005.

9. ABAQUS Version 6.6 User Manuals. ABAQUS Inc., 2006.

10. S.K. Itapu and K.H. Muci-Küchler, "Visualization of Material Flow in the Refill Friction Stir Spot Welding Process" SAE Paper 2006-01-1206.

11. A. Tartakovsky, G. Grant, X. Sun, and M. Khaleeel, "Modeling of Friction Stir Welding process using Smooth Particle Hydrodynamics." SAE Paper 2006-01-1394.

12. MIL-HDBK-5H: Military Handbook, Metallic Materials and Elements for Aerospace Vehicle Structures, December 1998.

EFFECT OF HEAT TREATMENT ON MICROSTRUCTURE, MECHANICAL PROPERTIES AND COMPOSITION VARIATION ACROSS THE INTERFACE FOR THE FSW 6061 Al ALLOY WELDMENTS

Karanam Bhanumurthy, Nitin Kumbhar and Beant Prakash Sharma

Materials Science Division, Bhabha Atomic Research Centre, Mumbai - 400 085, India

Keywords: FSW, 6061 Al alloy, Electron probe microanalysis (EPMA),
Post weld heat treatment (PWHT)

Abstract

Friction stir welding (FSW) is essentially a solid-state joining process and this technique is extensively used for joining 6061 Al alloys. The objective of the present work is to study the microstructural changes, defect formation and mechanical properties during the FSW and also after post weld heat treatment (PWHT). FSW experiments were carried out on 5 mm thick sheets with variable tool rotation speed and traverse speed to optimize the process parameters.

Mechanical properties of the as welded specimens were superior to the parent material in the O-condition. Concentration profiles of Al, Mg and Si across the weld zone showed nearly homogeneous distribution and there was oxygen pickup at the top surface of the welded zone. Substantial recovery of microhardness and mechanical properties were noticed after solutionizing at 530° C and aging at 160° C for different annealing periods. For most of the PWHT specimens, fracture occurred on the retreating side and the fractured surface showed dimple structure.

Introduction

Friction stir welding (FSW) is an emerging technology for joining Al alloys. This process has distinct advantages over conventional fusion welding and the weldments made by this technique show superior mechanical properties. The simplicity of the process, coupled with fair degree of reproducibility has lead to its application in advanced manufacturing processes [1-3]. Most of the earlier experimental work on Al base alloys was devoted to the designing of tool, optimization of process parameters and the recent work is aimed at correlation of experimentally observed microstructure with the mechanical properties and simulation studies [4-7]. Microstructural characterization of FSW of precipitate-hardened 7075-T651 Al alloy was studied by Rhodes et al. [8], 6063 Al alloy by Sato et al. [9,10] and these studies indicated that post weld heat treatment (PWHT) substantially improved the mechanical properties of the weldments.

PWHT studies on FSW 6061 Al alloy have attracted considerable interest recently [11-12]. Krishnan [11] has studied the effect of PWHT on the properties of the FSW joints and suggested that the ageing treatments have produced higher hardness in the nugget zone, which resulted in the failure of these joints during bend tests. Recently, Lee et al. [12] have indicated that the hardness values depend on the distribution of precipitates and dislocation density in the weld zone and not on the grain size. Most of the research activity on 6061-T6 Al alloy is related to the evaluation of the microhardness, mechanical properties and their changes during PWHT [11-15]. Limited work is reported on the variation of chemical composition of the alloying elements across the weld zone and composition changes after PWHT [16].

The objective of present studies is to optimize the process parameters for FSW of 6061-O Al in a narrow window based on earlier studies [11-14] and to examine the influence of PWHT on the microstructure and mechanical properties. In addition, an attempt has been made to analyze the nature and variation of alloying elements across the weld by electron probe microanalyser (EPMA).

Experimental Work

<u>Material and Welding Process</u>

The material used in the present study was 6061-O aluminum alloy plate of dimensions 30 cm x 5 cm x 0.5 cm. The chemical composition and mechanical properties of this alloy are listed in Table I (a) and (b). FSW trials were carried out using a vertical milling machine. Two plates of aluminum alloy were abutted and welded in a narrow window of rotation feed {Tool rotation speed (rpm)/ Tool traverse speed (mm/min)} ranging from 7.1 to 22.2 rotations/mm (rotation speed: 710-1400 rpm and traverse speed: 63-100 mm/min). The dimensions of the tool are given in Table II. The tool tilt was kept at 2° in the forward direction for all welding runs. In the initial stages the quality of the joints were assessed on the basis of die-penetration and X-ray radiography tests. These non-destructive tests helped to minimize the number of runs for optimizing the process parameters in the selected narrow window. All the welded specimens, which passed the initial die-penetration tests and X-ray radiography tests, were taken for detailed metallography and mechanical testing. In addition, specimens were given PWHT schedule consisting of solutionizing at 530° C for 0.5 hrs and quenching in water. Subsequently these solution treated samples were aged at 160° C for 1, 4, 8, 12 and 18 hrs. The following designations are used to identify the specimens. The specimen welded at a rotation feed 17.5 rot/mm is designated as W-17.5 and the specimen after solutionizing at 530° C for 0.5 hrs and aging at 160° C for 18 hrs is designated as PWHT-17.5-18.

Table I.

(a) CHEMICAL COMPOSITION (wt%) of 6061 Al ALLOY USED IN THE PRESENT STUDIES.

Si	Fe	Cu	Mg	Mn	Ti	Zn	Ca	Al
0.6	0.33	0.18	0.92	0.06	0.02	0.03	0.2	Balance

(b) MECHANICAL PROPERTIES of 6061 Al in O and T6 CONDITION.

Condition	UTS (MPa)	0.2 % Y_s (MPa)	Elongation (%)	Hardness (VHN)
O	125	55	24	38
T6	310	275	12	115

Table II. DIMENSIONS OF TOOLS USED FOR FSW.

Shoulder Φ (mm)	Pin Φ (mm)	Pin length (mm)
15	5	4.8

Metallographic Preparation

Cross sections of the welded specimens were taken using a low speed diamond saw. These specimens were prepared for EPMA using standard metallographic technique. The polished surfaces were etched in a solution consisting of 10 ml HF + 15 ml HNO_3 + 25 ml HCl +50 ml H_2O to reveal the microstructure. The microstructure was also observed by optical and scanning electron microscope (SEM).

Electron Probe Microanalysis

A Cameca SX100 electron probe microanalyser equipped with three wavelength dispersive spectrometers was used to analyze Al (K_α), Si (K_α), Mg (K_α) and O (K_α) across the welded zone. A well focused stabilized electron beam of energy 20 KeV at 20 nA was used for analysis. Quantitative analysis of the samples was done on point-to-point scale at an interval of 25 to 100 μm to cover all the regions of the welded zone. Quantification was restricted only to the major elements Al, Si and Mg. Oxygen pick up was studied only on the top surface of the weld zone. Pure elemental standards Al, Si and Mg were used for calibration and standard PAP correction scheme [17] was used to arrive at the concentration from the intensity values.

Microhardness and Mechanical Properties

Vickers hardness measurements were made on the cross section perpendicular to the welding directions, using indenter with a load of 50 g for a 10 sec dwell time.
Standard tensile specimens having gauge length of 25 mm and gauge width of 6 mm were machined from the welded specimen plates by keeping tensile axis perpendicular to the welding direction. The gauge length of the tensile specimen extended form the stirred zone in to the parent material and a minimum of three tests were done to arrive at the consistent value. Mechanical properties were evaluated by screw driven machine at a strain rate of 10^{-3} sec^{-1}. The fracture surfaces were examined by SEM.

Results and Discussions

Digital low magnification images for the W-17.5 and PWHT-17.5-18 specimens are shown in Fig. 1(a) and (b) respectively. Several back scattered electron (BSE) images and optical micrographs were taken along the horizontal line B-B' and vertical line A-A' as shown in Fig. 1 and few of these micrographs are shown in Fig. 1(c) to (f). Fig 1(c) and (d) are back scattered electron (BSE) images showing the microstructure at the top and bottom in the nugget zone along the line A-A' respectively for the W-17.5 specimen. Fig. 1(e) is the BSE image in the nugget region at the position A and Fig. 1(f) is the optical micrograph of the parent material for the PWHT-17.5-18 specimen. It is clear from these micrographs that the welded region is free from defects. Based on the detailed metallographic studies, three regions could be identified as i) nugget zone (NZ) ii) thermo-mechanical and heat affected zone (TMAZ) and iii) parent material and these regions are marked in Fig. 1(a) and 1(b). These regions particularly NZ, TMAZ and the interface between TMAZ and parent material are practically indistinguishable in Fig. 1(b). The stirred region is more of inverted rising wave on the advancing side and of inverted diminishing wave on the retreating side. The stirred region narrows down from top to bottom. The shape and the size of the stirred zone are well discussed for 6063 Al and 7075 Al alloys [5,10]. In the case of 7075 Al alloys, the shape of the stirred zone was more of elliptical in nature and this shape strongly depended on the tool geometry and the thermal properties of base material that is being welded. In the present experiments the shape of the stirred zone remained nearly the same for all the process parameters.

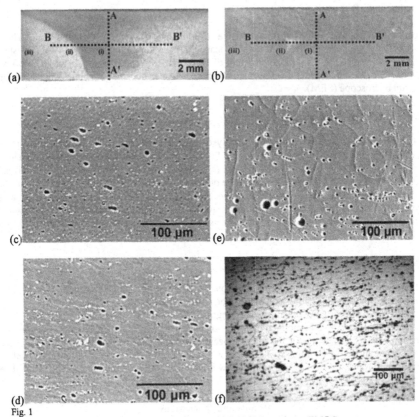

Fig. 1
(a) Digital images taken at low magnification for as welded Al 6061 specimen (W-17.5).
(b) Digital images taken at low magnification for post weld heat-treated Al 6061 specimen (PWHT-17.5-18).
(c) BSE image in the nugget corresponding to region (A) in Fig. 1(a) for W-17.5 specimen.
(d) BSE image in the nugget corresponding to region (A') in Fig. 1(a) for W-17.5 specimen.
(e) BSE image in the nugget zone corresponding to region (A) in Fig. 1(b) for PWHT-17.5-18 specimen.
(f) Optical micrograph in the base material of PWHT-17.5-18 specimen corresponding to region (iii) in Fig. 1(b).

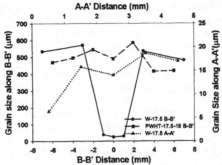

Fig. 2 Plot of grain size measurements taken along the A-A' and B-B' lines marked in Fig. 1(a) and (b) for W-17.5 and PWHT-17.5-18 specimens.

382

The grain size variation across B-B' (Fig. 1(a), (b)) for both W-17.5 and PWHT-17.5-18 specimens is shown in Fig. 2. For W-17.5 specimen the grain size decreases along B-B' and is around 15-20 μm in the nugget zone. Fine grain size in the nugget zone is due to the combination of plastic deformation and frictional heat occurring locally. The microstructure evolved in the stirred zone is essentially due to the process of dynamic recrystallization [18,19]. The average grain size for PWHT-17.5-18 specimen across the welded zone (B-B' in Fig. 1(b)) increases and the size of the grains in the nugget zone is about 500 μm and this value is an order of magnitude larger than the grain size in the nugget zone for as welded specimen (W-17.5). The stirred zone has experienced considerable amount of deformation during the stirring action of the tool and this must have resulted in the formation of large defects in this zone. During PWHT, rapid recrystallization occurs preferably at high defect density areas and this process of static recrystallization must be responsible for substantial increase of the grain size in the nugget zone. Similar change in the microstructure of the nugget zone after PWHT is reported for 6063 Al [9, 10] and 6061-T6-Al alloy [11,12].

The grain size variation across A-A' (Fig. 1(a)) is not significant for W-17.5 specimen. The grain sizes at positions A and A' in Fig. 1(a) is 5 μm and 20 μm respectively. The grain size distribution at top and bottom positions of the nugget zone has been studied for 7075 Al alloy [8] and also for 6063 Al alloy [20]. In the case of 7075 Al alloy the grain size obtained at the top of the nugget zone was marginally large and these results are contrary to the present observation. In the case of 6063 Al alloys, practically no difference was observed in the grain size. The temperature profile from top to bottom, thermal conductivity of material that are being joined and that of blanking plate used for holding the specimens are primarily responsible for the grain size variations in the nugget zone. In the present case high carbon steel was used as a blanking plate and its thermal conductivity is inferior to 6061 Al alloy. This may have resulted in marginally larger grains in the bottom.

Composition Variation Across the Weld Zone

Typical composition profiles taken across the region B-B' (Fig. 1(a) and 1(b)) for W-17.5 and PWHT-17.5-18 specimen are shown in Fig. 3(a) and (b) respectively. The distribution of Al, Mg and Si for both these specimens is nearly uniform. Though there is certain degree of inhomogeneity in terms of grain size in the nugget zone, there is practically a high degree of chemical homogeneity in the stirred region. The discontinuities observed in the concentration profiles indicate the formation of intermetallics. Two types of intermetallics i) aluminum rich $Al_xSi_y(Fe)$ and ii) magnesium rich $Mg_xSi_y(Al)$ could be identified from these profiles. It is observed that the size of these precipitates in the nugget zone is small compared to those found in the matrix and this reduction possibly could be due to the stirring action of the tool.

X-ray line scans of O (K_α), Mg (K_α), Si (K_α) and Al (K_α) taken from top to about 60 μm (along A-A' in Fig. 1(a) and (b)) inside the weld zone for W-17.5 and PWHT-17.5-4 specimens are shown in Fig. 3(c) and (d) respectively. Care was taken to avoid the flash regions of the stirred zone. The initial zero intensities for all the elements correspond to the cold set resin material used for mounting the sample.

There is oxygen pickup up to a depth of about 20 μm and this resulted in the change in the local chemical composition of the alloying elements Mg and Si. Recent studies on FSW of 6111 Al alloy show that the highest temperature reached during welding is close to 0.94 T_s [21] where T_s is the solidus temperature of the alloy in K. Assuming similar relation in the present studies, the temperature of 530°C could have been seen at the top of the weld, resulting in pickup of oxygen from the surroundings.

383

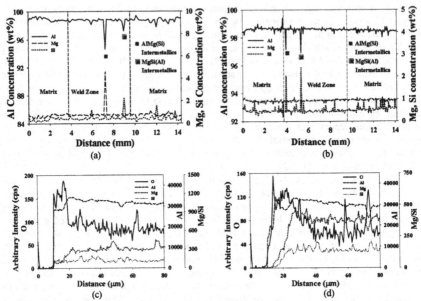

Fig. 3
(a), (b) Concentration profile across along the line B-B′ (Fig. 1(a) and (b)) in W-17.5 and PWHT-17.5-18 specimens.
(c), (d) Line scan of O (K_a), Mg (K_a), Si (K_a) and Al (K_a) taken along the line A-A′ (Fig. 1(a) and (b)) from top to about 60 μm for W-17.5 and PWHT-17.5-18 specimens.

Microhardness

Fig. 4 shows the microhardness profiles taken across the regions B-B′ as shown in Fig. 1(a) and (b) for W-17.5 and PWHT-17.5-18 specimens respectively. In addition, two horizontal lines are also plotted in Fig. 4 for the parent material in O and T6 conditions for comparison.

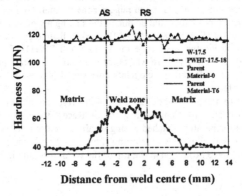

Fig. 4 Microhardness profile along the line B-B′ (Fig. 1(a) and (b)) for W-17.5 and PWHT-17.5-18 specimens.
(AS is advancing and RS is retreating sides of the weld direction)

384

The hardness value in the nugget zone is maximum for W-17.5 specimen and is around 65 VHN and this value gradually comes down on either side of the nugget zone. This reduction in the hardness values on the retreating side occurs over a larger length compared to the advancing side. The present studies showed a marginal hardening in the nugget zone compared to the parent material. This maximum hardness of 65 VHN in the nugget zone is close to the lowest value reported in the stirred zone for 6061-T6 Al alloy [12,13]. Similar behavior of hardness profiles was obtained for other rotation feeds and also for different PWHT. It may be mentioned here that the hardness values depended on the condition in which the parent material was received. Marginal increase in the hardness value in the nugget zone could be due to the on-set of formation of precipitates and also could be due to the presence of high defect density in the nugget zone.

The hardness values after PWHT reach a value of 110 VHN and there is practically no variation across the welded zone (Fig. 4). In the present investigation, TEM studies were not done. Based on earlier studies on 6061 [12] and 6063 [9] Al alloys, it may be inferred that the increase in the hardness values could be due to reprecipitation of fine precipitates during PWHT.

Mechanical Properties

Tensile properties of all the welded specimens with variation of rotation feeds (rotations/mm) are shown in Fig. 5(a). In this figure 0.2 % offset yield strength is used as the yield strength (Y_s). In the chosen narrow window of process parameters, the values of UTS, Y_s and elongation (%) do not vary significantly. In fact all the specimens failed away from the nugget zone. The UTS and Y_s values for the welded specimens are much larger compared to the parent material at O-condition (Table I(b)). The maximum elongation value for the welded specimens is around 18 % and this value is about 25 % less than the parent material.

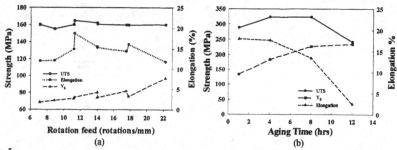

Fig. 5
(a) Plot of mechanical properties of specimens at varying rotation feeds.
(b) Plot of mechanical properties of PWHT specimens aged for different periods.

In order to study the behavior of mechanical properties on ageing, the specimens processed at 17.5 rotation feed (rotations/mm) was taken for PWHT for different duration. The mechanical properties of PWHT-17.5 specimen aged at different duration are shown in Fig. 5(b). It can be seen from Fig. 5(b) that ageing at 4 hrs shows superior mechanical properties compared to those of the parent material (Table I(b)) and these specimens also failed away from the weld zone. As discussed in section (c) superior mechanical properties after PWHT could be due to homogeneous distribution of precipitates in the weld zone [22].

Normally FSW of most of the aluminum alloys are carried out in the aged condition (T6 condition). This always requires higher tool forces and optimization of process parameters needs a larger window. In addition, during welding softening takes place in the nugget zone resulting in poor mechanical properties [13-14]. In order to restore all the mechanical properties, generally

385

PWHT in T6 condition may be required. The present studies suggest that experiments carried out under O-condition have specific advantages of FSW in a narrow window and also at lower rotation feeds. Further, PWHT may be carried out to restore most of the mechanical properties.

Fractography

Fracture of the as welded and also PWHT specimens mostly occurred on the retreating side and at the lowest hardness values. A few as-welded specimens at rotation feed of 22 rotations/mm failed at the advancing side. The fractographs taken at nearly same magnification for W-17.5 and also for PWHT-17.5 specimens aged for 4, 8 and 12 hrs are shown in Fig. 6(a) to (d) respectively. Fractograph corresponding to as welded specimen (Fig. 6(a)) shows several flat regions. In fact this specimen showed a lower hardness (Fig. 4) compared to the PWHT specimens. All PWHT specimens showed superior mechanical properties (Fig. 5(b)) and the fractured surfaces show the presence of fine dimple structures, a clear evidence of ductile mode of fracture (Fig. 6(b), (c) and (d)).

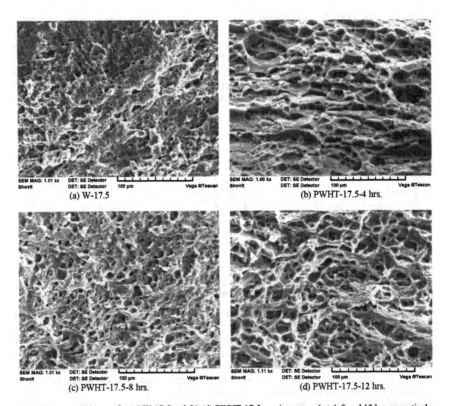

(a) W-17.5

(b) PWHT-17.5-4 hrs.

(c) PWHT-17.5-8 hrs.

(d) PWHT-17.5-12 hrs.

Fig. 6 Fractograph images for (a) W-17.5 and (b)-(d) PWHT-17.5 specimens aged at 4, 8 and 12 hrs respectively.

Conclusions

a) Microstructural inhomogeneity exists in the FSW specimens. There is large variation of grain size and microhardness across the weld zone. PWHT substantially reduces these inhomogeneities.

b) In general FSW does not result in chemical inhomogeneity in a scale of 1 μm. There is oxygen pickup resulting in the change of local chemical composition of the alloying elements at the surface. FSW on the O-condition has specific advantages in terms of optimizing the process parameters.

c) Mechanical properties substantially improve during PWHT and at an optimized heat treatment schedule, these mechanical properties are superior to those of the parent material. Fracture mostly occurred on the retreating side and the fracture surface show dimple structure.

References

1. G. Liu et al., "Microstructural aspects of the friction-stir welding of 6061-T6 aluminum," *Scripta Materialia*, 37 (1997), 355-361.

2. W. M. Thomas and E. D. Nicholas, "Friction stir welding for the transportation industries," *Materials and Design*, 18 (1997), 269-273.

3. R. S. Mishra and Z. Y. Ma, "Friction stir welding and processing," *Materials Science and Engineering R*, 50 (2005), 1-78.

4. H. Fujii et al., "Effect of tool shape on mechanical properties and microstructure of friction stir welded aluminum alloys," *Materials Science and Engineering A*, 419 (2006), 25-31.

5. M. Boz and A. Kurt, "The influence of stirrer geometry on bonding and mechanical properties in friction stir welding process, *Materials and Design*, 25 (2004), 343–347.

6. J. D. Robson et al., "Modeling precipitate evolution during Friction Stir Welding of aerospace Al alloys," *Materials Science Forum*, 519-521 (2006), 1101–1106.

7. W. B. Lee et al., "Effects of local microstructures on the mechanical properties in FSW joints of a 7075-T6 Al alloy," *Z. Metallkunde*, 96 (2005), 940-947.

8. C.G. Rhodes et al., "Effects of friction-stir welding on microstructure of 7075 aluminum," *Scripta Materialia*, 36 (1997), 69-75.

9. Y. S. Sato and H. Kokawa, " Distribution of tensile property and microstructure in friction stir weld of 6063 Aluminum," *Metallurgical and Materials Transactions*, 32A (2001), 3023-3031.

10. Y. S. Sato et al., "Microstructural evolution of 6063 Aluminum during friction stir welding," *Metallurgical and Materials Transactions*, 30A (1999), 2429-2437.

11. K. N. Krishnan, "The effect of post weld heat treatment on the properties of 6061 friction stir welded joints," *Journal of Materials Science*, 37 (2002), 473-480.

12. W. B. Lee et al., "Effect of PWHT on behaviors of precipitates and hardness distribution of 6061 Al alloy joints by friction stir welding method," *Material Science Forum*, 449-452 (2004), 601-604.

13. S. Lim et al., "Tensile behavior of friction-stir-welded Al 6061-T651," *Metallurgical and Materials Transactions*, 35A (2004), 2829-2835.

14. W. B. Lee, Y.M. Yeon and S. B. Jung, "Mechanical properties related to microstructural variation of 6061 Al alloy joints by friction stir welding," *Materials Transactions*, 45 (2004), 1700-1705.

15. H. Liu et al., "Tensile properties and fracture locations of friction-stir-welded Al 6061-T6 aluminum alloy," *Journal of Materials Science Letters*, 22 (2003), 1061-1063.

16. I. Shigematsu et al., "Joining of 5083 and 6061 aluminum alloys by friction stir welding," *Journal of Materials Science Letters*, 22 (2003), 353-356.

17. J. L. Pouchou and F. Pichoir, *Microbeam analysis* (California: San Fransico Press, 1985).

18. K. Tsuzaki, H. Xiaoxu and T. Maki, "Mechanism of dynamic continuous recrystallization during superplastic deformation in a microduplex stainless steel," *Acta Materialia*, 44 (1996), 4491-4499.

19. L.E. Murr, H. K. Shih and C. S. Niou, " Dynamic recrystallization in detonating tantalum shaped charges: A mechanism for extreme plastic deformation," *Materials Characterization*, 33 (1994), 65-74.

20. Y. S. Sato et al., "Parameters controlling microstructure and hardness during friction-stir welding of precipitation-hardenable aluminum alloy 6063," *Metallurgical and Materials Transactions*, 33A (2002), 625-635.

21. A. Gerlich, P. Su, T. H. North, "Tool penetration during friction spot welding of Al and Mg alloys," *Journal of Materials Science*, 40 (2005), 6473-6481.

22. L. E. Murr, G. Liu, J. C. McClure, "A TEM study of precipitation and related microstructures in friction-stir-welded 6061 aluminum," *Journal of Materials Science*, 33 (1998), 1243-1251.

INVESTIGATION OF LASER DEPOSITION OF HIGH TEMPERATURE REFRACTORY PIN TOOLS FOR FRICTION STIR WELDING

Bharat K. Jasthi[1], Aaron C. Costello[2], William J. Arbegast[3], Stanley M. Howard[1]

[1]Department of Materials and Metallurgical Engineering
[2]Additive Manufacturing Laboratory
[3]Advanced Materials Processing and Joining Laboratory
South Dakota School of Mines and Technology
501 E. St. Joseph St., Rapid City, SD 57701 USA

Keywords: Friction stir welding, Laser deposition, Pin tools

Abstract

High temperature pin tools used for friction stir welding (FSW) of higher temperature materials than aluminum are expensive and often fail prematurely. In this investigation an attempt was made to develop high temperature refractory pin tools using laser deposition via 3-KW Nd:YAG laser. Powder materials including WC-Co, Ni-Tung60 and CCW+ were deposited onto MP159, H13 and WC-Co shank materials. Pin tools made with WC-Co pin on WC-Co substrate exhibited thermally induced stress cracking in the substrate and failed during plunge trials conducted in 0.25-in Ti-6Al-4V plates. Initial trial depositions show good bonding characteristics of CCW+ on MP159 and H13 substrates. Metallurgical analysis of the laser depositions are analyzed and reported.

Introduction

Friction stir welding demonstrated potential for welding high melting temperature materials such as ferrous alloys [1], titanium alloys [2], oxide dispersion strengthened (ODS) alloys [3], and other iron-nickel alloys [4]. Development of pin tool for friction stir welding of these high temperature materials is a major challenge. An ideal pin tool should have high toughness, good strength, excellent wear resistance, and be chemically inert. Polycrystalline cubic boron nitride (PCBN) pin tools appear to meet all the requirements of a high temperature pin tools. However, premature cracking and the cost of the pin tool limits its use. Laser powder deposition is a possible way to produce near net shaped and functionally graded materials suitable for pin tools.

Laser powder deposition is an additive manufacturing technology that can build components with the help of a computer-aided-design (CAD) model in a layer-by-layer fashion. In a laser deposition process, a laser beam is focused on to the substrate to create a melt pool. Powder particles are then injected into the laser focal zone and melted [5]. The solidification rates are very high and controlled by the movement of the laser beam. The near-net-shape nature of this process eliminates the machining and also greatly reduces the time for manufacturing the pin tools. Various combinations of metals and cermets can be claded to obtain the optimum combination of properties. However, porosity and residual stresses in the material may cause serious problems.

Previous work by Ouyang, Mei, and Kovacevic demonstrated that WC-based functionally graded pin tools can be produced successfully using a one-step laser cladding process [6]. In this investigation, powder materials including WC-Co, Ni-Tung60 and CCW+ were deposited onto MP159, H13 and WC-Co shank materials.

Experimental

A 3-KW Nd:YAG laser cladding system was used to deposit the working end of the pin tools. Figure 1 shows a setup for direct laser deposition. Powder materials such as WC-Co (WC-85%, Co-15%), Ni-Tung60 (WC-60%, Ni-40%) and CCW+ (Cobalt based-Carpenter patented product) were used for the laser deposition. The substrate materials considered for this study were MP159, H13 and WC-Co. WC-Co was deposited onto WC-Co substrate and Ni-Tung60 was deposited onto MP 159 substrate. Similarly CCW+ was deposited onto MP 159 and H13 substrates. All the depositions were made in an inert argon atmosphere to minimize atmospheric oxygen contamination. Process parameters such as the laser power, travel speed, and powder feed rate were optimized by running initial trial depositions. Both visual inspection and metallographic evaluation was performed on these depositions to determine the quality of the clad and also to develop a process parameter window. All the depositions were mounted and polished with fine diamond paste. Electrolytic etching was performed on all the specimens to reveal the microstructure. A laser deposited WC-Co pin tool was plunge tested on Ti-6Al-4V titanium plates at 300 rpm to evaluate the effectiveness and performance of the deposition.

Figure 1. Laser deposition setup.

Results

Figures 2-7 shows the micrographs of WC-Co, CCW+, and NiTung60 depositions on selective substrates. Figure 8 shows the cracks and delamination that were formed during the trial deposition of WC-Co powder on WC-Co substrate. Figure 9a shows the complete pin tool with the selected processing parameters with WC-Co on WC-Co substrate. Figure 9b shows the macrograph of the broken WC-Co pin tool that failed during the plunge testing in Ti-6Al-4V plate.

390

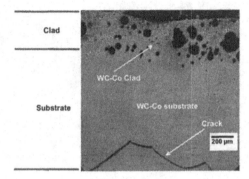

Figure 2. Micrograph of laser deposited WC-Co clad on WC-Co substrate.

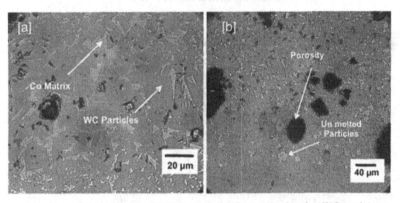

Figure 3. Micrograph of WC-Co clad; (a) Co-matrix showing WC particles, (b) unmelted particles and porosity in the clad.

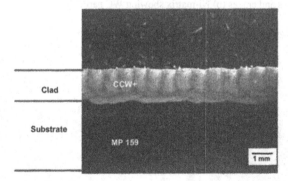

Figure 4. Macrograph of laser deposited CCW+ clad on MP159 substrate.

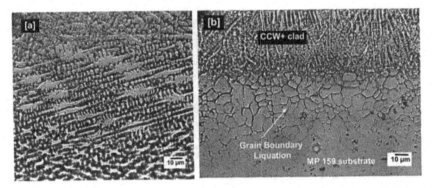

Figure 5. Micrographs of CCW+ clad on MP159 substrate; (a) dendritic structure of clad (b) grain boundary liquation in MP159 substrate.

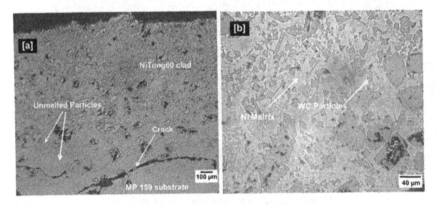

Figure 6. Micrograph of NiTung60 clad on MP159 substrate; (a) interface showing unmelted particles and cracks (b) Ni-matrix showing WC particles.

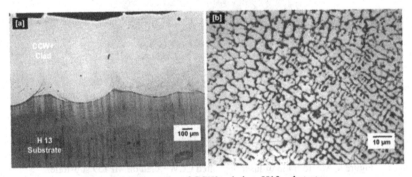

Figure 7. Micrographs of CCW+ clad on H13 substrate; (a) interface between CCW+ clad and H13 substrate, (b) grain boundaries decorated with metal carbides in clad.

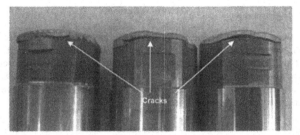

Figure 8. Schematic view of the WC-Co clad on WC-Co substrates showing cracks in the substrates.

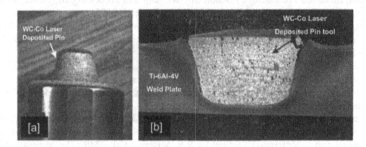

Figure 9. WC-Co laser deposited pin tool;
(a) before plunge test, (b) cross-sectional view of broken pin tool in Ti-6Al-4V plates.

Discussion

Almost all of the WC-Co depositions cracked during deposition. The reason for the cracking is likely thermal stresses involved during the deposition. Metallurgical examination of the broken pin tool revealed the presence of many pre-existing cracks in the WC-Co shank. It is interesting to note that these cracks appear to initiate some distance from the deposition-substrate interface within the substrate. This suggests high subsurface thermally induced tensile stress. A significant quantity of layered porosity and partially-melted powder particles was observed within the clad areas indicating insufficient energy input or inadequate hatching during deposition. Pre-heating the shank before deposition may reduce thermal gradients and thermal cracking.

Deposition of NiTung60 on MP159 substrate produced small cracks in the clad. Metallurgical examination revealed the presence of some cracks and unmelted particles in the clad. Deposition of CCW+ on MP159 substrate showed good bonding characteristics. However, metallurgical examination revealed the grain boundary liquation in the substrate material. The effect of liquation on strength of the pin tools is unknown. Deposition of CCW+ on H13 steel showed good bonding characteristics. Metallurgical examination revealed no crack formation.

Conclusions

All of the WC-Co depositions on WC-Co substrates cracked during laser deposition. Initial trial depositions have shown good bonding characteristics of CCW+ on MP159 and H13 substrates. CCW+ deposition on MP 159 showed grain boundary liquation in the substrate. From the preliminary investigation, CCW+ deposition on H13 substrate appears promising for friction stir welding of high temperature materials.

Acknowledgments

This research was sponsored by the Army Research Laboratory and was accomplished under Cooperative Agreement Number DAAD19-02-2-0011. The views and conclusions contained in this document are those of the authors and should not be interpreted as representing the official policies, either expressed or implied, of the Army Research Laboratory or the U.S. Government. The U.S. Government is authorized to reproduce and distribute reprints for government purposes notwithstanding any copyright notation hereon.

References

1. B. M. Tweedy, W. J. Arbegast, and C. D. Allen, " Friction Stir Welding of Ferrous Alloys Using Induction Preheating", Friction Stir Welding and Processing-III, Edited by K.V. Jata, et al, TMS (The Minerals, Metals and Materials Society), 2005.
2. R. L. Goetz and K. V. Jata, " Modeling Friction Stir Welding of Titanium and Aluminum Alloys", Friction Stir Welding and Processing, K. V. jata etal, ed., ISBN 0-87339-502-6, pp. 35, 2001.
3. B. K. Jasthi , S. M. Howard, W. J. Arbegast, G. J. Grant, S. Koduri, and D. R. Herling, "Friction Stir Welding of MA 957 Oxide Dispersion Strengthened Ferritic Steel", Friction Stir Welding and Processing-III, Edited by K.V. Jata, et al, TMS (The Minerals, Metals and Materials Society), pp. 75, 2005.
4. Bharat K. Jasthi, Stanley M. Howard, Casey D. Allen, William J. Arbegast, "Effects of Friction Stir Welding on the Coefficient of Thermal Expansion of Invar 36", Friction Stir Welding and Processing-IV, Edited by R.S. Mishra, et al, TMS (The Minerals, Metals and Materials Society), 2007.
5. Kai Zhang, Weijun Liua, Xiaofeng Shang, "Research on the processing experiments of laser metal deposition shaping", Optics & Laser Technology -39, 549–557, 2007.
6. J. H. Ouyang, H .Mei, and R. Kovacevic, "Rapid prototyping and characterization of a WC-(NiSiB alloy) ceramet/tool steel functionally graded material (FGM) synthesized by laser cladding", Rapid Prototyping of Materials; Columbus, OH; USA; 7-10 Oct. 2002. pp. 77-93. 2002.

LIQUID METAL EMBRITTLEMENT OF MP-159 PIN TOOLS

Chuck Standen[1], Bharat K. Jasthi[2], Dana J. Medlin[2], William J. Arbegast[3]

[1] Spearfish High School, Spearfish, SD 57783, USA
[2]Department of Materials and Metallurgical Engineering
[3]Advanced Materials Processing and Joining Laboratory
South Dakota School of Mines and Technology
501 E. St. Joseph St., Rapid City, SD 57701 USA

Keywords: Friction Stir Welding, Liquid Metal Embrittlement, MP-159

Abstract

During a recent friction stir weld (FSW) development program with aluminum-based alloys, pin tools made from MP-159 experienced premature failures. Subsequent analysis of the pin tool fracture surfaces revealed an intergranular fracture surface morphology indicating that the pin tools become embrittled during service. It was initially theorized that the MP-159 alloy experienced liquid metal embrittlement (LME) from exposure to a low temperature eutectic aluminum-based compound during FSW. In this study stressed and unstressed samples of MP-159 were placed in a molten aluminum alloy (5083) to determine the susceptibility of MP-159 to liquid metal embrittlement. Traditional metallography, SEM and EDX were used to evaluate the samples after exposure to molten aluminum for various lengths of time.

Introduction

Liquid metal embrittlement (LME) is the phenomenon where the fracture stress of a solid metal is reduced due to contact with a liquid metal [1]. According to Stoloff [2], there are four types of liquid metal embrittlement. Type I is the instant failure of a solid metal, occurs when in contact with certain liquid metal components under an applied tensile stress or residual stress. This is the most common type of LME. Type II is the delayed failure, occurs when the liquid metal is in contact with the metallic component and the liquid slowly penetrates in to the grain boundaries of the metal causing decohesion of the grain boundaries. Type III is the disintegration of the solid metal, occurs when in contact with the specific liquid metal due to mass transfer (atomic diffusion) of the metallic component. An applied residual stress is not required for Type III LME. And finally, Type IV is the embrittlement of the solid metal when in contact with the liquid metal caused by high temperature corrosion reactions.

MP-159 is a cobalt-based alloy that exhibits excellent strength, high wear resistance, and high temperature toughness. Recent use of this pin tool resulted in premature failures during FSW of Al-5083 alloy. Al-5083 alloy is rich in magnesium (~ 4.9 wt %), which makes this alloy vulnerable to low temperature eutectic phases during friction stir welding. Furthermore, the binary phase diagram of aluminum and magnesium suggests the formation of a liquid phase at a temperature of about 450°C with magnesium concentrations of 35 wt% [3]. Previous research has shown that temperatures reach up to 550°C during FSW of aluminum-5083 alloys [4]. In this paper an attempt has been made to determine the susceptibility of MP-159 to liquid metal embrittlement in liquid Al-5083.

395

Experimental

The investigation of the MP-159 failure was accomplished in two steps. Step one involves testing specimens of MP-159 machined from 5.97-mm diameter bar stock in the unstressed condition and step two is the testing of stressed specimens in molten Al-5083. Specimens of 4.3 mm long and 5.97 mm diameter of the MP-159 stock were used in the unstressed test phase of the project. Al-5083 alloy was heated in a Lindberg electric furnace and melted in an alumina crucible. The temperature of the molten aluminum was maintained at 750 ±2°C during the entire experiment. The experimental plan was to place the MP-159 unstressed specimens in molten aluminum for 1, 8, and 24-hours, and then removed them for analysis. However, due to oxidation of the molten aluminum, one specimen was removed after one hour and the remaining two specimens were removed after eight hours. The specimens were prepared for metallurgical analysis by cutting them along the vertical axis and then mounting, polishing, and etching them with standard metallographic techniques.

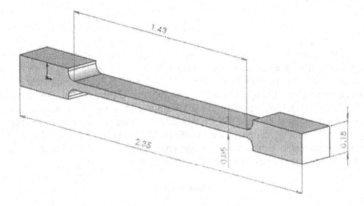

Figure 1. Schematic view of MP-159 sample for the stress test (dimensions in inches)

To test the MP-159 material in a stressed condition, specimens were machined to fit in a tool which held the specimen at each end and used a screw to put the specimen in tension on one side of the sample. The schematic view of the MP-159 specimen, which is used for stressed test, is shown in Figure 1. The machined MP-159 sample was subjected to tension by advancing the screw in the tension device, as shown in Figure 2. The amount of stress developed in the specimen was determined by measuring maximum deflection in the specimen with a micrometer and calculating the stress using standard bending moment equations. The entire assembly was immersed in the molten Al-5083 alloy that is maintained at a temperature of 750°C. The specimens were immersed for three hours and 15 minutes. The test conditions for the stress tests were summarized in Table I. After removing the specimens from the crucible, they were allowed to cool to room temperature and metallography was preformed on the specimens.

SEM-EDX analysis was performed to identify the chemistries of the phases formed at the interface between the MP-159 and liquid Al-5083. Area fraction and grain size measurements were calculated using an image analyzer to determine the quantity and distribution of the precipitates formed at/near the interface.

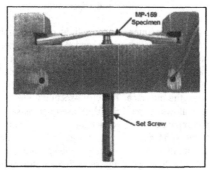

Figure 2. Tension tool for stressing MP-159 specimen.

Table-I. Load Conditions for MP-159 stressed tests

	Test 1	Test 2
Maximum Deflection	3.183 mm	3.023 mm
Calculated Stress	1,971.9 MPa	1,599.9 MPa
% of Yield	99.5%	80.0%

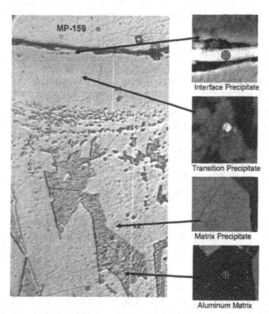

Figure 3. Various precipitates formed at the interface between the MP-159 and aluminum.

Results and Discussion

Examination of the unstressed specimens under a light metallography showed that previously squared corners of the MP-159 samples were rounded, the cross-section of the specimens had decreased, and precipitates had formed in the solidified aluminum matrix surrounding the test specimens. The precipitates that formed at the interface are termed as interface precipitates, transition precipitates and matrix precipitates. Figure 3 shows the various precipitates formed and their location relative to the interface. The cross-sectional view of the unstressed specimens are shown in Figure 4, and shows a significant decrease in cross-sectional area for the specimen soaked for 8 hours when compared with specimen soaked for 1 hour. Subsequent examination of the test specimens with the SEM detailed the changes taking place in the MP-159 specimen through the solidified aluminum matrix surrounding the specimen. A primary observation from the SEM was the compositional change of the solidified aluminum grains surrounding the MP-159 specimen. The SEM-EDX analyses of the specimens are shown in Table-II. The data show that the compositional chemistries in the precipitates change with distance from the MP-159 interface. The amount of cobalt and nickel present in the precipitates decrease with distance from the specimen while the amount of chromium in the precipitates increases.

The size and shape of the precipitates change with time of exposure to the molten aluminum. An image analysis evaluation of the one hour and eight hour test specimens was performed to measure the average precipitate size formed near the MP-159 interface, as shown in Table-III. These data show that the average precipitate size decreased and the number of precipitates increased when comparing the one hour and eight hour specimens. The eight hour specimen formed two precipitate transition zones described as the "inner area" and the "outer area" with each having distinctively different microstructural features, as shown in the SEM backscattered images of Figure 5. These images clearly show a difference in precipitate size and distribution due to the two dissimilar soak times.

Table-II. Composition of Precipitates and Solidified Aluminum Matrix in Weight%

	Interface Precipitate		Transition Precipitate		Matrix Precipitate		Al- Matrix	
1-hr Sample	Al	54.28%	Al	63.62%	Al	70.30%	Al	95.48%
	Cr	8.21	Cr	7.70	Cr	14.45	Cr	0.09
	Co	15.71	Co	13.60	Co	1.97	Co	0.18
	Ni	10.63	Ni	6.23	Ni	0.90	Ni	0.22
	Ti	2.82	Ti	1.16	Ti	1.41	Ti	0.16
	Fe	4.30	Fe	4.09	Fe	1.15	Fe	---
	Mo	3.57	Mo	2.56	Mo	6.48	Mo	0.19
	Mn	---	Mn	0.76	Mn	2.79	Mn	0.38
8-hr Sample	Al	52.08%	Al	70.61%	Al	70.69%	Al	97.34%
	Cr	10.31	Cr	14.63	Cr	14.35	Cr	0.57
	Co	18.19	Co	1.89	Co	2.82	Co	0.26
	Ni	10.63	Ni	1.22	Ni	0.77	Ni	0.15
	Ti	1.63	Ti	1.90	Ti	1.54	Ti	0.12
	Fe	5.10	Fe	1.18	Fe	1.23	Fe	---
	Mo	2.93	Mo	5.56	Mo	5.97	Mo	0.25
	Mn	---	Mn	---	Mn	2.76	Mn	0.79

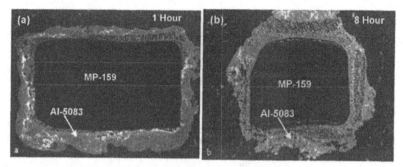

Figure 4. Macrographs of the unstressed MP-159 specimens soaked in liquid aluminum for (a) one hour (b) eight hours. There is a significant decrease in the MP-159 specimen size with increased soak time in molten aluminum.

Table-III. Precipitate Size and Distribution in Unstressed MP-159 Specimens

	1-Hr. Specimen	8 –Hr. Specimen	
		Inner Area	*Outer Area*
Area Fraction (% Precipitates)	73.2%	41.0%	36.3%
Precipitate Size (ASTM G.S. Number)	13.5	12.4	9.33

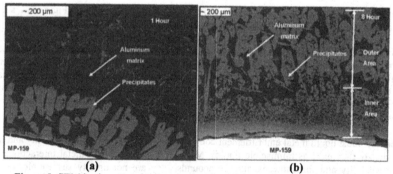

(a)   **(b)**

Figure 5. SEM backscattered micrographs of precipitates formed at interface of the (a) one hour (b) 8 hour specimens. The two images are at two different magnifications.

Table-IV. Precipitate Size and Distribution in the Stressed MP-159 Specimens

	15-Min. Sample	3-Hr. Sample
Area Fraction (% Precipitate)	36.2%	47.6%
Precipitate Size (ASTM G.S. Number)	14.3	10.9

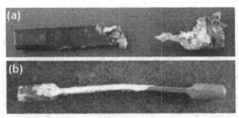

Figure 6. Stressed MP-159 after soaking in molten Al-5083 (a) three hours and
(b) 15 minutes in molten aluminum. There is significant material loss after three hours of
soaking in molten aluminum

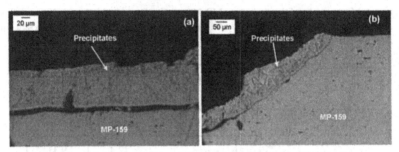

Figure 7. Micrographs of stressed MP-159 after (a) 15 minutes and
(b) three hours in molten aluminum.

After soaking for 3 hours in molten aluminum, the stressed specimen (Figure-6a) showed significant material loss when compared to the stressed specimen soaked for 15 minutes (Figure-6b). The specimen which was immersed in molten aluminum for three hours showed complete dissolution in the highest stressed portion of the specimen, where as the specimen immersed for 15 minutes showed a partial reduction in the cross-sectional area. Figure 7 shows the interface microstructures from the same two stressed specimens and reveals microstructures similar to those found in the unstressed specimens described previously. The size and distribution of the precipitates at the interface of the two stressed MP-159 specimens are shown in Table-IV and show that longer soak times result in larger precipitate size formation.

Generally, most liquid metal embrittlement couples have a very limited solid solubility and in the liquid state they exist as immiscible liquids. More over, if solid-liquid metal interactions have good solubility and form intermetallic compounds; they are not usually susceptible to liquid metal embrittlement [1]. In this current investigation intermetallic compounds are forming near the interface between the MP-159 specimens and the molten aluminum alloy giving additional evidence that the liquid metal embrittlement reaction is not occurring.

Conclusions

The results from this investigation show that a liquid metal embrittlement reaction between the MP-159 pin tool specimens and the molten aluminum alloy (5083) are not the likely cause for premature failure of the pin tools. The formation of intermetallic compounds at the interface between the MP-159 specimens and the molten aluminum alloy agree with previous research on other alloy combinations that liquid metal embrittlement is unlikely to occur. In addition, metallographic examinations did not reveal any liquid metal penetration into the grain boundaries of the MP-159 specimens or grain boundary decohesion typically seen in liquid metal embrittlement couples. Additional research into the premature failure of the MP-159 pin tools is needed.

Acknowledgements

The research presented in this paper was sponsored by the Center for Friction Stir Processing, which is a National Science Foundation I/UCRC. The research work was performed under project number CFSP06-AMP-04. The views and conclusions contained in this paper are those of the authors and should not be interpreted as representing the official policies, either expressed or implied, of the center, the industrial members of the center or the National Science Foundation.

References

1. ASM Handbook, Vol 1, 10th edition, Properties and Selection: Irons, Steels, and High Performance Alloys, p 718.
2. N.S. Stoloff, Liquid Metal Embrittlement, Surfaces and Interfaces II, Syracuse University Press, p 157-182, 1968.
3. ASM Handbook, Vol 8, Metallography, Structures, and Phase Diagrams, 8th Edition, ASM-International.
4. T. Hashimoto, S. Jyogon, K. Nakata, Y.G. Kim, M.Ushio, in: Proceedings of the First International Symposium on Friction Stir Welding, Thousand Oaks, CA, USA, June 14-16, 1999.

STIR ZONE TEMPERATURES DURING FRICTION STIR PROCESSING

K. Oh-ishi[1,2], A.P. Zhilyaev[1,3], S. Swaminathan[1], C.B Fuller[4], B. London[5], M.W. Mahoney[4], and T.R. McNelley[1]

[1]Naval Postgraduate School, Monterey, CA 93943, USA
[2] National Institute for Materials Science, Tsukuba 305-0047, Japan
[3]Centro National de Investigaciones Metalúrgicas (CENIM), 28040 Madrid, SPAIN (On leave from Institute for Metals Superplasticity Problems, Ufa, 450001 Russia)
[4] Rockwell Scientific Company, Thousand Oaks, CA 91360
[5]California Polytechnic State University, San Luis Obispo, CA 93407

Keywords: Friction Stir Processing, NiAl Bronze, Temperature Measurements

Abstract

Local peak temperatures have been estimated by analysis of stir zone microstructures in a NiAl bronze material subjected to friction stir processing and compared to embedded thermocouple and optical pyrometry measurements. The peak temperatures measured using the different techniques are in good agreement and ranged from 940°C to 1020°C.

Introduction

Friction stir processing (FSP), an allied process of friction stir welding (FSW), has been used in the surface modification of as-cast components [1-4]. FSP and FSW are solid-state techniques wherein friction due to rotation of a non-consumable tool on a work piece surface induces intense local shear deformation and adiabatic heating in the material. Severe but localized deformation due to flow of material around the tool results in refinement of the microstructure. In FSW, this flow also produces coalescence and formation of a weld; in both FSP and FSW the tool shoulder forges the processed material and prevents upward flow [5]. The thermomechanical cycles of FSW/P involve rapid transients and steep gradients in strain, strain rate, and temperature, and peak temperatures experienced by the material in the stir zone have not been reliably determined by direct measurement during experiments.

In the present work, FSP has been applied to as-cast NiAl bronze to achieve localized modification and control of microstructures in near-surface layers of large marine components. Direct thermocouple and optical pyrometry measurements have been made for selected combinations of processing parameters to assess the peak temperatures associated with this process. These measurements have been compared to temperature estimates based on quantitative optical microscopy analysis of stir zone microstructures.

Detailed descriptions of the complex physical metallurgy of NiAl bronze have been presented elsewhere [6-9]. Briefly, during cooling from the melt at nominal rates of 10^{-3} °Cs^{-1}, the bcc β phase forms initially and then transforms to the fcc primary α phase with a Widmanstätten morphology beginning at about 1030°C. At 930°C, nucleation of globular κ_{ii}, which is nominally Fe$_3$Al with a DO$_3$ structure, starts in the β phase. When the temperature reaches 860°C, fine κ_{iv} precipitates, also nominally Fe$_3$Al, begin to form in the α phase. The remaining β decomposes by the eutectoid reaction β → α + κ_{iii} at about 800°C, resulting in the formation of a lamellar constituent. The κ_{iii} is nominally NiAl with a B2 structure; proeutectoid κ_{iii} may exhibit a globular morphology, or may form by epitaxy on the globular κ_{ii}. When the cooling rates are increased to around ~10^0 °Cs^{-1}, or greater, the rapid cooling suppresses the

403

eutectoid reaction and the Widmanstätten morphology of the α as well as bainitic and martensitic products of β phase decomposition may become apparent. Thus, slowly cooled as-cast material is comprised of a primary terminal fcc α solid solution, a eutectic constituent, and various dispersoids. This material undergoes a phase transformation upon heating, such as during FSP, and the bcc β phase begins to form at 800°C. During subsequent cooling at high rates various β phase transformation products may become apparent. By determining the volume fractions of the β phase transformation products at various stir zone locations the local peak temperatures experienced can be estimated. This requires the assumption that the severe concurrent straining during FSP results in equilibrium microstructures upon heating to peak temperature. With this assumption, initial investigations indicated that peak temperatures of ~930°C were attained in the stir zone during FSP of a NiAl bronze material [9].

Experimental Procedure

The FSP was conducted on NiAl bronze plates approximately 30 × 15 cm (12 × 6 in) in size and 19.0 mm (0.75 in) in thickness; typical composition data are given in Table 1. The processing was conducted using Densimet® tools having a 12.7 mm pin with a step-spiral design. Tool rotation rates were either 600 or 1000 rpm and the traversing rate was 2 ipm (50.4 mm min^{-1}).

Table1: Composition (wt. pct.) data for NiAl bronze.

Element	Cu	Al	Ni	Fe	Mn	Si	Pb
Min-max	(min)79.0	8.5-9.5	4.0-5.0	3.5-4.5	0.8-1.5	0.10(max)	0.03(max)
Nominal	81	9	5	4	1	-	-
Typical	81.2	9.39	4.29	3.67	1.20	0.05	<0.005

The direct thermocouple (TC) measurement was accomplished using a sheathed type K thermocouple 1.6 mm (0.0625 in) in diameter and inserted into a hole drilled into the bottom center of the plate prior to FSP, as illustrated in Fig. 1. The depth of the hole allowed the TC junction to be located 6.35 mm (0.25 in) below the top surface of the plate and along the intended line of the FSP traverse. On the bottom of the plate, a channel was also machined in the transverse direction from the location of the TC hole to the side of the plate for the thermocouple to bend 90° into a direction perpendicular to the FSP travel direction. The diameter of the drilled hole was sufficiently small for the TC sheath to contact the side as well as the end of the hole. The TC sheath was readily bent around the radius from the hole to the channel while the TC end maintained contact with the bottom of the hole. The channels were just wide enough for a press fit of the TC sheath, and this helped keep the TC in place. Finally, both sides of the channel were lightly peened to lock the TC sheath in place. The optical pyrometer (Wahl Model DH552) was calibrated using a black-body source in conjunction with a Hart Scientific calibrator. The pyrometer was calibrated at a temperature of 950°C, based on values from microstructure analysis [9]. During FSP, the pyrometer was focused on a location on the NiAl bronze surface just in the wake of the pin tool and care was taken to assure that 'flash' from the processing did not interfere with the pyrometer measurement.

After FSP, the processed material was sectioned on the transverse plane and prepared for optical microscopy using standard methods. Samples were etched in a two-step process involving, first, immersion for 1 – 2 s in a solution of 40 ml water – 40 ml ammonium hydroxide – 2 ml hydrogen peroxide (30%), rinsed in water, followed by immersion for 1 –2 s in a solution of 60 ml water – 30 ml phosphoric acid – 10 ml hydrogen peroxide. Etched samples were examined using bright-field illumination in a Jenaphot 2000 equipped with a digital imaging

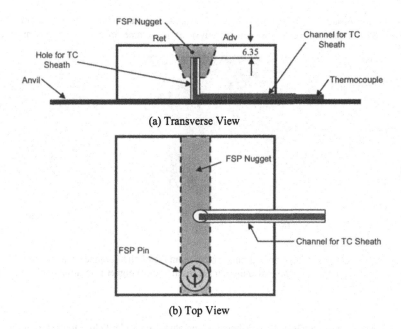

(a) Transverse View

(b) Top View

Figure 1: Thermocouple placement in a NiAl bronze plate. (a) Transverse and (b) top view.

system. The volume fraction of β transformation products was recorded at different locations in the stir zone. The measured volume fraction was converted to a temperature based on a formula[1] derived from experimental studies of annealing alone followed by quenching into either air or water, or of hot rolling with reheating between passes for temperatures of 800 – 1000°C. Full details of the latter analysis will be presented in a subsequent report.

Results and Discussion

Fig. 2(a) shows a representative montage of optical micrographs from the transverse plane of the NiAl bronze after FSP at 600 rpm and 2 ipm (50.4 mm min⁻¹). The tool traversing direction was into the plane of the image and the advancing and retreating sides are indicated. The tool pin is shown in the inset in this image. An image at higher magnification from a location 6.35 mm from the top of the stir zone is shown in Fig. 2(b). This micrograph shows a microstructure consisting of horizontally elongated bands of Widmanstätten α that has formed in a dark-etching matrix of other β transformation products. These bands reflect locations that experienced transformation to β during the FSP thermomechanical cycle, and are interspersed with bands comprising of fine, equiaxed α grains. The latter have formed by recrystallization of primary α that did not transform to β during the FSP thermomechanical cycle. The volume fraction of β transformation products at this location, which is 6.35 mm below the plate surface, is consistent with local heating to ~930°C during processing. Further down in the stir zone the microstructure is comprised of mostly deformed and recrystallized α grains with a lower volume fraction of β transformation products. At the interface between the stir zone and base material, the microstructure is made up of elongated and deformed α grains and dark appearing β

---

[1] $T = 248V_\beta + 758$; T: temperature in °C and V_β: volume fraction of β.

transformation products. The variation in local peak temperature with depth in the stir zone, determined from quantitative microstructure analysis of the volume fraction of β transformation

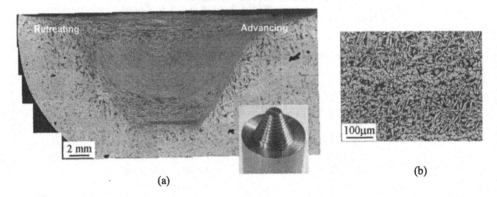

(a)

(b)

Figure 2: (a) Representative optical microscopy montage of NiAl bronze subjected to FSP at 600 rpm and 2 ipm using the tool shown in the inset; and (b) higher resolution optical image from a depth of ~ 6.35 mm.

products, is shown in Fig. 3(a). The predicted temperatures are highest close to the top of the stir zone (980°C); this may reflect additional local heating produced by the tool shoulder. The peak temperature initially decreases with depth in the stir zone and then increases again at depths corresponding to the 'onion ring' flow patterns at locations 4 – 6 mm below the plate surface. Within the standard error, the temperature is still ~980°C at a depth of 4 mm. Below 6 mm, the peak temperatures decrease relatively steeply to a value of ~840°C at a depth of 10 mm. Fig. 3(b) shows the temperature readout for the TC in NiAl bronze processed at 1000 rpm and 2 ipm (50.4 mm min⁻¹). Despite displacement of the thermocouple from its initial position, the peak temperature measured is ~1020°C. Finally, the optical pyrometer recorded a surface temperature of 944°C when the tool was at the location of the embedded thermocouple.

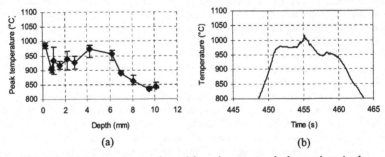

(a)

(b)

Figure 3: (a) Temperature estimated by microstructural observations in the center of the stir zone of NAB subjected to FSP at 600 rpm at 2 ipm and (b) temperature measured using thermocouple at a depth of 6.35 mm in the center of the stir zone of NAB subjected to FSP at 1000 rpm and 2 ipm.

Conclusions

Table 2 summarizes the peak temperatures obtained by these three techniques. The temperatures predicted by microstructure based estimates and direct observations generally agree well. Clearly, the microstructural gradient through the thickness of the friction stir processed zone is a consequence of a temperature gradient inherent to FSP. The dependence of the distribution of peak temperature on processing parameters will be the subject of a future report. Finally, all three measures of peak temperature are consistent with the absence of melting during FSP of this material.

Table2: Peak temperatures recorded by different techniques.

Technique	Microstructure	Thermocouple	Optical Pyrometry
Peak Temperature	930°C (at ~6.35 mm) 900°C-980°C (surface to 6 mm depth) 840°C-900°C (6mm-10mm depth)	1020°C (at ~6.35 mm)	944°C (surface)

Acknowledgements

The as-cast NiAl bronze materials were provided by Mr. William Palko and Mr. Steven Fielder of the Naval Surface Warfare Center - Carderock Div, Bethesda, MD. The assistance of Mr. William Bingel at the Rockwell Scientific Corporation is gratefully acknowledged. Funding for this work was provided by the Defense Advanced Research Projects Agency (DARPA), Dr. Leo Christodoulou program monitor.

References

1. Z.Y. Ma, R.S. Mishra and M. W. Mahoney, *Friction Stir Welding and Processing II*, K.V. Jata, M.W. Mahoney, R.S. Mishra, S.L. Semiatin and T. Lienert, eds., TMS, Warrendale, PA, 2003, pp. 221-30
2. M.W. Mahoney, W.H. Bingel, S.R. Sharma, and R.S. Mishra, Materials Sci. Forum Vols, 426-432 (2003) pp. 2843-2848.
3. W.A. Palko, R.S. Fielder, and P.F. Young, Materials Sci. Forum Vols, 426-432 (2003) pp. 2909-2914.
4. L. Christodoulou, W. Palko, and C. Fuller, *Friction Stir Welding and Processing III*, K. V. Jata, M.W. Mahoney, R.S. Mishra, and T.J. Lienert, eds., TMS, Warrendale, PA, 2005, pp. 57-66.
5. W.M. Thomas, E.D. Nicholas, J.C. Needham, M.G. Murch, P. Templesmith and C.J. Daws, G.B. Patent Application No. 9125978.8, December, 1991; U.S. Patent No. 5460317, October, 1995
6. P. Brezina, Int. Met. Rev. 27 (1982) 77.
7. E.A. Culpan and G. Rose, J. Mater. Sci. 13(1978) 1647.
8. F. Hasan, A. Jahanafrooz, G.W. Lorimer and N. Ridley, Metall. Trans. A 13(1982) 1337.
9. K. Oh-ishi and T. R. McNelley, Metall. Mater. Trans. A., 35A (2004) 2951.

FRICTION STIR PROCESSING OF D2 TOOL STEEL FOR ENHANCED BLADE PERFORMANCE

Carl D. Sorensen[1], Tracy W. Nelson[1], Scott M. Packer[2], Charles Allen[3]

[1]Brigham Young University, 435 CTB, Provo, UT 84602 USA
[2]Advanced Metal Products,2320 North 640 West, West Bountiful, UT 84087, USA
[3]Knives of Alaska, Inc., 3100 W. Airport Drive, Denison, TX 75020, USA

Keywords: Friction Stir Processing, Blade Testing, Microstructural Modification, D2 Steel

Abstract

Friction stir processing (FSP) has been applied to blade blanks made from D2 steel to improve blade performance. D2 blanks of moderate hardness (30-40 HRC) are subjected to FSP using a convex scrolled shoulder, step spiral pin (CS4) tool made from PCBN. The resulting microstructures are hard, tough, and corrosion resistant. Hardness on the order 65-68 Rockwell C is achieved consistently. Grain sizes are on the order of a micron or less. The chromium content remaining in the matrix is higher than that in quenched and tempered D2. Knife blades manufactured from FSP D2 steel exhibit up to a 10 fold increase in edge life over traditional thermomechanically processed and heat treated D2 blades. Methods for repeatably testing blade edge performance are presented, along with a microstructural analysis of the FSP D2.

Introduction

Friction Stir Welding has received significant attention over the past few years as a revolutionary joining process. It produces a fine wrought microstructure with properties significantly better than the cast microstructure of fusion processes. More, recently the friction stirring process has been applied to materials where no joint is involved, as a means of changing the microstructure on a local scale. This application is termed Friction Stir Processing (FSP).

Because FSP is a relatively new metal working process, there are many applications yet to be explored. With the advent of new FSP tooling, applications in steels and other high softening temperature metals are becoming feasible. One of the most recent applications of FSP has been in processing of D2 tool steel for application to knives and other types of blades. This paper discusses preliminary results from this application.

Previous Work

FSP for Microstructural Modification.

FSP for microstructural modification has gained increasing interest during the last six years. A derivative of Friction Stir Welding [1], FSP locally alters the microstructure of a metal by the application of extreme plastic deformation under pressure and temperature. As a result, highly refined microstructures with enhanced mechanical and physical properties have been reported [2-5].

Properties obtained through FSP often exceed those obtainable by traditional metal working processes. Mahoney and co-workers successfully applied FSP to high-strength, low-density aluminum alloys in order to achieve thick section superplastic properties [2,3]. They demonstrated elongations greater than 600% in 5mm thick 7050-T7 without localized necking. Similarly, aerospace components which had been processed via FSP exhibited more complete forming at half the pressure and time than traditional superplastic material [6]. Previously, superplastic characteristics have been obtainable only in materials less than 2 mm thick because of limitations in the sheet production process.

Mishra and co-worker have successfully demonstrated increased mechanical properties in aluminum castings processed via FSP. The FSP material showed tensile strengths and elongations almost double those of the casting [4].

Blade Testing

The leading organization for standardizing test methods for blades is the Cutlery and Allied Trades Research Association (CATRA). CATRA has defined a number of standard tests and associated testing equipment for use in testing commercial blades. Many of these tests have become ISO or EU standards. For testing durability of knife edges, CATRA developed the CATRA Edge Retention Tester (ERT)[7]. This equipment is used to perform cutting tests according to ISO 8442.5-2004, which is specified for food preparation knives. The ERT tests the ability of a knife edge to cut through standard media, consisting of silica-impregnated paper. The paper is cut by means of knife strokes of fixed length and velocity, under the effects of a constant normal force. Typically, the test is run with a normal force of 50 N, with a stroke of 40 mm and a speed of 50 mm/s. When necessary, the media is lifted off the knife edge, and advanced automatically to provide new media to cut.

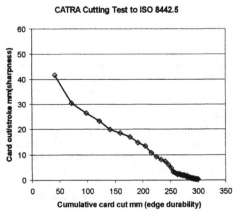

Figure 1: The CATRA Edge Retention Tester, along with a sample data plot from ERT Testing (from [7]).

The ERT measures two important data values for each blade tested. The initial cutting performance (ICP), calculated as the total media cut in the first three strokes, measures the initial sharpness of the blade. The total card cut (TCC), calculated as the total media cut during a test of 60 strokes, measures the life of the blade edge.

410

For extremely sharp blades such as razors and scalpels, CATRA has developed a different machine for measuring blade sharpness, called the Razor Edge Sharpness Tester (REST)[7]. The REST machine pushes a blade a fixed distance into a silicone medium similar to weatherstripping. During the test, the force required to push the blade into the material is recorded. In contrast to the ERT test, there is no slicing or motion parallel to the blade edge. The REST value is recorded as the peak force during the cutting cycle, measured in N. The lower the peak cutting force, the sharper the blade.

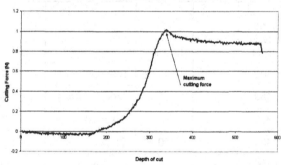

Figure 2: The CATRA Razor Edge Sharpness Tester in action, along with a typical REST plot (from [8]).

D2 Steel

D2 is an air-hardenable high carbon, high chromium cold work tool steel [9]. Primary strengthening is attributed to the high carbon (C), chromium (Cr) and vanadium (V) content. The high carbon contributes to higher hardness in the martensite that forms during quenching from the austenitic temperature. Cr and V slow the transformation kinetics of the austenite to pearlite, increasing the hardenability. Cr and V also act as secondary hardening agents by forming carbides thought the microstructure.

411

Typical applications of D2 tool steel include tool and dies for cutting, forming and shaping of materials. However, the D-series tool steels also exhibit excellent wear resistance because of their higher carbon and vanadium, making them suitable materials for knife blades. Table 1 shows the specified composition for alloy D2 [9].

Table 1 Nominal composition of D2 tool steel.

Steel/Comp.	C	Cr	Mn	Si	Ni	Mo	V
D2	1.4-1.6	11.0-13.0	0.6 max	0.60 max	0.30 max	0.70-1.20	1.10 max

Despite the high Cr content, D2 is not considered "stainless". D2 is not considered stainless because much of the chromium in D2 is tied up in the form of chromium carbides, which form during cooling from the austenitic temperature in the range of 650-800C. The carbides obtain their high Cr levels by absorbing Cr from the surrounding steel. This Cr depletion greatly reduces corrosion resistance; in the steel surrounding the carbides the Cr concentration is less than 12%. Thus, D2 is not considered stainless steel.

FSP Methods

FSP was performed on 6.5mm thick D2 steel with nominal composition shown previously in Table 1. Two different heat treatments were used in this investigation. Both were austenitized and quenched then tempered to Rc 30 and 40. Plates were cut 15cm in width and 60 cm in length. The processed sides of the plates were ground and cleaned with acetone prior to processing.

Table 2: Parameters used in DOE

Weld Side Next to Blade Edge	Spindle Speed, RPM	Feed, IPM	Hardness	Pin	Blade IDs
Retreating	300	3	40	Stepped Spiral	2-1, 2-2
Retreating	450	3	30	Stepped Spiral	4-1, 4-2
Advancing	300	5	30	Stepped Spiral	5-1, 5-2
Advancing	450	5	40	Stepped Spiral	7-1, 7-2
Advancing	300	3	30	Tri-flat	1-1, 1-2
Advancing	450	3	40	Tri-flat	3-1, 3-2
Retreating	300	5	40	Tri-flat	6-1, 6-2
Retreating	450	5	30	Tri-flat	8-1, 8-2

FSP was performed using two different tool designs. Both tools had a convex scrolled shoulder that was 25.4 mm in diameter and a convex radius of 90 mm. The diameter at the base of the pin was 6mm, with a pin half-angle of 15 degrees, and a pin length of 2.2 mm. One tool had a simple truncated cone pin and the other featured a step spiral.

FSP was performed at 300 and 450 RPM and 3 and 5 IPM (1.3 and 2.1mm/s). During processing, the center of the tool was 12 mm from the edge of the plates. The edge of the knife blade was placed between the center of the processed zone and the edge of the plate. For some blades, the advancing side of the processed zone was closest to the edge of the plate; for others the retreating side was closest to the edge. All plates were processed under position control. Table 2 shows the parameters for each of the blades produced in this study.

Transverse samples were removed from each parameter set. These were mounted, polished, etched, and examined via optical microscopy (OM) for weld quality and microstructure. Each sample was etched with 10% Nital for up to 8 minutes. Optical microscopy images were examined to determine grain structure and carbide distribution in the processed zone.

Vickers microhardness was performed on each set of conditions. Microhardness tests were performed with an automatic microhardness tester using a Vickers diamond indenter, a 300 gram load, and an 8 second dwell. A line trace along the plate centerline through the processed zone was performed on each sample.

Blade Testing Methods

Following microhardness testing, a waterjet cutter was used trim the plates to the intended blade edge at a distance of 3 mm from the process zone centerline. Microhardness testing showed that the hard zone would be included in the blade if the cut were made at this location. The blades were then ground with a bevel angle of 20 degrees to an edge thickness of 0.75 mm (0.030 in). The final sharpening was performed on an abrasive belt using a fixture to hold a constant edge angle, followed by a cardboard wheel to remove the wire edge.

The blades created by the process described above were then tested using a combined test sequence based on CATRA testing principles. Blade sharpness was measured using the CATRA REST tester. Blade durability was measured by a test similar to the CATRA ERT test, but using a different cutting medium.

The ERT-style testing used in this study used manila rope ¾ inch in diameter. Rope was used instead of CATRA test media because the CATRA test media was too aggressive, and not indicative of the typical use of outdoor knives.

After REST testing, the sharpened blade was placed in the ERT tester. 20 strokes were made on the ERT tester. The blade was removed and tested in the REST tester. The blade was returned to the ERT tester for 40 strokes, then REST tested. The cycle of 40 strokes on the ERT followed by REST testing was repeated until the blade was too dull to shave.

Results and Discussion

Microstructure Evaluation

Grain Size Reduction. As shown in the photomicrographs in Figure 3, the grain size of the FSP D2 (Figure 3 b and c) has been reduced significantly over that of the base metal (Figure 3 a). The grain size of the base metal (traditionally processed) D2 ranges from 5 to 15 μm. Figure 3 b and c show that the grain size for the FSP D2 is significantly smaller than that of the as-received material. Grain sizes of the FSP D2 range from submicron to a few μm. In addition to grain size, carbides appear to be smaller on average compared to those in the base metal.

413

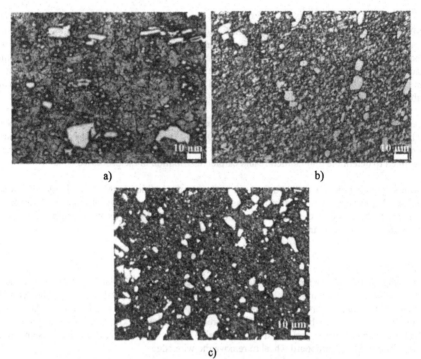

a)

b)

c)

Figure 3: Optical micrographs of FSP D2 showing changes in grain size. a) Normally heat-treated D2, b) FSP at 600 RPM and 4 in/min, and c) FSP at 250 RPM and 4 in/min

Figure 3 also demonstrates that FSP processing parameters can also have a significant effect on grain size. The microstructure shown in Figure 3 b was processed at 650 rpm and 4 in/min. The grains in this image appear to be between 1 and 5 μm in size. At 250 RPM and 4 in/min (Figure 3 c) the grains are so fine that, even at 1000X, the interior of the grain cannot be resolved.

Transmission electron microscopy (TEM) was used to analyze the grain size of the FSP D2 at 250 RPM and 4 IPM. Figure 4 shows a TEM micrograph of this material at a magnification of 80,000X. The average grain size in this figure is about 500 nm.

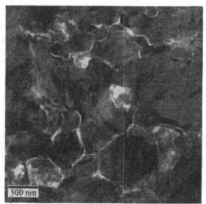

Figure 4: Transmission electron micrograph of FSP D2 at 250 RPM and 4 in/min showing typical grain sizes of 500nm.

Grain sizes of this magnitude in tool steels are unobtainable by traditional metal working processes. Perhaps the most promising candidate would be powder metallurgy (PM) processing. PM processing may be capable of attaining grain sizes of this magnitude, but the strength and toughness of the material is severely compromised by the presence of oxide inclusions and voids inherent to the process. As an example, a photomicrograph of S30V, a powder metallurgy processed steel, is shown in Figure 5. The grain sizes here are on the order to 5-10 μm, which is about 30% smaller than traditional D2, and approximately 5 times larger than the coarsest FSP D2. Although this process leads to smaller grains than traditional D2, the void and oxide content (seen as darker region in Figure 5) result in very poor fracture toughness.

Figure 5: Photomicrograph of powder metallurgy processed S30V steel.

Greater Levels of Alloying Elements in the Matrix In addition to the reduction in grain size, FSP D2 gets a second bump in hardness from the addition of chromium (Cr) and carbon (C) to the martensite matrix. Under the elevated temperature and strain rates present during FSP, the Cr rich carbides in the D2 steel begin to dissolve, with the Cr and C diffusing into the austenite matrix. During post-processing cooling, these elements are retained in solution, thus increasing the steel's hardenability and martensite hardness. This leads to strength not typically reached in traditional D2 processing.

415

The increased level of Cr in the matrix is demonstrated by the increased corrosion resistance of the FSP D2. The material processed by FSP exhibits excellent corrosion resistance when subjected to a nitric acid solution as shown in Figure 6. The FSP region (in the middle of the figure) shows no indication of attack by nitric acid implying that the processed zone is "stainless". In contrast to the undisturbed FSP region, the base metal is severely attacked by the etchant, leading to a dark grey color.

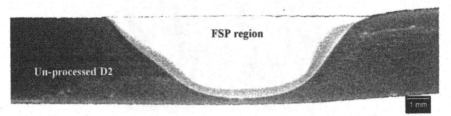

Figure 6: Photomacrograph of FSP D2 steel etched with 10% Nitric acid in methanol.

Microhardness Results

Figure 7 shows typical microhardness results for the FSP zone in the D2 blades. As can be seen, there is no reduced hardness associated with the HAZ. The hardness in the stir zone is uniformly high, with some scatter that is mostly due to the presence of carbides in the microstructure. The peak hardness in the stir zone is over 1000 HV, and the minimum hardness is the stir zone is over 900 HV. Hardnesses of this magnitude are comparable to cemented carbide, and significantly higher than those usually seen in D2.

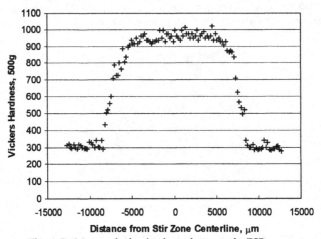

Figure 7: Measured microhardness data near the FSP zone.

Sharpness Testing

The primary functional test for the sharpness of an outdoor knife is the ability to shave with the knife. A series of D2 blades was tested in this sequence, coupled with a shaving test. Following each REST test, the blade was tested to see if it would shave the arm hair of the operator. The REST value and the shaving ability of the blade were recorded and used to determine the REST

416

value that constituted the loss of shaving sharpness. Over this testing, the average minimum REST value that led to a loss of shaving was 3.07 N. Accordingly, 3 N was chosen as the limit for shaving sharpness.

Figure 8 shows the performance of four different blade steels in the ERT test using manila rope. All blades had identical edge geometry. As can be seen, the FSP D2 blade (called Friction Forged[TM] in the legend) cuts 50% more rope on the first stroke than S30V, and 80% more than S90V and traditionally processed D2. This higher initial sharpness lasts throughout the entire testing cycle. One way of determining blade durability is to measure the total amount of rope cut before the blade reaches a minimum sharpness level. Using a 0.25 inches/stroke sharpness value for comparision, we see that the FSP blade falls permanently below 0.25 at 200 inches of rope cut. In contrast, the S30V reaches this sharpness at 40 inches, S90V reaches this sharpness at 15 inches, and D2 reaches this sharpness at 5 inches. By this measure, FSP D2 is five times as durable as S30V and 40 times as durable as traditionally processed D2.

Edge Retention

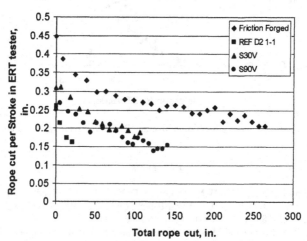

Figure 8: Rope cut (ERT sharpness) per stroke as a function of total rope cut for various blade materials.

Figure 9 shows total inches of rope cut as a function of REST value. As mentioned previously, a REST value of 3 is the mean value of the loss of shaving sharpness. D2 loses its shaving sharpness at less than 5 inches of rope cut. S30V loses shaving sharpness at 100 inches of rope cut. S90V loses shaving sharpness at 140 inches of rope cut. FSP D2 retains shaving sharpness at over 250 inches of rope cut. The FSP blade has significant advantages in edge retention by this measure as well.

Conclusions

FSP has been successfully used to increase the durability of knife blade edges. The grain size in the processed zone was greatly reduced from the base metal, with a typical grain size of 500 nm. The FSP zone also had higher concentrations of chromium and carbon in the martensite, as indicated by the stainless nature of the stir zone. The additional Cr and C also contribute to high hardness in the stir zone. The hardness in the FSP zone exceeded 900 HV. This high hardness allows the formation of an exceptionally sharp edge. It also results in an extremely durable knife edge, with the edge up to 40 times as durable as traditionally processed D2.

Cutting Life

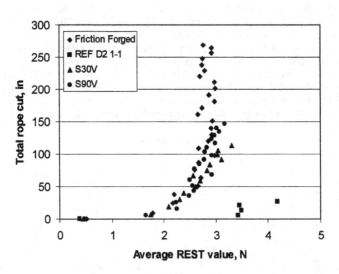

Figure 9: Total rope cut as a function of average REST value for various blade steels.

References

1. The Welding Institute; TWI; W.M. Thomas, E.D. Nicholas, J.C. Needham, M.G. Murch, P. Temple-Smith, C.J. Dawes, PCT World Patent Application WO 93/10935. Field: 27 Nov. 1992 (UK 9125978.8, 6 Dec. 1991). Publ: 10 June 1993.
2. Z. Y. Ma, R. S. Mishra and M. W. Mahoney, Superplastic deformation behavior of friction stir processed 7075Al alloy, *Acta Materialia*, **50** (2002) 4419.
3. M.W. Mahoney, R.S. Mishra, T.W. Nelson, (2002) "Friction Stir Processing Creates Aluminum Alloy Superplasticity," *Industrial Heating*, February, pp. 31-33.
4. The Effect of Friction Stir Processing on 5083-H321/5356 Al Arc Welds: Microstructural and Mechanical Analysis" by Christian Fuller and Murray Mahoney
5. J.Q. Su, T.W. Nelson, and C.J. Sterling, (2003) "A New Route to Bulk Nanocrystalline Materials", *Journal of Materials Research*, 18(8), pp 1757-1760.
6. Murray W. Mahoney, "Friction Stir Welding and Processing: A Sprinter's Start, A Marathoner's Finish", *Proceedings of the 7th international Conference on Trends in Welding Research*, pp 233-240. April 2005.
7. Cutlery and Allied Trades Research Association, http://www.catra.org/products/testing/knives/knives_aet.htm, visited October 2006.
8. Cutlery and Allied Trades Research Association, http://www.catra.org/products/testing/knives/sharpness_tester.htm, visited October 2006.
9. ASM Handbooks, Volume 1, Properties and Selection: Irons, Steels, and High Performance Alloys, Tenth Edition, 1994.

THE RELATIONSHIP BETWEEN FRICTION STIR PROCESS (FSP) PARAMETERS AND MICROSTRUCTURE IN INVESTMENT CAST TI-6AL-4V

Adam L. Pilchak[1], Z. Tim Li[2], James J. Fisher[2],
Anthony P. Reynolds[3], Mary C. Juhas[1], James C. Williams[1]

[1]The Ohio State University, Department of Materials Science and Engineering,
2041 North College Rd, Columbus, OH 43210
[2]Edison Welding Institute, 1250 Arthur E. Adams Drive, Columbus, Ohio 43221
[3]University of South Carolina, Department of Mechanical Engineering, 300 Main Street,
Columbia, SC 29208

Keywords: Friction stir processing, thermo-mechanical processing, Ti-6Al-4V, microstructure

Abstract
Titanium alloy castings have a coarse, fully lamellar microstructure which results in lower yield strength and high cycle (HCF) fatigue life than comparable wrought products. This study has demonstrated that the microstructure at the surface can be modified using FSP to eliminate the coarse as-cast microstructure. FSP can transform the as-cast microstructure into a range of microstructures, all of which have smaller length scales than the as-cast structure. This ranges from very fine (1 μm to 2 μm diameter) equiaxed microstructures to fine lamellar structures with 20 μm to 40 μm prior β grains. Such small length scale microstructures are difficult to produce in wrought products. We have used a variety of characterization techniques to investigate these microstructural changes. The changes will be illustrated and discussed in terms of the FSP parameters. The mechanism for the observed microstructure evolution will be described.

This work is supported by the Office of Naval Research under contract N00014-06-1-0089.

Introduction
Friction stir processing (FSP) is a relatively new thermo-mechanical processing technique based on the principles of friction stir welding [1] (FSW). During FSP, there is no joint or weld created, instead, the tool is used to locally refine the microstructure in specific regions by in situ mechanical working. FSP is illustrated schematically in Figure 1. A non-consumable tool is rotated and plunged into the workpiece. The tool generally consists of a small diameter pin (usually 0.20 inch to 0.30 inch) [2-4], sometimes with features such as flutes or threads on it that is concentric with a larger diameter shoulder (typically near 1 inch). When the pin and the shoulder contact the workpiece frictional heat is generated which locally softens and deforms the workpiece. The rotation of the tool moves material from the leading edge of the tool to the trailing edge. Concurrent with deformation, recrystallization occurs to refine the microstructure in the stir zone (SZ) [5-7]. There are many types of FSW and FSP tools in use and the exact dimensions and material composition are often proprietary information and not reported in the open literature. This makes comparison of seemingly similar studies difficult since the tool geometry has a significant effect on material flow [8].

Since no joint is created during FSP, the depth of penetration can be reduced to only process the near surface region of the material. In the present study we have used shallow plunge depths to modify the coarse lamellar microstructure on the surface of investment cast and hot isostatic pressed (HIP) Ti-6Al-4V plate. Investment casting is used to make large structural castings with complex geometries, however, solidification shrinkage can lead to pores. Hot isostatic pressing involves applying isostatic pressure during at an elevated temperature below the β transus followed by slow cooling. This process has been shown to close and heal internal pores, which

419

are often sites for fatigue crack initiation. It has been reported [9] that high cycle fatigue (HCF) strength of Ti-6Al-4V castings increased after HIP compared to the as-cast condition. Although HIP increases fatigue strength, further improvements can be made by thermo-mechanical processing to reduce the long slip lengths associated with the coarse, fully lamellar colony structure. Reducing the effective slip length delays fatigue crack initiation. FSP can be used to refine the microstructure on the surface of a casting, which is where most fatigue cracks initiate, while leaving the coarse lamellar microstructure in the remainder of the component. A review by Lütjering and Gysler [10] reported the effect of α morphology on fatigue crack growth (FCG) rates in Ti-6Al-4V. At R = 0.1, in air at room temperature and $\Delta K = 8$ MPa√m the FCG rate of coarse lamellar titanium was an order of magnitude lower ($da/dN = 10^{-9}$ m / cycle) than an equiaxed microstructure with an average α grain size of 2 μm ($da/dN = 10^{-8}$ m / cycle). The fatigue strength of the fine equiaxed microstructure and coarse lamellar microstructure was approximately 625 MPa and 490 MPa, respectively. Therefore, FSP can produce a microstructure on the surface of the component that has improved fatigue crack initiation resistance while maintaining a structure with higher FCG resistance in the remainder of the component.

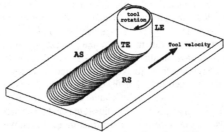

Figure 1. Schematic of friction stir processing showing the tool rotation and travel direction which designates the advancing side (AS), retreating side (RS), leading edge (LE) and trailing edge (TE).

Experimental Procedure

Two investment cast Ti-6Al-4V plates with dimensions of approximately 175 x 15 x 1.5 cm were provided by Precision Cast Parts Company. The plates were hot isostatically pressed at 1000 atm (15,000 psi) at 896 °C (1645 °F) for two hours and slowly cooled followed by a two hour anneal at 845 °C (1553 °F) and slow cooling. The plates were grit blasted, chemically milled 1.25 mm and finally mechanically milled to provide flat, parallel surfaces for FSP. One plate was processed at the University of South Carolina (USC) with a W-25%Re tool. The second was processed at Edison Welding Institute (EWI) using a different tungsten alloy tool. An image of the tool used to process the EWI plate is shown in Figure 2.

Figure 2. Tungsten based tool prior to friction stir processing at EWI.

The tool used at USC had a shoulder diameter of 19.05 mm, a pin diameter of 5.08 mm and a pin length of 1.27 mm. The processing parameters used in this study are shown in Table I. The USC

420

plate was processed using load control, so the down force reported was constant. Conversely, the EWI plate was processed in depth control. Therefore, the down force reported for the EWI plate is an average taken from the entire processing distance of 8.9 cm (3.5 inches) not including the spikes that occur during plunging. The plunge depth that is reported for the USC plates is an estimate based on measurements taken in the scanning electron microscope (SEM). Since load control was used, this could vary slightly along the entire length of the processed zone. Argon gas was used to prevent oxidation of the workpiece as well as the tool during processing of both USC and EWI plates.

Table I. FSP parameters used in this study.

Sample ID	tool travel speed, cm/min (in/min)	tool rotation speed, rpm	Down force, kN (lbf)	Plunge depth, mm (in)
USC-2	5.1 (2)	100	16.5 (3700)	~1.8 (0.07)
USC-4	10.2 (4)	100	38.7 (8700)	~1.8 (0.07)
EWI-2	5.1 (2)	150	15.2 (3418)	2.54 (0.10)
EWI-4	10.2 (4)	150	19.6 (4405)	2.54 (0.10)

Transverse samples for microstructure examination were removed by sectioning with an abrasive water jet. The samples were mounted in electrically conductive bakelite and ground with progressively finer SiC paper followed by 3 μm and 1 μm diamond compound. Final polishing was done using 0.05 μm colloidal silica in a vibratory polisher. A FEI Sirion SEM was used to characterize the microstructure in the samples. Texture data was obtained using a Phillips/FEI XL-30 ESEM equipped with a TSL electron backscatter diffraction (EBSD) detector for orientation imaging microscopy (OIM). Both microscopes are equipped with energy dispersive x-ray detectors and field emission sources. Backscattered electron images were collected from the as-polished specimens. A Scintag XDS-2000 x-ray diffractometer with a Cu Kα source operating at 45 kV and 25 mA with a scan rate of 0.2 degrees per minute was used for x-ray diffraction (XRD) measurements.

Results and Discussions

The base material microstructure was essentially the same for both the USC and EWI processed plates. It consisted of 1 mm to 1.5 mm diameter prior β grains bounded by grain boundary α. The interior of the β grains contained a coarse, fully lamellar microstructure. The α colonies were on the order of several hundred microns and the width of the α laths was approximately 5 μm. Hot isostatic pressing appeared to close all of the internal porosity. The base material was also assumed to be free of residual strain since the cooling rate from above the β transus was slow and the plate was subsequently annealed.

Figure 3. Typical macro-etched cross sections of the (a) USC plate and the (b) EWI plate.

A macro etched transverse cross section of both the USC and EWI plates are shown in Figure 3. These were prepared using Kroll's reagent and scanned with a flatbed scanner. The light etching parabolic regions near the top are the stir zones and the coarse prior β grains are seen in the remainder of the sample. Higher magnification SEM observations revealed that a transition zone [11] (TZ) was also present between the SZ and the heat affected zone of the base material. This area is similar to what is commonly called the thermo-mechanically affected zone in Al alloys. It is also worthwhile to note that the plunge depth was greater on the EWI plate compared to the USC plate.

The SZ for both processing conditions on the USC plate consisted of 1 μm to 2 μm diameter equiaxed primary α grains with a volume fraction of approximately 0.70. Between the primary α grains was a fine α + β transformation product with acicular α plates approximately 0.05 to 0.075 μm wide. A representative micrograph is shown in Figure 4.

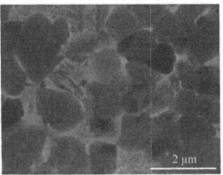

Figure 4. Backscattered SEM micrograph of the stir zone in USC-4.

Electron backscatter diffraction data was used to assess the degree of microtexture in the stir zone. The data for USC-4 is presented in Figure 5a as pole figures. Figure 5b shows that the maximum degree of preferred orientation is less than 2 times random. This shows that there is essentially no microtexture in the equiaxed α grains in the SZ. Greater than 95% of the grain boundaries had a misorientation angle exceeding 15°. The misorientation distribution is shown in Figure 5c. High misorientation angle between adjacent grains is indicative of a recrystallization process. Although results are only shown for USC-4, a similar lack of microtexture was also observed in USC-2.

In general, the diameter of equiaxed α grains that are formed by recrystallization of a lamellar microstructure corresponds with the width of the α laths before deformation [12]. In the present study, however, the primary α grains in the SZ are about one fifth of the width of the α laths in the base material. This observation can be explained by the temperature gradient that precedes the friction stir tool. The temperature rise requires that the volume fraction of β increase at the expense of the α laths, which become thinner by preferential lateral shrinkage. The amount of deformation and heat generation increases moving from the TZ into the SZ. Therefore, observations of the α laths in the TZ provide some insight to the state of the base material as it begins to be deformed. The width of the α laths in the TZ is approximately 1.5 μm to 2 μm for both processing conditions (Figure 6), which is more consistent with the diameter of the equiaxed α grains in the SZ. Orientation contrast in backscattered SEM images, for example Figure 6, implies that subgrain formation has occurred, although a TEM study is necessary to confirm this observation.

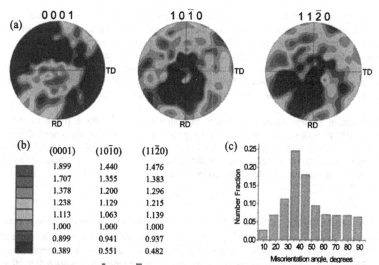

(a) 0001   10\overline{1}0   11\overline{2}0

(b)

	(0001)	(10\overline{1}0)	(11\overline{2}0)
	1.899	1.440	1.476
	1.707	1.355	1.383
	1.378	1.200	1.296
	1.238	1.129	1.215
	1.113	1.063	1.139
	1.000	1.000	1.000
	0.899	0.941	0.937
	0.389	0.551	0.482

(c)

Figure 5. (a) $(0001)_\alpha$ $(10\overline{1}0)_\alpha$ $(11\overline{2}0)_\alpha$ pole figures, (b) legends and (c) the misorientation distribution corresponding to the EBSD data from USC-4.

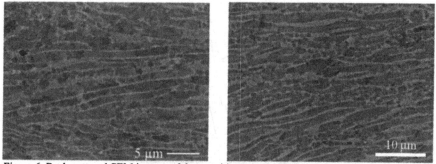

Figure 6. Backscattered SEM images of the transition zone in USC-4. The α lath width is refined compared to the base material.

Several mechanisms for the conversion of lamellar microstructures to equiaxed microstructures were reported by Peters et al. [12] and Weiss et al. [13]. These mechanisms can explain the observations of equiaxed α grains in the transition zone and can be extended to explain the microstructure in the SZ. The separation of α laths into discrete equiaxed grains can occur by the formation of subboundaries or shear bands across α laths followed by penetration of β along the α/α boundary. Higher angle α/α boundaries correspond with higher energy boundaries and as a result the β phase penetrates these high angle boundaries more readily. Penetration of the β phase occurs to balance interfacial energies between the α/α boundary and the α/β boundaries. The aligned recrystallized α grains resemble the original lamellar structure in the transition zone (Figure 6b). The lamellar arrangement, however, is not preserved in the SZ. In the present case, subgrain or shear band formation ahead of the tool and penetration of the β phase into α/α boundaries could sever the α laths into the equiaxed microstructure we observe in the SZ. The continued deformation and stirring imposed by the tool could cause further rotation of the equiaxed α grains to form high angle α/α boundaries. This is similar to the mechanisms based on continuous dynamic recrystallization that have been proposed for Al-Li-Cu [5] and 7050-T651 [6]. Jata and Semiatin [5] reported that dislocation glide gave rise to gradual rotation of subgrains present in the base material, however the recrystallized grains in the SZ were smaller than the subgrains in the base material suggesting that there is an intermediate step. Su et al. [6] elaborated on the continuous dynamic recrystallization mechanism to include the formation of fine subgrains by dynamic recovery. This occurs ahead of the tool and subsequent deformation causes rotation of the finer scale subboundaries to high angle boundaries.

X-ray diffraction results suggest that there is residual strain in the α phase in the SZ, as evidenced by both peak shift and peak broadening [14]. The experimentally observed broadening (B_{exp}), that is broadening beyond the integral breadth of the peak and any instrumental broadening effects, can be related to the amount of stored strain (ε) by (1), where d is interplanar spacing and θ is the location of the peak maximum [14].

$$\varepsilon = \frac{\Delta d}{d} = -\frac{B_{exp}}{2\tan\theta} \qquad (1)$$

Figure 7 shows the $(1\bar{1}20)_\alpha$ peaks for USC-2, USC-4 and the base material. The base material in this study was assumed to be strain free, so instrumental broadening effects can be ignored since the breadth of the peak for the base material would also contain any instrumental broadening effects. The slower tool velocity resulted in less peak broadening compared to the faster condition. Slower travel speed corresponds to hotter processing conditions which could allow more time for recovery to occur.

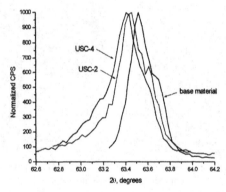

Figure 7. $(1\bar{1}20)_\alpha$ peak for USC-2, USC-4 and base material.

EWI

The microstructure in the SZ of the plate processed at EWI was considerably different than the plate processed at USC. The temperature during processing exceeded the β transus and the deformation was sufficient to recrystallize the β phase. The recrystallized prior β grain size was measured after FSP at depths of 0.5 mm and 1.5 mm in the center of the SZ and was similar for both tool travel speeds (Table II). Upon cooling from above the β transus, grain boundary α nucleated at prior β grain boundaries and there was subsequent nucleation and growth of α colonies. Grain boundary α and the α lamellae were both on the order of 1 μm wide. An example of the SZ for both processing conditions is shown in Figure 8.

Table II. Recrystallized prior β grain size for EWI reported in $\mu m^2/\mu m^3$.

	0.5 mm depth	1.5 mm depth
EWI-2	0.0528	0.0443
EWI-4	0.0527	0.0475

The SZ in the present study is very similar to that observed by Lienert et al. [15], Ramirez and Juhas [16] and Karogal [17] who reported FSW of mill-annealed and β-annealed Ti-6Al-4V. Our observations support the claim that the microstructure in the SZ is independent of the starting material microstructure and is instead dependent on the processing parameters [16]. On the other hand, the transition zone microstructure is dependent on the starting material microstructure. The TZ of the EWI plate had 1 µm to 2 µm equiaxed α grains (Figure 9) similar to those observed in the entire SZ of the USC plate. The volume fraction of transformed β was significantly higher in the EWI plate than in the USC plate indicating higher processing temperatures. The presence of retained α in the TZ confirms that this region did not exceed the β transus since there is no thermodynamic barrier to the dissolution of α upon heating. The most likely mechanism for the formation of the equiaxed α grains is similar to those in the USC plate. Deformation imposed by the compressive force or rotation of the tool introduces dislocations and causes buckling of the lamellae in the base material. The dislocations arrange into subgrains in the laths and there is subsequent penetration of the β phase into the α/α boundary causing either complete or partial severing of the α lath depending on the energy of the boundary. Again, a TEM study is necessary to confirm this mechanism, but it is quite similar to what has been reported in the literature for the conversion of lamellar microstructure to equiaxed microstructures [12, 13]. Upon cooling, the β phase in the TZ of the EWI plate transformed to a lamellar α + β structure.

Figure 8. The stir zone in (a) EWI-2 and (b) EWI-4.

The deeper plunge depth and higher tool rotation rate used in processing the EWI plate compared to the USC plate led to higher heat input during processing and a maximum temperature that clearly exceeded the β transus. Deformation occurred above the β transus which substantially refined the prior β grain size. The time spent above the β transus must have been relatively short considering that β grain growth is rapid and the refined β grain size is on the order of 30 µm.

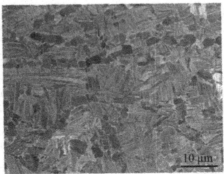
Figure 9. The transition zone in EWI-2 on the retreating side.

Conclusions

Two types of microstructures were observed as a result of friction stir processing an investment cast Ti-6Al-4V plate. A shallow plunge depth led to the formation of equiaxed α grains on the order of 1 μm. The EWI plate had a deeper plunge depth, wider pin and higher rotational speed, all corresponding with higher heat input, which resulted in maximum temperature exceeding the β transus and plastic strains which refined the prior β grain size from greater than 1 mm in diameter to approximately 30 μm in diameter.

The process variables examined were plunge depth, down force and travel speed. Among these, varying the tool travel speed seemed to have an insignificant effect on the diameter of the equiaxed α grains in the USC plate as well as the prior β grain size in the EWI plate. A slower tool travel speed resulted in less residual strain in the stir zone in the USC plate.

Acknowledgements

The author's acknowledge the financial support of The Office of Naval Research with Dr. J. Christodoulou as project monitor as well as Precision Cast Parts Company for providing the investment castings.

References

1. W.M. Thomas, E.D. Nicholas, J.C. Needham, M.G. Murch, P. Templesmith, C.J. Dawes, G.B. Patent Application No. 9125978.8 (December 1991).

2. B. London et al. "Friction Stir Processing of Nitinol," *Friction Stir Welding and Processing III*, TMS, San Francisco, CA, Feb. 2005.

3. A. Stahl and C.D. Sorensen, "Experimental Measurements of Load Distributions On Friction Stir Weld Pin Tools," *Friction Stir Welding and Processing III*, TMS, San Francisco, CA, Feb. 2005.

4. Y. Li et al. "Friction Stir Welding of TIMETAL 21S," *Friction Stir Welding and Processing III*, TMS, San Francisco, CA, Feb. 2005.

5. K.V. Jata and S. L. Semiatin, "Continuous Dynamic Recrystallization During Friction Stir Welding of High Strength Aluminum Alloys," *Scripta Mat.*, 43 (2000), 743-749.

6. J.-Q. Su et al., "Microstructural investigation of friction stir welded 7050-T651 aluminum," *Acta Materialia*, 51 (2003), 713-729.

7. B. Heinz and B. Skrotzki, "Characterization of a friction-stir-welded aluminum alloy 6013," *Metall. Mater. Trans. B* 33 (6) (2002), 489-498.

8. R.S. Mishra and Z.Y. Ma, "Friction stir welding and processing," *Mat. Sci. Eng. R*, (50) 2005, 1-78.

9. D. Eylon and R. Boyer, "Titanium Alloy Net-Shape Technologies," *in Proc. Int. Conf. Titanium and Aluminum*, Paris, Feb. 1990.

10. G. Lütjering and A. Gysler, "Fatigue," *Titanium Science and Technology, Procedings of the Fifth International Conference on Titanium, Munich, Germany, 1984.*

11. A.L. Pilchak, M.C. Juhas, and J.C. Williams: "Microstructural Changes Due to Friction Stir Processing of Investment Cast Ti-6Al-4V," *submitted to Met. Trans. A.*, 8/2006.

12. M. Peters, G. Lütjering, and G. Ziegler, "Control of Microstructures of (α + β)-Titanium Alloys," *Zeitschrift für Metallkunde*, (74) 1983, 274-282.

13. I. Weiss et al., "Modification of Alpha Morphology in Ti-6Al-4V by Thermomechanical Processing," *Met. Trans. A*, (17A) 1985, 1935-1947.

14. B.D. Cullity, *Elements of X-ray Diffraction 2nd ed.*, (Reading, MA: Addision-Wesley Publishing Company, 1978), 286-287.

15. T.J. Lienert et al., "Friction Stir Welding of Ti-6Al-4V Alloys," in: Proceedings of the Joining of Advanced and Specialty Materials III, (Materials Park, OH, ASM International, 2001), 160-167.

16. A.J. Ramirez and M.C. Juhas, "Microstructural Evolution in Ti-6Al-4V Friction Stir Welds" *THERMEC'2003, Mat. Sci. For.*, (426-432) 2003, 2999-3004.

17. N. U. Karogal, "Microstructural Evolution in Friction Stir Welding of Ti-6Al-4V," (M.S. Thesis, The Ohio State University, 2002).

FRICTION STIR PROCESSING OF A CAST WE43 Mg ALLOY

T. A. Freeney[1], R. S. Mishra[1], G. J. Grant[2] and R. Verma[3]

[1]Center for Friction Stir Processing, Materials Science and Engineering, University of Missouri,
Rolla, MO 65409, USA
[2]Pacific Northwest National Labs, Richland, WA 99356, USA
[3]General Motors R&D Center, Warren, MI 48090, USA

Keywords: Friction Stir Processing, WE43, cast magnesium, grain size

Abstract

A heat treatable sand cast WE43 magnesium alloy was friction stir processed (FSP) at 3 heat indices to determine the effect on microstructural and mechanical properties of the processed region. In addition, the sequence of solutionizing and aging heat treatment was investigated to maximize the processing efficiency, which is defined as the percent increase of FSP nugget yield strength over base yield strength. The maximum processing efficiency achieved for the casting was 125%, when a yield strength of 250 MPa was achieved in the FSP + T5 heat treatment condition. Compared to a yield strength of 202 MPa in the cast + T6 condition, this increase in mechanical properties is an attractive prospect for the local enhancement of Mg-Y-RE sand castings.

Introduction

Cast magnesium alloys are becoming increasingly popular for use as structural components in the automotive and aerospace industries. Due to the growing need among these industries to lower fuel consumption and satisfy emission guidelines the use of magnesium to replace components currently made from steel, aluminum, and polymer composites has become an attractive route [1,2]. The use of casting processes in these applications are beneficial due to the ease of creating complex shapes and internal cavities, which can ultimately result in the reduction of total part cost for assembly [3]. The benefit of magnesium casting is highlighted by exceptional fluidity, fast cycle time due to low latent heat, and good die life. State of the art magnesium alloys offer high specific strength at room and elevated temperatures, good creep and corrosion properties, good castability, exceptional machineability and dampening characteristics [4]. Despite all benefits of the casting process, mechanical properties can vary significantly due to solidification issues such as coarse interdendritic phases, large grain size, and interdendritic microporosity [5,6].

FSP has been proven to be an effective technology for the production of a fine grained, homogenous, porosity free microstructure in cast aluminum alloys, A356 and A319 [6,7]. In the studies by Ma et al. [6,8,9] on A356 a significant increase in mechanical properties, strength, ductility and fatigue life in the nugget region was reported. In magnesium casting FSP may be a viable process for local mechanical property improvement as well as for the repair of casting defects. FSP has been demonstrated on magnesium-aluminum-manganese (AM) and magnesium-aluminum-zinc (AZ) casting alloys AM60B, AM50, AZ91/6, AZ91D4, AZ95D/5 [4,10,11,12]. The aluminum-manganese and aluminum-zinc alloy series, such as in the previous studies are relatively inexpensive due to the relatively low cost of the alloying additions and ease of melt handling [13]. WE43 is a high strength casting alloy developed by Magnesium Elektron with excellent high temperature strength retention, corrosion resistance and long term temperature stability up to 250°C. However, the use of rare earth elements in this alloy (Y, Nd) as well as the attention to melt practices that must be taken makes the alloy more costly [14]. Cost plays an important role when deciding to repair or scrap a casting. In the case of expensive castings, repair and modification processes like FSP can become justifiable from a cost perspective.

Experimental Approach

In this study, the as received material was a sand cast WE43 component having a nominal composition of Mg-4.0Y-2.25Nd-1.0RE-0.6Zr (in wt. %The component was sectioned into six approximately 25mm thick pieces which were machined so the backing and processing surfaces were parallel. Single pass FSP was performed on all samples such that adequate time was given in between runs for the backing plate and tool to return to room temperature. The processing tool made from MP159 had a 12 mm shoulder with a conical threaded pin. For mechanical testing, mini-tensile samples with 4.2 mm gage length and 1 mm gage width were machined from the nugget oriented with gage length being entirely in the nugget through transverse section. Tensile samples were ground down to a final thickness of 0.5-0.55 mm and given a 1μm polished finish before testing. Tensile testing was done at a strain rate of $1\times10^{-3}s^{-1}$.

Since heat treatable alloys are rarely used in the as cast condition, the effects of heat treatment on the nugget properties are important. To understand the effect of heat treatment on microstructural and mechanical properties the sequence of heat treatment was varied before and after FSP according to the experimental schedule shown in Table I.

Table I Summary of processing parameters and heat treatment conditions.

Tool Rotation Rate (rpm)	Tool Traverse Speed (mm/min)	Heat treatment schedule			
		Cast + FSP	Cast + FSP +T6	Cast + FSP +T5	SS + FSP +T5
400	102	x	x	x	x
700	153	x	x	x	-
900	127	x	-	x	x

The T6 treatment for WE43 casting involved solutionizing (SS - 520°C for 8h), warm water quench (60-80°C), and ageing (250°C for 16h). The T5 treatment involved aging at 250°C for 16h. The microstructure was studied by mounting, mechanically polishing and etching of the nugget cross section with a mixture of acetic acid, nitric acid and ethanol. Images for microstructural analysis were taken with the use of optical microscopy (OM) and scanning electron microscopy (JEOL 840A); analysis was done with ImageJ software package.

Results and Discussion

Fig. 1 shows SEM images of the polished as received WE43 casting. The microstructure consists mainly of α-magnesium dendrites and β-yttrium-neodymium rich lamellar interdendritic eutectic regions, marked 'a'. The needle-like precipitates of similar composition to the interdendritic phase are located at grain boundaries and sparsely through the matrix, labeled 'b'. The α-Zr cored structure which is a characteristic phase in many Mg-Zr alloys reported by [15,16,17] and are thought to be formed through a peritectic reaction upon solidification [17], is marked 'c'. Current understanding is that, zirconium contributes to very effective grain refinement by two primary mechanisms: constitutional undercooling due to a high growth restriction factor [15] and grain nucleation due to magnesium and zirconium having the same crystal structure and similar lattice parameters [16].

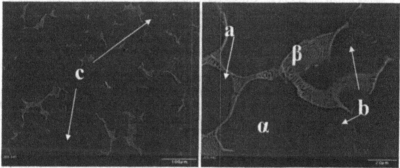

Fig. 1. SEM images of as-cast WE43. Apart from α-Mg, three other phases were identified through EDS: interdendritic lamellar RE (Y-Nd) rich eutectic phase marked 'a', needle precipitate of RE (Y-Nd) composition marked 'b', and α-Zr rich cored structure marked 'c'.

In the cast + T6 condition, the coarse β phase is significantly reduced when compared to the as-cast material. The α-zirconium clusters also partially dissolve and precipitate into a more regular spacing which can be seen in the OM image of the cast + T6 material (Fig. 2). The cast + T6 microstructure had an average grain size of 80±5 μm determined through mean linear intercept technique.

431

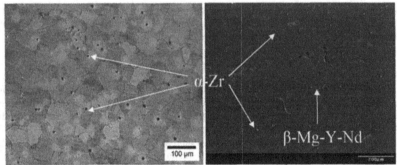

Fig. 2. OM (left) and SEM image (right) of cast + T6 condition. The location of the α-Zr clusters is apparent and is generally centrally located in a grain or on the boundary. SEM image shows incomplete dissolution of β phase and solidification defects.

Fig. 3 shows macro images of the cross section for the three FSP conditions. Fig. 4 shows the microstructures taken from the center of the nugget for the 3 different heat inputs in the T5 condition. The grain sizes in order of increasing heat input were 3.9±0.32μm, 5.6±0.32 μm and 7.1±0.48 μm. This is a significant refinement over the T6 grain size of 80±5 μm. The microstructures show significant difference in the fraction of dissolution of β phase and the corresponding precipitation in the matrix upon aging. At the lowest heat input dissolution was moderate, and the undissolved β phase could easily be seen on the grain boundaries. At the middle, heat input dissolution was almost complete and intragranular precipitates were observed. At the highest heat input the β phase completely dissolves and re-precipitates on grain boundaries as a large intergranular connected phase. The present results indicate an optimum heat input from the standpoint of β phase dissolution and re-precipitation.

In agreement with investigations done on FSP 1050Al and 7075Al-T651 [18], the peak temperature of the FSW/P thermal cycle is the primary contributor to grain size in WE43. Charit et al. [19] investigated the grain size of aluminum alloys 7075 and 2024 during FSW. Their study found no monotonic increase in grain size versus pseudo-heat index over a very broad heat input range. Yet for similar heat input, the grain sizes achieved in 2024Al seemed to correlate fairly well with the results from this study. In Fig. 5(a) the correlation between the average grain size and pseudo heat index is shown. A logarithmic relationship was found to exist between grain size and pseudo heat index,

$$\text{Mean Grain Size} = 5.2 * \log (\text{Rotation Rate}^2 / \text{Traverse Rate})-12.7 \qquad \text{(Eq.1)}$$

The grain size distribution in the center of the nugget for the three heat indices was analyzed to determine how grain distribution was effected by FSP (Fig. 5(b)). It was found that grain size distribution followed a log-normal distribution typical of a recrystallized material. The grain size of a material affects the mechanical behavior of that material in service situations. To predict the behavior of a material an average grain size is generally insufficient.

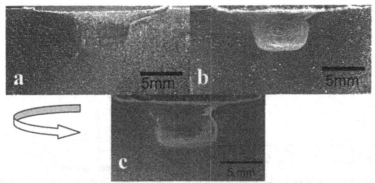

Fig. 3. Macro image of nugget geometry and onion ring structure for a) 400 rpm 102 mm/min, b) 700 rpm 153 mm/min and c) 900 rpm 127 mm/min. Advancing side is on the right for all images.

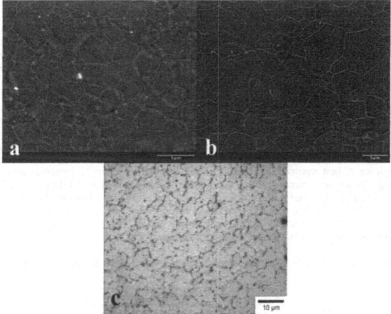

Fig. 4. SEM (a,b) and OM (c) image of nugget grains at respective increasing heat inputs of 400 rpm 102 mm/min, 700 rpm 153 mm/min and 900 rpm 127 mm/min in the T5 condition. Grain size increases with increasing heat input, respectively 3.9 μm, 5.6 μm and 7.1 μm. Bright white particles in SEM Image are α-Zr, and light gray particles are β phase. In OM α-Zr are black, and dark gray particles are β phase.

433

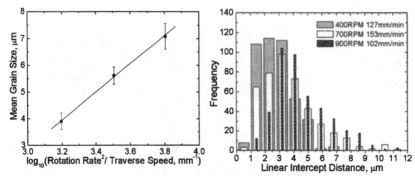

Fig. 5(a). Variation of mean linear grain size in the central region of the nugget with pseudo heat index. (b) A plot of frequency variation with intercept distance in the central region of the nugget for different heat indices.

Rhodes et al. postulated through quench studies of 7050Al that the fine grained FSP microstructure was the result of nucleation and growth [20]. In the post FSP microstructure the mean grain size at each heat index was found to be skewed toward the smaller grains. As the peak temperature and thermal cycle in the processed region increase, grain growth occurs more rapidly, and smaller grains are consumed shifting the mean of the distribution away from the median value.

In order to maximize the strength of the parent material, a T6 heat treatment involving a solutionizing and aging step is performed. These results will serve as a baseline comparison for further discussion (Table II). Although a T6 heat treatment is necessary to strengthen the base casting, the fine grained FSP microstructure is unstable at the solutionizing temperature (Fig. 6), resulting in a significant coarsening of grains. Due to the instability of the nugget microstructure, a sequence of heat treatment must be developed in order to maximize properties across the casting. The sequence investigated in this study involved solutionizing (SS) the casting prior to processing (FSP), then aging after processing. These results have been compared to cast + T6, as FSP, and FSP + T5. The processing condition that provided best strength was 700 rpm-153 mm/min. The microstructure that corresponded to this parameter had the combination of highest volume precipitation of β in the aged sample and fine grain size.

Table II Summary of mechanical properties for the as cast and cast + T6 heat treatments.

Tensile Properties	Sample Condition	
	As Cast	Cast + T6
UTS (MPa)	181 ± 9.6	261.3 ± 5.2
0.2 %YS (MPa)	136.8 ± 14	202 ± 6
%Elongation	1.9 ± 0.6	5

Fig. 6. OM of WE43 FSP nugget in T6 condition illustrating excessive grain growth caused by the instability of the fine grained microstructure at solutionizing temperature of 520°C.

This indicates that in WE43, dissolution, precipitation and subsequent coarsening of the β phase has a more significant impact on strength than the grain size variation due to grain growth. Fig 7 gives the tensile properties of all heat treatment steps and the variation with heat input. From Fig. 7(a), it can be noted that the UTS does not vary much with heat input and aging only contributes to a slight increase. Also noted is that solutionizing prior to processing imparts a slight increase in UTS. Fig. 7(b) shows a large variation in yield strength due to aging effects. The lower heat input exhibits the lowest strength due to the incomplete homogenization of second phase particles. It can be seen that solutionizing the casting prior to processing at this heat index has increased the YS by 15 MPa. The middle heat input shows the same trend, yet achieves a higher strength due to more complete homogenization because of the increased thermal cycle during processing. For the highest heat index it can be argued that the processing temperature is so high that coarsening of precipitates at grain boundaries depletes intragranular precipitates (Fig. 4c). Fig. 7(c) shows a substantial increase in % elongation from the lowest heat input to the middle heat input, but seems to reach a plateau of 25 %. For both tensile and yield strength the properties across heat index rise and fall, yet the elongation increases across heat index due to the increase in uniformity and distribution of equiaxed grains.

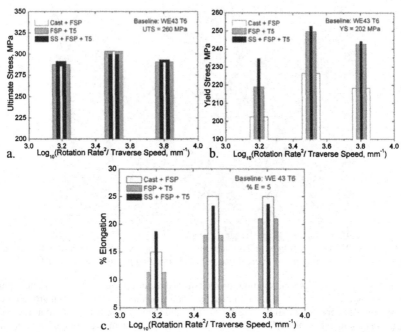

Fig. 7. Variation of nugget mechanical properties with processing thermal input. Note that the grain sizes for three increasing heat indices are: 3.9±0.32μm, 5.6±0.32 μm and 7.1±0.48 μm.

Conclusions

1. Friction stir processing of a cast WE43 magnesium alloy resulted in significant grain refinement and dissolution of the β-Mg-Y-Nd phase.
2. The impact of the dissolution of β phase had more influence on strength than the variation in grain size due to FSP thermal cycle.
3. A solutionizing step prior to FSP and subsequent aging resulted in the best mechanical properties due to increased homogenization.
4. A maximum ultimate tensile strength of 303 MPa, 0.2% yield strength of 253 MPa and a elongation of 17% were achieved with processing parameters of 700 rpm and 153 mm/min. These values resulted in a 125% processing efficiency when compared to the cast + T6 condition.

References

1. J. Aragones et al., "Development of the 2006 Corvette Z06 Structural Cast Magnesium Crossmember," *SAE International* 2005-01-0340.
2. S. Schumann and H. Friedrich, "Current and Future use of Magnesium in the Automobile Industry," *Materials Science Forum*, 51(2003), 419-422.

436

3. N. Li et al., 2006 Ford GT Magnesium Instrument Panel Cross Car Beam," *SAE International* 2005-01-0341.
4. M. Santella et al. "The Use of Friction-Stir Technology to Modify the Surfaces of AM60B Magnesium Die Castings," *JOM* May 2006. V. 58 No.5.
5. B.J. Coultes, J.T. Wood, G.Wang, R. Berkmortel, "Mechanical Properties and Microstructure of Magnesium High Pressure Die Castings," *TMS* Magnesium Technology 2003, 45-50.
6. Z.Y. Ma, S.R. Sharma, and R.S. Mishra. "Effect of multiple pass friction stir processing on microstructure and tensile properties of a cast aluminum-silicon alloy," Scripta Mater., 54 (2006) 1623-1626.
7. M. L. Santella. T. Engstrom, D. Storjohann, and T. Y. Pan, Scripta Mater., 53 (2005) 201.
8. Z.Y. Ma, S. R. Sharma, R. S. Mishra, and M. W. Mahoney, Mat. Sci. Forum 426-432 (2003) 2891.
9. Z.Y. Ma, S. R. Sharma, and R. S. Mishra, "Effect of friction stir processing on the microstructure of cast A356 aluminum," Mat. Sci. A: 433 (2006) 269-278.
10. R. Johnson, Mater. Sci. Forum, 419–422 (2003) 365.
11. W.B. Lee, Y.M. Yeon, and S.B. Jung, Mater. Sci. Tech. 19 (2003) 785.
12. G. Kohn, S. Antonsson, and A. Munitz, in: S.K. Das (Ed.), Automotive Alloys 1999, TMS, 2000, pp. 285–292.
13. P. Lyon, "New Magnesium Alloy for Aerospace and Specialty Applications," Magnesium Technology 2007, TMS, 2007, pp.311-315.
14. P.Lyon, "Elektron WE43," Foundry Processing, Phoenix 1998
15. Ma Qian, D. Graham, L. Zheng, D. H. StJohn and M. T. Frost, Mat. Sci. & Tech., 19 (2003) 156-162.
16. Z. K. Peng et al. "Grain Refining Mechanism in Mg-9Gd-4Y alloys by Zirconium," Mat. Sci. & Tech., 21 (2005) 722-726.
17. D. H. StJohn, M. Qian, M. A. Easton, P. Cao, Z. Hildebrand. "Grain Refinement of Magnesium Alloys," Metall. & Mat. Trans. A, 36A (2005) 1669.
18. R.S. Mishra, and Z.Y. Ma, in: A.G. Cullis and S.S. Lau (Eds), "Friction Stir Welding and Processing: A Review Journal," Mat. Sci. 50 (2005) 1-78.
19. Y.C. Lee, A. K. Dahle and D.H. StJohn: Metall. & Mater. Trans. A, 31A (2000) 2895-2906.
20. C.G. Rhodes, M.W. Mahoney, et al. "Fine-grain Evolution in Friction-Stir Processed 7050 aluminum", Scripta Materialia 48 (2003) 1451-1455

FRICTION STIR MICROSTRUCTURAL MODIFICATION OF INVESTMENT CAST F357

S. Jana[1], R. S. Mishra[1], H. N. Chou[2] and D. R. Herling[3]

[1]Center for Friction Stir Processing, Materials Science and Engineering, University of Missouri,
Rolla, MO 65409, USA
[2] The Boeing Company, St. Louis, MO 63166, USA
[3]Pacific Northwest National Laboratory, Richland, WA 99356, USA

Keywords: Aluminum alloys, friction stir processing, casting modification

Abstract

A hypoeutectic Al-Si alloy has been friction stir processed in this study using various run parameter combinations. Tensile test results indicate at least three times improvement in ductility value over as-cast T6 condition because of refinement in Si particle size. Si particle size and shape has been quantified and correlated with mechanical properties. Tool rotation rate seems to have the most significant effect on properties. Higher tool rotation rate resulted in more uniform and homogeneous microstructure though some anomaly is observed at very high tool rotation rate.

Introduction

In today's world of energy crisis, aluminum alloys are gaining more and more importance because of their high specific strength and moderate cost. One of the commercially important aluminum casting alloy is F357 because of its good castability, corrosion resistance and high strength and it is placed into the group of "premium casting alloys" [1]. The microstructure of as-cast F357 alloy consists of primary aluminum dendrites with coarse acicular eutectic Si particles in the interdendritic region which causes limited ductility of about 3 % [1].

In a cast aluminum alloy, most important factors responsible for mechanical properties are (1) Grain size and shape (2) secondary dendrite arm spacing and (3) size and distribution of second phase particles and inclusion [1]. Out of these, first two factors depend on cooling rate and therefore on a particular casting technique. Control of inclusion size and distribution can again be done by controlling cooling rate. Rapid cooling will result in a finer distribution of constituent particles [1]. On the other hand size and shape of second phase particles e.g. primary Si particles can be changed by using chemical modifiers. Chemical modification is achieved by the addition of certain elements such as calcium, sodium, strontium, and antimony [1]. The other alternative is thermal modification which can be done by exposing the casting to high temperature. Solution heat treatment can result in spherodization and coarsening of Si particles [2, 3]. None of the available methods can redistribute Si particles uniformly throughout the matrix which would be essential in order to achieve good mechanical properties, especially ductility.

Recently, friction stir processing technique was applied to modify microstructure of a sand-cast A356 alloy which resulted in break up and dispersion of acicular Si particles resulting in enhanced mechanical properties [4-8].

439

The principle of friction stir processing (FSP), a development based on friction stir welding (FSW) [9], is quite simple. A rotating tool with pin and shoulder is first plunged into a single piece of material and then traversed along the desired path to create a processed zone. FSP causes severe plastic deformation and extensive material flow in the processed zone. This helps in refinement and mostly uniform distribution of Si particles in the processed zone.

In this study, FSP is applied to investment cast sheets of F357 alloy, a higher strength version of A356, to see its effect on mechanical properties.

Experimental Procedure

Commercial investment cast plates of alloy F357 having nominal composition of 7.0 Si-0.6 Mg-Al (bal.) (in wt. pct.) were used for this study. The alloy was received in the shape of investment cast boxes without any heat treatment. The boxes were cut to obtain individual sheets of ~ 3.30 mm thickness. The sheets were friction stir processed using a tool having conical pin with steps (Fig.1). Processing parameters used for this study are listed in Table I.

Fig. 1 Image of the tool used.

Table I. Process parameters used for the present study

Tool rotation speed, rev/sec	Tool traverse speed, mm/sec	Heat index [Tool rotation speed^2 (rev/min) / Tool traverse speed (in/min) x 10^4]	Pitch (Tool traverse speed/ Tool rotation speed) μm
37	0.98	215	26
32	0.64	253	20
42	2.33	113	56
27	2.33	46	86
16	3.67	11	229

Microstructural characterization of the as-processed sheets was done using optical microscopy in the transverse section. As polished samples without any etching were investigated as it led to better delineation of Si particles and other phases present in the alloy. Image J software was used to carry out size and aspect ratio analysis of Si particles. Each Si particle in the micrograph was approximated to nearest ellipse. The size of individual Si particle was defined in terms of equivalent diameter, D ($D = (d_{maj}d_{min})^{1/2}$), where d_{maj} and d_{min} refers to major and minor axis of the ellipse. The aspect ratio of the particles was defined as d_{maj} / d_{min}.

To investigate, the effect of heat treatment on properties, part of the processed material was solution treated at 540 °C for 8 hours followed by hot water quenching. Aging was done at 170 °C for 3 hours after the solution treatment. The heat treatment procedure conformed to handbook T6 specification [1].

For mechanical property evaluation, mini tensile samples of 4.3 mm gage length and 1.0 mm gage width were machined using a desktop CNC. The samples were obtained in the longitudinal direction with gage center along the processing centerline. The locations of the samples were at a depth of 1.5 – 2.0 mm from top surface. After machining, individual tensile samples were ground to 3 μm surface finish and a final thickness of ~ 1.0 mm. Tensile tests were conducted at room temperature using a computer controlled, custom built mini tensile machine employing an initial strain rate of 1×10^{-3} s $^{-1}$. Mini tensile testing was done for material in as-cast, as-cast + T6, as-FSP and FSP + T6 condition to compare the properties. Fracture surface of tensile samples were investigated in SEM. Also, longitudinal cross section of the failed samples were mounted, polished and observed using optical microscope.

Results and Discussion

Fig. 2a-2d shows optical micrographs of the parent material in as-cast and T6 heat treated conditions. As expected, the as-cast structure consisted of primary Al dendrites with eutectic Si particles in the interdendritic region. The average secondary dendrite arm spacing (SDAS) was calculated to be around 100 μm. Apart from Si particle, presence of π phase, an intermetallic compound rich in Fe, was also observed in the interdendritic region (Fig. 2b). As the material was given T6 heat treatment, the long acicular Si particles got rounded because of exposure to high temperature during solution treatment (Fig. 2c). Also some precipitation could be observed along the grain boundaries in T6 condition (Fig. 2d).

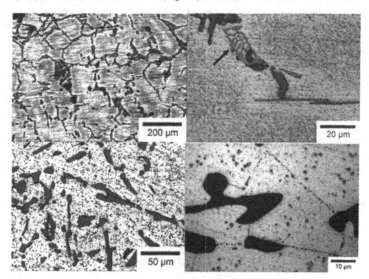

Fig. 2 Micrographs of parent material in various conditions; (a) dendritic pattern of Al with Si particles in interdendritic region, (b) Presence of Fe-rich π phase as shown by arrow, (c) Si particles getting rounded after T6 treatment, and (d) grain boundary precipitation after T6 treatment.

In Fig. 3 macro image of processed zone cross-section is shown with the tool outline superimposed on it. The processed zone had a hemispherical shape with advancing side on the right and retreating side on the left. Specimens for mechanical property evaluation as well as microstructural quantification were obtained from the region marked by white dashed line.

Fig.3 Macro image of the processed zone cross section. The tool outline is shown to highlight the relative sizes of pin and nugget.

In Fig. 4, a montage of the processed zone cross section is shown to give a better view of the material flow. It was observed that there is a distinct boundary on the advancing side between processed and unprocessed material. At higher magnification, a very thin layer of thermo-mechanically affected zone (TMAZ) was noticed along the boundary where Si particles were partially broken (Fig. 5a). Inside the nugget region, an elliptical area was observed (Fig. 4) where Si particles were much finer and mostly equiaxed as compared to TMAZ. In earlier reports [5, 6], it was shown that lower tool rotation rate (300-500 rpm) leads to basin shaped nugget with wide top region. Increasing tool rotation speed to 900 rpm changes the nugget to an elliptical shape together with decrease in size. It also led to generation of onion ring patterns. Apart from that, a macroscopically visible banded pattern was also reported for some run parameters combinations. Present results show a basin shaped nugget (Fig. 3) observed for all tool rotation speeds i.e. 960–2500 rpm. Onion ring pattern was also observed in the present study (Fig. 5b). When run parameter combinations of the present study are compared against the earlier reports, it was noticed that tool rotation speed is much higher in the present case while tool traverse speed is comparable. So it seems onion ring pattern is related to tool rotation rate directly while shape of the nugget must depend on some other parameter. As noted earlier, maximum refinement of Si particles took place inside the elliptical area. This region is asymmetric with reference to the centerline of the run, remaining more shifted towards advancing side. Towards the retreating side, dendritic pattern was observed. Si particles in this region were again partially broken as compared to unprocessed zone but primary Al grains remained mostly unchanged (Fig. 5c).

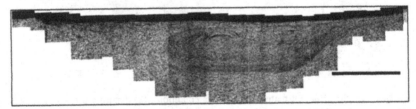

Fig. 4 Montage of the processed zone cross section.

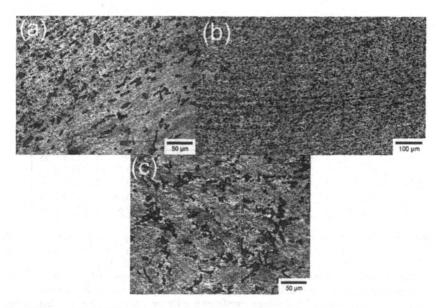

Fig. 5 Micrographs of different region within the processed zone; (a) TMAZ along the advancing side, (b) onion ring pattern in the nugget, and (c) dendritic morphology in the retreating side.

In Fig. 6, representative micrographs show Si particle size and shape in as-FSP and FSP + T6 conditions. FSP resulted in considerable breakdown of Si particles in the nugget area as well as reduction in aspect ratio. In T6 condition, the morphology of Si particles was even better because of rounding of sharp corners. The distribution of the particles in the nugget zone was quite uniform.

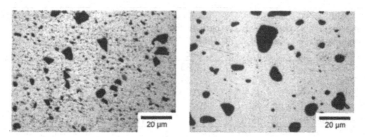

Fig. 6 Micrographs from nugget region showing Si particle size and shape, (a) as-FSP and (b) FSP + T6

Table II summarizes the average particle size and aspect ratio of the particles. It can be noted from Table II, average particle size remains quite uniform for all the processing conditions, though there is a slight increase in the particle size when processed at lower tool rotation rate. The particle size in T6 condition did not alter much. T6 heat treatment reduced the aspect ratio as sharp corners were rounded. Tool rotation speed is one of the most important criteria for Si particle break up as reported by others [4-6]. No significant trend can be observed in the present data (Table II) though it appears that the average particle size and aspect ratio is minimum at 1950 rpm.

Table II. Average particle size and shape analysis of Si particles for different run conditions in nugget area.

Run parameter combinations (rpm;ipm)	Average particle size, (as processed) μm	Average particle size, (FSP + T6) μm	Average aspect ratio, (as processed)	Average aspect ratio, (FSP + T6)
2236; 2.32	3.46 ± 2.41	3.36 ± 1.74	2 ± 0.6	1.63 ± 0.48
1950; 1.5	3.13 ± 2.32	3.91 ± 1.99	2.14 ± 0.72	1.58 ± 0.45
2500; 5.5	3.48 ± 2.56	3.9 ± 2.4	2.18 ± 0.74	1.7 ± 0.51
1600; 5.5	3.41 ± 2.32	3.51 ± 2.15	2.02 ± 0.64	1.66 ± 0.49
964; 8.68	3.68 ± 2.93	3.61 ± 2.21	2.07 ± 0.66	1.72 ± 0.56

The size of the largest particles as well as average of top 10 % of particle population for each processing condition was also determined. Coarser particles will act as preferential crack nucleation sites. Therefore, information about larger particles could help in obtaining better correlation between microstructure and properties. The values of largest and average size of top 10 % particle population are listed in Table III while Table IV reports the aspect ratio values.. The largest particle size did not change significantly with the process parameters. Even when average values were considered the process parameters do not seem to have significant influence, though lower tool rotation rate again resulted in larger particles. From Table IV, it can be noted that largest particles were more elongated in shape which is evident from their high aspect ratio. T6 treatment led to decrease in aspect ratio due to rounding of sharp corners. Aspect ratio of top 10 % particles (Table IV) is observed to be higher than average aspect ratio (Table II) in both as-FSP and FSP + T6 condition. This would be detrimental towards achieving better properties.

Table III. Largest and average size of top 10 % particle population

Run parameter combination (rpm;ipm)	Largest particle size, (as-FSP) μm	Largest particle size, (FSP + T6) μm	Average particle size (top 10 %), (as-FSP)	Average particle size (top 10 %), (FSP + T6)
2236; 2.32	22.5	12.83	9.04 ± 2.78	7.31 ± 1.68
1950; 1.5	21.75	15.33	8.65 ± 2.6	8.51 ± 1.72
2500; 5.5	17.96	17.99	9.52 ± 2.68	9.49 ± 2.32
1600; 5.5	22.76	16.78	8.83 ± 2.65	8.66 ± 2.28
964; 8.68	25.55	14.01	10.67 ± 3.38	8.86 ± 1.85

Table IV. Aspect ratio of largest and top 10 % of particle population

Run parameter combination (rpm;ipm)	Aspect ratio of largest particle (as-FSP)	Aspect ratio of largest particle (FSP + T6)	Mean aspect ratio of top 10 % particle (as-FSP)	Mean aspect ratio of top 10 % particle (FSP + T6)
2236; 2.32	4.6	3.91	3.2 ± 0.34	2.75 ± 0.35
1950; 1.5	5.22	3.88	3.57 ± 0.54	2.62 ± 0.39
2500; 5.5	5.74	4.76	3.72 ± 0.51	2.82 ± 0.41
1600; 5.5	4.82	4.48	3.34 ± 0.41	2.71 ± 0.39
964; 8.68	5.05	4.19	3.38 ± 0.48	2.94 ± 0.43

Mechanical properties of as processed and heat treated samples obtained by mini tensile testing are summarized in Tables V and VI. It is very clear that FSP led to significant improvement in mechanical properties over the unprocessed alloy. Strength values improved in the processed condition and most importantly, percent elongation to failure improved significantly for the FSP alloy. Elongation to failure remained quite uniform for processed and heat treated samples for all the run conditions. This implies a wide window over which FSP can be performed. Combined improvement in strength and ductility usually enhances the fatigue properties as it has already been reported [10]. Cross section metallography of the fractured tensile samples showed that larger particles did act as preferential crack nucleation sites. Fig. 9 shows one such fractured Si particle. Fractography of failed tensile samples in SEM revealed cleavage mode of fracture in as-cast alloy (Fig. 10a). Dimpled rupture pattern was observed in as-FSP alloys which is consistent with the high ductility values obtained (Fig. 10b). As-FSP fracture samples when observed at higher magnification revealed Si particle located inside most of the dimples. Each individual Si particle fractured by cleavage mode, but as their size was small, surrounding matrix could elongate much more and therefore resulted in much higher ductility.

Table V. Mechanical properties of the processed alloy.

Run parameter combination (rpm; ipm)	As cast	2236; 2.32	1950; 1.5	2500; 5.5	1600; 5.5	964; 8.68
Yield strength (MPa)	116	137	107	118	119	128
Tensile strength (MPa)	134	187	173	197	163	164
Elongation (%)	1.0	21.5	26	33	11	14.3

Table VI. Mechanical properties of the processed and heat treated alloy.

Run parameter combination (rpm; ipm)	As cast + T6	2236; 2.32	1950; 1.5	2500; 5.5	1600; 5.5	964; 8.68
Yield strength (MPa)	203	245	218	250	235	250
Tensile strength (MPa)	230	294	258	293	275	295
Elongation (%)	2.75	11.5	11.5	11	12.6	8.0

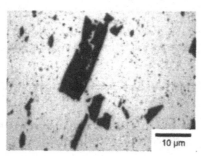

Fig. 9 Cross section micrograph of failed sample, a cracked Si particle can be seen.

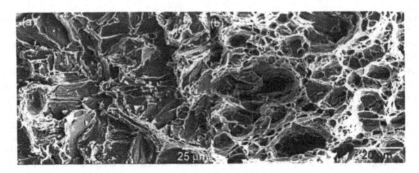

Fig. 10 Fractographs of mini tensile samples, (a) as-cast alloy showing cleavage morphology, and (b) as-FSP alloy showing dimpled rupture morphology.

Conclusion

Friction stir processing led to 3-4 times improvement in ductility value and around 50 MPa increment in both yield strength and ultimate tensile strength over the unmodified alloy in T6 condition. The improvement in ductility was even better when alloy was not heat-treated. This result conforms to results achieved for A356 [4-8, 10], a lower strength version of the present alloy. Lower tool rotation rate resulted in coarser Si particles and smallest particle size was obtained at an intermediate tool rotation speed. In general, for all the run parameter combinations studied, there was significant improvement in mechanical properties which implies a broad processing range for FSP.

Acknowledgement

This work was performed under the NSF-IUCRC for Friction Stir Processing and the additional support of NSF, GM and Friction Stir Link for the UMR site is acknowledged. This report was prepared as an account of work sponsored by an agency of the United States Government. The views and opinions of authors expressed herein do not necessarily state or reflect those of the United States Government or any agency thereof.

References

1. ASM handbook, vol. 2, Properties and Selection: Nonferrous Alloys and Special-Purpose Materials.

2. F. Paray, J E Gruzleski, "Microstructure-Mechanical Property Relationships in a 356 Alloy. Part I. Microstructure," *Cast Met.* Vol. 7, no. 1 (1994), 29-40.

3. G. Atxaga, A. Pelayo, A.M. Irisarri, "Effect of microstructure on fatigue behavior of cast Al – 7Si –Mg alloy," *Mater. Sci. Technol.*, 17 (2001), 446 – 450.

4. Z .Y. Ma, S. R. Sharma, R. S. Mishra, M. W. Mahoney, "Microstructural modification of cast aluminum alloys via friction stir processing," *Materials Science Forum.* Vol. 426-432, Part 4 (2003), 2891-2896.

5. Z .Y. Ma, R. S. Mishra, M. W. Mahoney, "Friction stir processing for microstructural modification of an aluminum casting," (Paper presented at Friction Stir Welding and

Processing II held at the 2003 TMS Annual Meeting; San Diego, CA; USA; 2-6 Mar. 2003), 221.

6. Z.Y Ma, S. R. Sharma, R. S. Mishra, "Effect of friction stir processing on the microstructure of cast A356 aluminum," *Mater. Sci. Engg A.*, 433 (2006), 269 – 278.

7. M. L. Santella, T. Engstrom, D. Storjohann, T. Y. Pan, "Effects of Friction Stir Processing on Mechanical Properties of the Cast Aluminum Alloy A356," *SAE Transactions: Journal of Materials & Manufacturing.* Vol. 114 (2006), 599-603.

8. M. L. Santella, T. Engstrom, D. Storjohann, T. Y. Pan, "Effects of friction stir processing on mechanical properties of the cast aluminum alloys A319 and A356," *Scripta Materialia.* Vol. 53 (2005), 201-206.

9. W. M. Thomas, E. D. Nicholas, J. C. Needham, M. G. Murch, P. Templesmith, C. J. Dawes, International Patent No. PCT/GB92/02203, (1991).

10. S. R. Sharma, Z. Y. Ma, R. S. Mishra, "Effect of friction stir processing on fatigue behavior of A356 alloy," *Scripta Materialia.* Vol. 51 (2004), 237-241.

CORROSION IN FRICTION STIR WELDED DISSIMILAR ALUMINUM ALLOY JOINTS OF 2024 AND 7075

Christian A. Widener[1], Jorge E. Talia[2], Bryan M. Tweedy[1], and Dwight A. Burford[1]

[1]National Institute for Aviation Research
[2]Department of Mechanical Engineering
Wichita State University
1845 Fairmount St., Wichita, KS 67260

Keywords: Friction stir, dissimilar alloys, 2024, 7075, artificial aging, corrosion

Abstract

Dissimilar aluminum alloy joints are often observed to be susceptible to corrosion, especially in 2XXX and 7XXX series alloys. Researchers have generally found this to be the case in friction stir welded joints as well; however, an initial temper and post-weld artificial aging combination has been identified which results in substantially improved exfoliation response when 0.125-inch 2024-T81 is joined to 7075-T73. The improved corrosion response is attributed to the overaged condition of both alloys which allows for a common post-weld artificial aging treatment and the placement of the 2024-T81 material on the advancing side of the joint. Samples were evaluated using optical microscopy, exfoliation, electrical conductivity, microhardness, and tension testing. While it was noted that the 7075-T73 material was preferentially attacked over the 2024-T81 material, no pitting or selective attack at the dissimilar interface was observed.

Introduction

A great advantage of friction stir welding (FSW) is its ability to join aluminum alloys previously unweldable by conventional fusion welding processes. Further, FSW has been shown to create strong joints between any combination of dissimilar aluminum alloys, whereas fusion welding techniques are very limited in their ability to join dissimilar alloys. Dissimilar aluminum alloys are often used in aerospace applications due the benefits of tailoring the properties of structures, e.g. maximizing strength in one location and fracture toughness, fatigue, or corrosion resistance in another. Before FSW can be more broadly applied to dissimilar alloy applications in aerospace and other industries, however, the nuances of joining dissimilar alloys need to be better understood, especially with regards to corrosion. To that end, two of aerospace's most common aluminum alloys, 2024 and 7075, have been investigated to lay the foundation for exploring trends in next generation overaged 2XXX and 7XXX alloys.

Background

Microstructure of Dissimilar Aluminum Alloys

The microstructure of dissimilar aluminum alloy joints is similar to that of each alloy on their respective portion of the joint; however, in a dissimilar alloy joint, there is the added complexity of a mixed zone (interface) in the weld nugget or dynamically recrystallized zone (DXZ). In the weld nuggets of dissimilar welds, the mixing of dissimilar materials are clearly seen using optical microscopy. Lee et al. [1] showed that the DXZ was dominated by the material that was placed on the advancing side of the weld, which in turn affected the microhardness and strength of the weld. Larson et al. [2] showed that although the dissimilar materials were intimately mixed, as evidenced by grains growing across dissimilar material boundaries in the nugget, the diffusional mixing between the two materials was limited to a very narrow zone at the boundary between the two materials. This was also concluded by Bala Srinivasan et al. [3], who felt that the inherent differences in chemistry observed within the nugget region clearly suggested that the dwell time at high temperature is insufficient for diffusion of elements during FSW.

Research by Nelson et al. [4] has shown, that in the case of 2024-T3, this alloy will become essentially stable after approximately 100-hours of natural aging, while 7075 will continue to naturally age at room temperature beyond 1,000-hours in the as-welded condition, as would be expected for 7075. Therefore, a degree of artificial aging is at least recommended to stabilize welds in 7075 and probably 7XXX series alloys in general. Gérard and Ehrström [5] reported the potential for welding 2XXX series alloys in the T8 condition to 7XXX series alloys, but made no mention of the effects of this type of approach on the corrosion resistance of the dissimilar couple.

Strength of Dissimilar Aluminum Alloys Joints

Few reports have been published on the performance of dissimilar joints between Al 2024 and Al 7075 or on dissimilar joints of any alloy, for that matter. In general, though, the strength of the joint has been found to be limited by the heat affected zone (HAZ) of the weaker material. Larson et al. [2] determined that in a dissimilar joint of Al 5083 and Al 6082, placing the material with the lower high temperature strength on the advancing side would more likely produce the best mechanical properties. Similarly, Lee et al. [1] concluded that when welding A356 aluminum to Al 6061, the highest strengths were achieved when the harder material was placed on the retreating side. Gérard and Ehrström [5] achieved similar results in an investigation of joints with dissimilar 7XXX series and 2XXX series alloys, noting that the joint always failed in the weaker material, and was slightly stronger with the weaker material on the advancing side. Cook et al. [6], looked at the effect of alternating advancing and retreating sides in a dissimilar joint of 2024-T3 and 7075-T73. Both joints performed well, but again reached the same conclusion. Cavaliere et al. [7] welded 0.100-inch 2024-T3 to 7075-T6 and reported the tensile and fatigue life data. Baumann et al. [8] also reported on 2024Al/7075Al joints produced in 1 inch thick joints with efficiencies greater than 80% of the 2024-T351 parent material properties with fractures occurring in the lower strength HAZ.

Corrosion of Dissimilar Aluminum Alloys Joints

There is even less data in the literature about corrosion related issues associated with dissimilar joints. One of the primary concerns with corrosion that may be related to the joining of dissimilar metals is the inducement of accelerated galvanic corrosion brought on by the coupling of materials with different corrosion potentials. Because different alloys have differences in their corrosion potentials, some extent of galvanic corrosion is possible and should be investigated. In the case of 2XXX series and 7XXX series alloys, it is expected that the 7XXX series will corrode anodically to the nobler 2XXX series alloy. As a result, the 2XXX series alloy would be protected cathodically since the 7XXX series corrodes preferentially. Cook et al. [6] also investigated the effect of alternating advancing and retreating sides in dissimilar welds of 2024-T3 and 7075-T73 on the exfoliation resistance of the FSW joint. They observed different behaviors on the top (weld side) and the bottom (anvil side) of the welds, with one side doing better than the other, depending on which material was on the advancing side of the joint. They also reported that the corrosion, where present, was worse in the 7075 material than in the 2024 material, which is expected due to the well-documented inferior resistance of 7XXX series to

450

exfoliation as compared to 2XXX series alloys. They concluded that the exfoliation resistance was maximized by placing 7075-T73 on the retreating side. This result is likely related to the temperature rise on the advancing side being slightly higher than on the retreating side during welding. Bala Srinivasan et al. [3] investigated the stress corrosion cracking resistance of 7075-T7351 joined to Al 6056 using slow strain rate testing in a sodium chloride environment. They found that only the TMAZ/HAZ interface on the 7075-T73 side of the joint was susceptible to SCC at strain rates in the order of 10^{-7} s^{-1}. Limited data was found in the literature about improving the corrosion resistance of dissimilar alloy joints through post-weld artificial aging (PWAA). On the contrary, Bauman et al. [8] reported on the difficulty of applying PWAA to dissimilar alloy joints due to the differences in treatment temperatures and times. Gérard and Ehrström [5], however, did investigate some applications of PWAA and reported the potential for welding 2XXX series alloys in the T8 condition to 7XXX series alloys, but did not discuss the effects on the corrosion resistance of the dissimilar couple.

Experimental Procedure

Butt welds were made between 0.125-in. thick sheets of 2024-T81 and 7075-T73 parallel to the rolling direction, using a standard FSW pin tool with an approximately 2:1 concave shoulder and a cylindrical threaded probe, reference Figure 1. The welds were made at 600 rpm and 8 ipm, under load control with a forging force of 2,150 lbs. The base material properties of the alloys welded are shown in Table I. Dissimilar alloy joints were produced alternating 2024-T81 on the advancing and retreating sides of the joint, and were then given post-weld artificial aging to enhance their resistance to corrosion.

0.1-in.

Figure 1. FSW Fixed Shoulder Pin Tool

Table I. Base Material Actual Compositions

Alloy	Composition								
	Si	Fe	Cu	Mn	Mg	Cr	Zn	Ti	Al
0.125-in 2024-T81	.07	.15	4.7	.63	1.6	.00	.06	.03	Bal.
0.125-in 7075-T73	.08	.28	1.6	.03	2.6	.19	5.5	.02	Bal.

Table II. Base Material Properties

Alloy	Tensile Strength (ksi)	Yield Strength (ksi)	Elongation (%)	Conductivity (%IACS)
0.125-in 2024-T81	72.1 (497 MPa)	66.3 (457 MPa)	8.0	39.0 - 39.7
0.125-in 7075-T73	73.1 (504 MPa)	61.1 (421 MPa)	11.1	39.1 - 39.8

Tension testing was performed per ASTM E-8. Yield strength calculations were based on a gauge length of 0.5-inch centered across the weld (incorporating both heat affected zones). Exfoliation testing was conducted per ASTM G34, with an exposure time of 96 hours. An exfoliation rating system was defined for FSW by assigning 0 points for N (no apparent attack),

1 point for EA (superficial attack), 2 points for EB (Moderate Attack), 3 points for EC (Severe Attack), and 4 points for ED (Very Severe Attack). Points were assigned for the parent material, HAZ, and weld nugget, so that the minimum score was 0 and the maximum was 12. Electrical conductivity testing was performed by a level-2 NDI technician using a hand-held Staveley Nortec 2000S Eddy Current tester. Measurements were the average of three measurements taken at each location across the friction stir weld starting on the advancing side on the root side of the weld. Microhardness measurements were taken using a Buehler Micromet® model 5103 microindentation hardness tester. Hardness measurements were taken at increments of 0.050-inch, at mid-thickness. Results are from one sample taken from a steady-state portion of the weld. Recorded values are the average of three measurements per location. The PWAA aging treatments and temper combinations which were investigated to find a beneficial heat treatment for 2024 and 7075 are shown in Table III.

Table III. Post-Weld Heat Treatments with Alloy and Temper Combinations Investigated for Exfoliation Corrosion Resistance

2024-T3	2024-T81	7075-T6	7075-T73	Dissimilar 2024/7075
Naturally Aged	Naturally Aged	Naturally Aged	Naturally Aged	Naturally Aged
24 hrs @ 225 °F	24 hrs @ 225 °F	24 hrs @ 225 °F	24 hrs @ 225 °F	24 hrs @ 225 °F
100 hrs @ 225 °F	100 hrs @ 225 °F	100 hrs @ 225 °F	48 hrs @ 225 °F	100 hrs @ 225 °F
9 hrs @ 355 °F	4 hrs @ 325 °F	9 hrs @ 355 °F	100 hrs @ 225 °F	2 hrs @ 325 °F
1.1 hrs @ 365 °F	1 hrs @ 375 °F	RRA - 8 hrs @ 320 °F	1 hr @ 325 °F	4 hrs @ 325 °F
2.3 hrs @ 365 °F	2 hrs @ 375 °F	RRA - 11 hrs @ 320 °F	2 hrs @ 325 °F	
4.5 hrs @ 365 °F	4 hrs @ 375 °F	RRA - 2 hrs @ 355 °F	4 hrs @ 325 °F	
9 hrs @ 365 °F		RRA - 3 hrs @ 355 °F	8 hrs @ 325 °F	
12 hrs @ 375 °F		9 hrs @ 355 °F	24 hrs @ 325 °F	
		12 hrs @ 375 °F		

Results

Tension

A bar chart showing the ultimate tensile strengths of the treated samples is shown in Figure 2. The results have also been compared to 0.250-inch thick naturally aged dissimilar joints of 2024-T3 (Adv) and 7075-T73 (Ret) tested by Cook et al. [6]. All of the values reported are for heat treatments that gave beneficial exfoliation corrosion resistance, with the exception of the data reported by Cook et al. [6]. The differences in ultimate tensile strength between all tested values was not significant. However, an apparent reduction in the elongation at failure between the 100-hours at 225°F samples and the 2- and 4-hours at 325°F samples was observed. The difference is attributed to the failure location which occurred in nugget for the 100-hours at 225°F samples and in the HAZ for the 2- and 4-hours at 325°F samples.

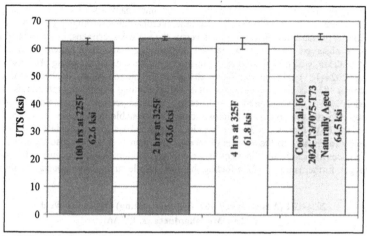

Figure 2. UTS of Dissimilar Welds of 2024-T81 (Adv) / 7075-T73 (Ret) plus PWAA Compared to Reported Data

<u>Microhardness</u>

The microhardness testing of the dissimilar joints of 2024 and 7075 clearly reveal the division between the higher strength 7075 material and the lower strength 2024 material. As shown in Figure 3, a clear demarcation exists between the 7075 and 2024 material. It can also be seen that the minimum microhardness was always in the heat-affected zone of the 2024 material; however, the microhardness appears to be lowest when 7075 is on the advancing side, which is agreement with tensile results and reports from other researchers.

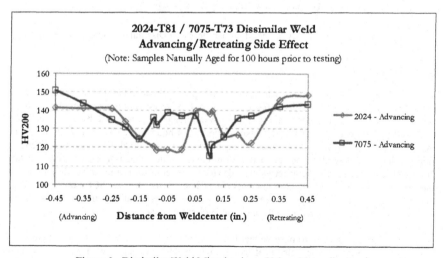

Figure 3. Dissimilar Weld Microhardness Values Naturally Aged

In Figure 4, the effect of post-weld aging on the microhardness of the dissimilar joint is evaluated. There is a clear difference between the microhardness readings on the retreating side (7075-T73) for the naturally aged specimens versus the PWAA specimens. Lower microhardness values were observed for the 7075 material in the nugget and HAZ when given a PWAA treatment at 325°F, which is most likely associated with precipitate coarsening. However, on the advancing (2024-T81) side, no significant change in the microhardness is observed due to the heat treatment. This is in strong agreement with earlier findings that the microhardness profile in as-welded 2024-T81 is not altered by short-term, lower-temperature, post-weld artificial aging treatments. Clearly, the precipitate microstructure is very stable in this case. It appears that by the end of the FSW process, 2024-T81 has already reached a stable precipitate structure that is subsequently insensitive to these types of thermal treatments. It should also be noted that the only differences in the microhardness of between 2- and 4-hours at 325°F occurs in the 7075 side of the nugget, further supporting the finding that their tensile strengths do not significantly differ either.

2024-T81 (Advancing) / 7075-T73 (Retreating) Dissimilar Weld
Vickers Microhardness vs. PWAA

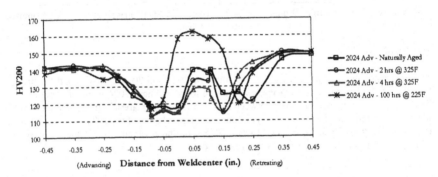

Figure 4. Effect of PWAA on the Microhardness of a Dissimilar Friction Stir Weld

Electrical Conductivity

The effect of friction stir welding with and without post-weld artificial aging on the electrical conductivity of friction stir welded 2024-T81 to 7075-T73 is shown in Figure 5 and Figure 6. In both cases, the electrical conductivity in the weld nugget is increased by post-weld heat treatments. The greatest improvement takes place with 4-hours at 325°F when 7075-T73 is on the advancing side; however, the effect is nearly the same when 2024-T81 is on the advancing side. It should be noted that the conductivities in the as-welded condition are the same whether 2024-T81 is on the advancing or retreating side (for the level of measurement resolution used in the tests). While SCC testing was not possible, it is expected that the SCC resistance of the dissimilar joint will be greatly improved by these treatments, especially on the more susceptible 7075 side of the joint, in part due to the increase in electrical conductivity [9,10].

Additionally, it is important to observe that both materials have nearly the same initial electrical conductivity. Since higher electrical conductivities are associated with a more general coarsening of precipitates , this means that the matrix surrounding those precipitates is more purely aluminum, thus reducing the potential differences between the two dissimilar alloys. Hence, one of the explanations for the unexpected corrosion resistance of the dissimilar alloy joint is that in the over-aged, stress-relieved condition, there is a greatly diminished corrosion potential between 2024 and 7075, a trend that could possibly apply to other 2XXX and 7XXX series alloys as well. In the as-welded condition, both alloys have a high degree of supersaturation and, as a result, a higher corrosion potential between them due to the differences in the concentrations of dissolved solutes. With increasing PWAA time, those differences in corrosion potential are reduced as precipitates form and coarsen in the matrix. This appears to be consistent with the fact that the degree of supersaturation, e.g. the high matrix solute content in non-overaged tempers, correlates well with the corrosion rates observed in the bi-alloy joints noted earlier.

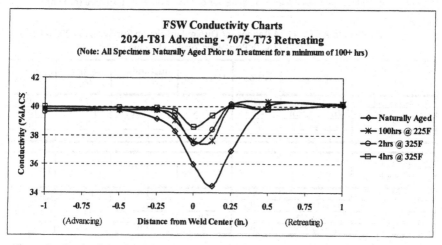

Figure 5. Conductivity Chart for Dissimilar FSW with 2024-T81 on the Advancing Side and 7075-T73 on the Retreating Side

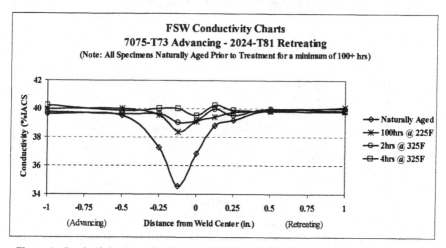

Figure 6. Conductivity Chart for Dissimilar FSW with 7075-T73 on the Advancing Side and 2024-T81 on the Retreating Side

<u>Exfoliation</u>

The results of the dissimilar welding exfoliation investigation on the top surface are summarized in Table IV and are shown graphically in Figure 7. Micrographs of the results of the are shown in Figure 8 and Figure 9. There has been an obvious improvement in the exfoliation resistance for the PWAA samples versus the naturally aged samples. No severe pitting is present in the HAZ on the 7075 retreating side, as was the case previously [6]. Each alloy is clearly distinguishable in each DXZ shown the micrographs in Figures 7 and 8, indicating that a definite boundary is retained following FSW. Exfoliation of the bottom surface (root) of the weld (not shown) exhibited the same basic corrosion behavior observed on the top surface (crown).

Table IV. Exfoliation ASTM G-34 Ratings for Dissimilar Friction Stir Welds of 2024-T81 to 7075-T73 with Post-Weld Artificial Aging

2024-T81 / 7075-T73 Dissimilar Joint	PWAA	Weld Zone	HAZ (Adv)	Base Material (Adv)	Base Material (Ret)	HAZ (Ret)
2024-T81 (Adv)	N.A.	N/EA	N/EA	N	EB	ED
"	100 hrs @ 225F	N/EA	N/EA	N/EA	EA/EB	EB
"	2 hrs @ 325F	N/EA	N/EA	EA	EB	EB
"	4 hrs @ 325F	N/EA	N/EA	EA	EA/EB	EB

2024-T81 / 7075-T73 Dissimilar Joint	PWAA	Weld Zone	HAZ (Adv)	Base Material (Adv)	Base Material (Ret)	HAZ (Ret)
7075-T73 (Adv)	N.A.	EC	ED	EB	EA/N	EA/N
"	100 hrs @ 225F	EA/EB	EB	EB	N/EA	N
"	2 hrs @ 325F	EC	EB	EB	EA	EA
"	4 hrs @ 325F	EB	EC	EB	EA/EB	EA

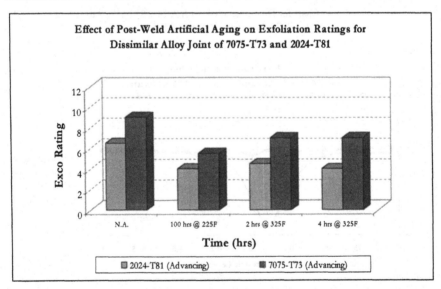

Figure 7. Exfoliation Numerical Ratings for 7075-T73 and 2024-T81 Dissimilar Welds with Post-Weld Artificial Aging

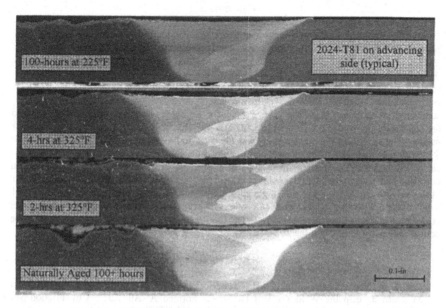

Figure 8. Dissimilar FSW Exfoliation Results: 2024-T81 Advancing (right) / 7075-T73 Retreating (left). Top: PWAA for 100-hours at 225°F; Middle Top: PWAA for 4-hours at 325°F; Middle Bottom: PWAA for 2-hours at 325°F; Bottom: Naturally Aged 100+ hours

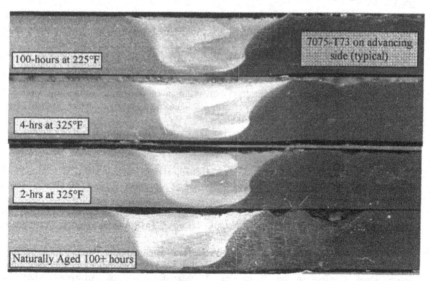

Figure 9. Dissimilar FSW Exfoliation Results: 7075-T73 Advancing (right) / 2024-T81 Retreating (left). Top: PWAA for 100-hours at 225°F; Middle Top: PWAA for 4-hours at 325°F; Middle Bottom: PWAA for 2-hours at 325°F; Bottom: Naturally Aged 100+ hours

Conclusions

An investigation of potential post weld heat treatments for dissimilar friction stir welded joints between 2XXX and 7XXX series alloys has been successfully evaluated. Through this study it was found that welding 2024 and 7075 in appropriate initial overaged tempers and then applying short-duration PWAA treatments can be used to restore the corrosion resistance of the joint. Joints were produced that maintained a high strength level and were found to achieve similar electrochemical potentials, as evidenced by the lack of galvanic attack and similar high electrical conductivities on either side of the joint. Previously this result was not thought attainable due to the apparently unavoidable effects of producing a galvanic couple when joining two dissimilar metals. The success of the present study was observed to be related to FSW 2024-T81 to 7075-T73 with the 2024-T81 material on the advancing side, followed by heat treating the joint for either 100-hours at 225°F or 2- to 4-hours at 325°F. With these treatments, the exfoliation corrosion resistance of the material associated with the joint was found to be largely restored. As observed with the corrosion of the single alloy joints, the corrosion resistance of the 2024-T81 side is superior to that of the 7075-T73 side; however, the difference is proportional to the corrosion behavior of the materials alone. The tensile strength of the joint was also found to be comparable to the strength of the 2024-T81 naturally aged specimens.

References

1. Lee, W.B., Yeon, Y.M., and Jung, S.B., "The Mechanical Properties Related to the Dominant Microstructure in the Weld Zone of Dissimilar Formed Al Alloy Joints by Friction Stir Welding," *Journal of Materials Science*, 38 (2003), 4183-4191.
2. Larsson, H., et al., "Joining of Dissimilar Al-alloys by Friction Stir Welding," *Proceedings of the 2nd International Friction Stir Welding Symposium*, Gothenburg, Sweden, 26-28 June, 2000.
3. Bala Srinivasan, P. et al., "Stress Corrosion Cracking Susceptibility of Friction Stir Welded AA7075-AA6056 Dissimilar Joint," *Materials Science and Engineering A*, 392 (2005), 292-300.
4. Nelson, T.W., Steel, R.J., and Arbegast, W.J., "Investigation of Heat Treatment on the Properties of Friction Stir Welds," ASM International Aeromat Conference Presentation, 2001.
5. Gérard, H. and Ehrström, J.C., "Friction Stir Welding of Dissimilar Alloys for Aircraft," *Proceedings of the 5th International Friction Stir Welding Symposium*, Metz, France 14-16 September, 2004.
6. Cook, R. et al., "Friction Stir Welding of Dissimilar Aluminum Alloys," *Friction Stir Welding and Processing III*, TMS Annual Meeting, San Diego, CA, 2-6 March, 2005, pp. 35-42.
7. Cavaliere, P., Cerri, E., and Squillace, A., "Mechanical Response of 2024-7075 Aluminum Alloys Joined by Friction Stir Welding," *Journal of Materials Science*, 40 (2005), 3669-3676.
8. Baumann, J.A. et al., "Property Characterization of 2024Al/7075Al Bi-Alloy Friction Stir Welded Joints," *Friction Stir Welding and Processing II*, TMS Annual Meeting, Warrendale, PA, 2003, pp. 199-207.
9. Widener, C.A. et al., "Evaluation of Post-Weld Heat Treatments to Restore the Corrosion Resistance of Friction Stir Welded Aluminum Alloy 7075-T73 vs. 7075-T6," *Thermec 2006*, International Conference on Processing & Manufacturing of Advanced Materials, Vancouver, Canada, July 4-8, 2006.
10. Widener, C.A. et al., "Investigation to Restore the Exfoliation Resistance of Friction Stir Welded Aluminum Alloy 2024," *Friction Stir Welding and Processing IV*, TMS Annual Meeting, Orlando, FL, Feb. 25 – Mar. 1, 2007.

INVESTIGATION TO RESTORE THE EXFOLIATION RESISTANCE OF FRICTION STIR WELDED ALUMINUM ALLOY 2024

Christian A. Widener[1], Jorge E. Talia[2], Bryan M. Tweedy[1], and Dwight A. Burford[1]

[1]National Institute for Aviation Research
[2]Department of Mechanical Engineering

Wichita State University
1845 Fairmount St., Wichita, KS 67260

Keywords: friction stir, 2024-T3, 2024-T81, artificial aging, corrosion

Abstract

In the as-welded condition, joints produced in aluminum alloy 2024-T3 by friction stir welding (FSW) have properties that are comparable to the parent material, including mechanical properties, stress corrosion cracking resistance, and damage tolerance. However, in the naturally aged condition they are highly susceptible to exfoliation corrosion. Therefore, a research program was undertaken to develop methods for restoring joint exfoliation corrosion resistance without significantly affecting other mechanical properties. The starting temper of the parent material and potential post-weld artificial aging (PWAA) treatments of butt-welds were investigated in sheets of 0.125-inch 2024-T3 and 2024-T81. The samples were given PWAA treatments for various time and temperature combinations. Thermal treatments were evaluated using optical microscopy, exfoliation, electrical conductivity, microhardness, tensile, and fatigue crack propagation (FCP) testing. It was demonstrated that the exfoliation resistance of friction stir welded Al 2024 joints may be restored through PWAA to the –T81 temper or when initially welded in the –T81 temper, followed by naturally aging.

Introduction

Aluminum alloy 2024-T3 is one of the most commonly used aluminum alloys in the aerospace industry because of its good balance of strength and low crack growth rates (in the –T3 temper). Friction stir welding (FSW) creates sound, defect free joints in 2024, but at the expense of its exfoliation resistance [1-4]. In general, aluminum alloys that contain appreciable amounts of soluble alloying elements, primarily copper, magnesium, silicon, and zinc, like 2xxx and 7xxx alloys, are potentially susceptible to exfoliation and stress corrosion cracking (SCC). A considerable amount of research has been conducted in order to develop corrosion-resistant tempers that still maintain desirable mechanical properties. However, one of the challenges of FSW is that it alters the temper of the base material across the weld zone. While some research has been conducted in order to find solutions to address corrosion in friction stir welds, more work is needed. Preferential corrosion of the FSW joint over the parent material is an important concern associated with broad implementation of this technology.

One established method for evaluating a material's corrosion resistance is an exfoliation test. Exfoliation corrosion testing determines a material's susceptibility to two types of corrosion: uniform attack and pitting corrosion. Uniform attack is the general attack of the material in a corrosive environment, often measured in average depth per unit time. Since the material is attacked evenly over its surface, the exfoliation test is the least destructive (if the attack does not occur at an accelerated rate). Pitting corrosion, on the other hand, is of great concern, since pits can grow well in advance of the general corrosion state of the material, creating large stress concentrations that can cause premature failure of a part under a variety of loading conditions. Pitting in wrought aluminum alloys primarily originates from a galvanic reaction between coarse precipitates and other constituent particles and the surrounding aluminum matrix. The pits form either by the dissolution of particles anodic with respect to the surrounding matrix or by the dissolution of the matrix adjacent to cathodic particles [5]. The challenge, therefore, is to determine if FSW in a given material causes it to be more susceptible to corrosion than the parent material and, if so, to develop methods to improve the corrosion response. One possible method is through post-weld artificial aging (PWAA).

Microstructure

In precipitation strengthened aluminum alloys a softening phenomenon is often observed in the heat-affected zone (HAZ) and thermo-mechanically affected zone (TMAZ) in FSW joints. The primary cause of this softening is the coarsening and transformation of precipitates. Genevois et al. [6], explained in detail the mechanism for softening in the HAZ for Al 2024-T3. In a follow-up study, Genevois et al. [7] reported on the effects of FSW on 2024-T6(T8). They carried out a quantitative investigation of changes that occur in the microstructure and mechanical properties of 2024-T351 given a post-weld artificial aging treatment of 10-hours at 375°F (which the authors referred to as T6, but could also be referred to as T8, since the starting temper was T351, and the PWAA time and temperature for T6 and T8 are almost the same – 10 vs. 12 hours at 375°F, respectively) and compared the results to that of material given the same T6(T8) aging treatment prior to welding. In their study, they found that the microstructure in the weld nuggets of 2024-T351 and 2024-T6(T8) are almost identical. However, they found that the two materials had opposite trends in the HAZ with respect to the distribution of GP zones and S′(S) precipitates. They also observed that when 2024-T351 is friction stir welded and then aged to T6(T8), the HAZ experiences the maximum coarsening of S′(S) precipitates and, therefore, a minimum strength value as compared to as-welded 2024-T351 and 2024-T6(T8).

It has also been shown that 2xxx series alloys, including 2024, will become essentially stable after approximately 100-hours of natural aging after welding [8]. As a result, PWAA is only required if it can have some other beneficial effect besides stabilizing the microstructure of the weld zone, such as improving corrosion resistance or relieving residual stresses.

Corrosion

One of the first and most complete evaluations of corrosion in friction stir welded 2024-T3 was conducted by Biallas et al. [1]. They looked at 1.6-mm and 4-mm thick 2024-T3 joined by FSW. They found evidence of selective attack of the weld zone during ASTM G-34 exfoliation testing, especially in the heat-affected zone, and susceptibility to intergranular attack in the weld nugget when immersed in an ASTM G-110 intergranular corrosion test solution. They found no susceptibility to stress corrosion cracking after 40 days of alternate immersion testing in an aqueous 3.5% NaCl solution (per ASTM G-44), using four-point loaded bent-beam specimens (ASTM G-39), loaded to a stress of 36.2 ksi (250 MPa). Hannour et al. [2] made a similar determination that the weld zone, especially the HAZ, was more sensitive to attack by corrosion as compared to the parent material. They used polarization curves to show that both the weld nugget and HAZ had high corrosion potentials, with the highest potential in the HAZ. Later, Davenport et al. [3] found that welding process parameters could affect whether the HAZ or nugget had the highest corrosion potential. The effect was due to changes in the thermal gradients and maximum temperatures caused by FSW, which in turn affected precipitate morphology. Kumar et al. [4] also reported the susceptibility of as-welded 2024-T3 to exfoliation corrosion for both 0.040-inch and 0.125-inch material, and recommended further investigation into methods to improve its corrosion resistance.

One of the reported ways to restore the corrosion resistance to 2024-T3 in the as-welded condition is through laser surface melting (LSM) treatments [3,9,10]. Through LSM, the surface can be homogenized, which eliminates variations in corrosion potential at the exposed surface, greatly reducing the resistance to corrosive elements. On the other hand, Corral et al. [11] did not find a severe sensitivity to corrosion for 2024 when it was friction stir welded in the T4 condition. They exposed test samples in 0.6M NaCl solutions for 20-hours and observed similar corrosion patterns in both the weld and parent materials. They also showed that both FSW and parent material have similar potentiodynamic polarization curves in de-aerated 0.6M NaCl to support their conclusions. Similarly, Li et al. [12] examined the stress corrosion susceptibility of two 2xxx series alloys in the –T8x condition, namely 2195-T8 and 2219-T87 in the as-welded condition. They found that both 2195 and 2219 possessed higher resistance to environmentally assisted cracking than their parent materials, with 2195 exhibiting the highest resistance to SCC.

Other authors have endeavored to investigate the effects of corrosion on fatigue life in 2XXX series and 2024 aluminum alloys. Alfaro Mercado et al. [13] compared the fatigue life of pristine and pre-corroded as-welded 2024-T3 coupons to pristine and pre-corroded parent material. Specimens were pre-corroded per the ASTM G-44 alternate immersion method and then left in laboratory air for 100, 250, and 1,000-hours. They reported a marked reduction in fatigue life for all of the pre-corroded samples, and demonstrated that the friction stir weld behaved comparably to the parent material in both pristine and pre-corroded conditions. They also compared experimental results to predicted values. In their model all corrosion damage was treated as a semi-elliptical surface crack with a depth equal to the deepest pit or intergranular attack and a width equal to the width of the corrosion area. Excellent agreement was achieved between their model and experimental results, demonstrating the ability to predict the behavior of corrosion damage using computational methods.

Experimental Procedure

Butt welds were made in 0.125-in. thick sheets of 2024-T3 and 2024-T81 parallel to the rolling direction, using the FSW pin tool shown in Figure 1, which has an approximately 2:1 concave shoulder and a cylindrical threaded probe. The welds were made at 600 rpm and 8 ipm, under load control with a forging force of 2,150 lbs. The base material properties of the alloys welded are shown in Table I.

0.1-in.

Figure 1. FSW Fixed Shoulder Pin Tool

Table I. Base Material Properties

Alloy	Tensile Strength (ksi)	Yield Strength (ksi)	Elongation (%)	Conductivity (%IACS)
0.125-in 2024-T3	67.7 (467 MPa)	46.8 (323 MPa)	17.7	30.1
0.125-in 2024-T81	72.1 (497 MPa)	66.3 (457 MPa)	8.0	39.0 - 39.7

Tension testing was performed per ASTM E-8. Yield strength calculations were based on a gauge length of 0.5-inch centered across the weld (incorporating both heat affected zones). Exfoliation testing was conducted per ASTM G34. An exfoliation rating system was defined for FSW by assigning 0 points for N (no apparent attack), 1 point for EA (superficial attack), 2 points for EB (Moderate Attack), 3 points for EC (Severe Attack), and 4 points for ED (Very Severe Attack). Points were assigned for the parent material, HAZ, and weld nugget, so that the minimum score was 0 and the maximum was 12. Electrical conductivity testing was performed by a level-2 NDI technician using a hand-held Staveley Nortec 2000S Eddy Current tester. Measurements were the average of three measurements taken at each location across the friction stir weld starting on the advancing side on the root side of the weld. Microhardness measurements were taken using a Buehler Micromet® model 5103 Microindentation Hardness tester. Hardness measurements were taken at increments of 0.050-inch, at mid-thickness. Results are from one sample taken from a steady-state portion of the weld. Recorded values are the average of three measurements per location. Testing to determine fatigue crack propagation (FCP) characteristics was conducted per ASTM E-647. FSW joints were tested at a loading ratio of $R = 0.5$, with an applied load of 1100 lbs, and a minimum of two samples per condition. FCP testing was conducted using the eccentrically loaded single-edge crack tension specimen ESE(T) with a width of 3.00-inches. The notch was placed in the weld nugget, against the welding direction, where crack growth rates in the FSW weld zone have been reported to be the highest.

Results

Tension

The effects of the PWAA heat treatments on ultimate tensile strength, versus the naturally aged condition for 2024-T3 and 2024-T81 are shown in Figure 2. For 2024-T3, a decrease in ultimate tensile strength of the heat treated samples was observed over the naturally aged samples. Between the various times at 365°F, tensile strength noticeably decreased with increasing time at temperature. An improvement in tensile strength is clearly observed by joining 2024 in the –T81 temper versus the –T3 temper followed by PWAA, which is consistent with the results reported by Genevois et al. [7].

Microhardness

The microhardness profiles that were obtained for the friction stir welds in 2024-T3 are shown in Figures 3. They display the classic FSW "W" profile, with the minimum hardness values occurring in the heat-affected zones. These results correlate well with tensile strength values, with friction stir welded 2024-T3 aged to T81 having both the lowest microhardness values and tensile strength. It can also be seen that 4.5 hours at 365°F was sufficient to increase the microhardness of the unaffected 2024-T3 material to –T81 parent values. Also, for the 2024-T3 samples, higher microhardness values in the weld zone were achieved with low-temperature artificial aging at 225°F for 100 hours, compared to the treatments at 365°F for 4.5 hours or more.

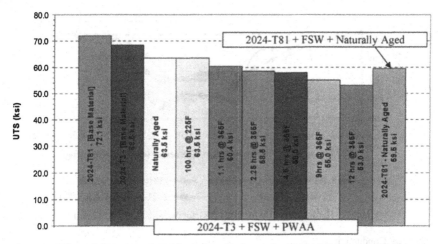

Figure 2. Effect of PWAA on the Ultimate Tensile Strength of 2024-T3 Compared to Parent Material and As-Welded 2024-T81

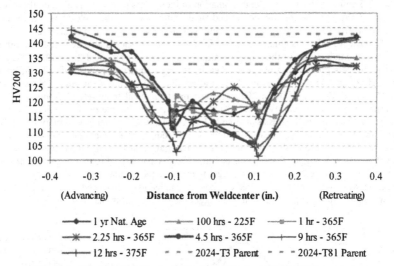

Figure 3. Hardness Profile for 2024-T3 for Various Thermal Treatments

The microhardness profiles that were obtained for the friction stir welds in 2024-T81 are shown in Figure 4. Low-temperature artificial aging (225°F) did not appear to have an effect on microhardness. When comparing as-welded 2024-T81 with 2024-T3 plus a T81 PWAA, however, there does appear to be an improvement in overall microhardness by welding initially in the T81 temper instead of applying a -T81 PWAA to material welded in the T3 temper. Again, this correlates well with tensile data, which showed that tensile and yield strength were improved by welding initially in the 2024-T81 temper versus welding initially in the T3 temper and applying PWAA to the T81 temper.

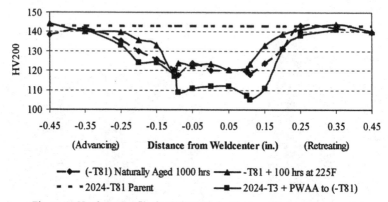

Figure 4. Hardness Profile for 2024-T81 for Various Thermal Treatments

Electrical Conductivity

Root side electrical conductivity plots of friction stir welded 2024-T3 and 2024-T81 are shown in Figures 5 and 6, respectively. An examination of these plots provides insight into the effect of FSW and PWAA on the microstructure and final tempers across the weld zone. With natural aging and artificial aging, the electrical conductivity in the weld zones can be seen to increase for 2024-T3; however, there is almost no effect for 2024-T81. It should also be noted that although the electrical conductivity of the nugget is almost the same in as-welded 2024-T3 or 2024-T81, the weld nuggets produced in material of each starting tempers responds radically different to post-weld aging. After 4-hours at 375°F, the conductivity in the weld nugget rises to about 37%IACS in 2024-T81, while in 2024-T3, an almost identical treatment (4.5-hours at 365°F) raises the electrical conductivity to 40%IACS. This indicates that the resulting microstructure in the initially –T81 material is less sensitive to PWAA than is the materially initially joined in a -T3 temper. When 2024-T3 is heat treated to the T81 temper, the response of the weld nugget is markedly different from that of the parent material, reaching a much higher electrical conductivity in the nugget than in the parent. This is partially due to the greater degree of supersaturation that exists in the weld nugget than in the parent material. There may also be an interaction with the increased number of sites for nucleation introduced by the FSW process, causing a more rapid aging response in the nugget than in the parent material.

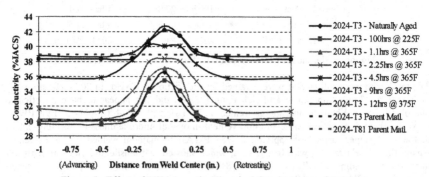

Figure 5. Effect of PWAA on the Electrical Conductivity of 2024-T3

464

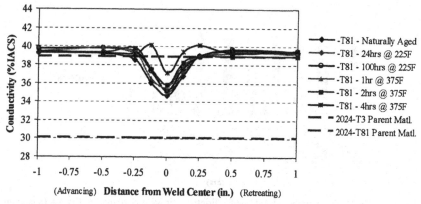

Figure 6. Effect of PWAA on the Electrical Conductivity of 2024-T81

Exfoliation

The only treatments that were found to restore the exfoliation resistance of the weld zone to initially 2024-T3 material were the 4.5-hour to 9-hour treatments at 365°F and 12-hours at 375°F. Heat treatments were based on a standard heat treatment for 2024-T81, which applies up to an additional 12-hours at 375°F to material in the T3 condition. The results are summarized in Table III and in Figure 7. The low-temperature treatments at 225°F resulted in almost no change from the naturally aged results. This was attributed to the rapid natural aging response of 2024.

Table III. Exfoliation ASTM-G34 Ratings for Friction stir welded 2024 with PWAA

Alloy	PWAA	Weld Zone	HAZ	Base Material
2024-T3	PARENT			EA/EB
"	Naturally Aged	ED	ED	N
"	24 hrs at 225F	ED	ED	EA
"	100 hrs at 225F	ED	ED	EA
"	1 hr at 365F	ED	ED	EC
"	2 hrs at 365F	EC	ED	EC
"	4 hrs at 365F	EA	EB	EB
"	9 hrs at 365F	N/EA	EA	EA
2024-T3	12 hrs at 375F	N	EA/EB	EA/EB
2024-T81	PARENT			EA
2024-T81	Naturally Aged	N	N	EA

Based on the observation that beneficial exfoliation results were achieved in welded material aged from T3 to T81, welding 2024 in the T81 condition was investigated. A number of heat treatments were initially investigated for 2024-T81 because it was expected that 2024-T81 would show a susceptibility to exfoliation corrosion in the as-welded condition. Surprisingly, 2024-T81 in the as-welded condition was found to have excellent exfoliation corrosion resistance. A comparison of micrographs from the 2024-T3 samples and the 2024-T81 naturally aged sample are shown in Figure 8. The best exfoliation resistance (lowest Exco Rating index) is achieved by the 2024-T81 sample, although the 2024-T3 exposed to PWAA for 9 hours at 365°F and 12 hours at 375°F are also considered very good. The results for 2024-T81 samples are summarized in Table IV. Some small pits were observed in the unaffected parent material, but no signs of corrosion were found in the weld or HAZ. The exfoliation behavior of the weld track surface root surface was observed to exhibit the same behavior as the top surface (crown) of the weld.

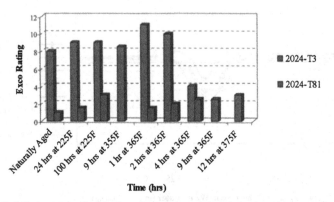

Figure 7. Exfoliation Numerical Ratings for Friction stir welded 2024 with PWAA (a decreasing Exco Rating index or number indicates an increasing exfoliation resistance)

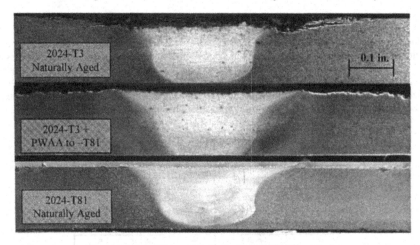

Figure 8. Comparison of 2024 Exfoliation Micrographs

Fatigue Crack Propagation

Figures 9 and 10 show the results from the fatigue crack propagation portion of this study on 2024 FSW samples welded in the -T3 condition. The FCP results for both the naturally aged samples and those treated for 24-hours at 225°F had superior FCP results to all other treatments. This is most likely a residual stress effect, as reported by Bussu and Irving [15], who found that when the residual stresses are relieved from FSW, the crack growth rates in the weld zone are almost identical to those in the base material (indicated in Figure 9). The 365-375°F PWAA treatments that were tested produced slightly increased crack growth rates as compared to that of the –T3 parent material. Also, it should be noted that a hold time of 1 hr at 365°F was sufficient to significantly alter crack growth rates in 2024-T3 FSW joints. For the 2024 samples, when tested against the direction of welding, cracks tended to quickly veer off the weld center into the base material, which could have been a microstructural interaction due to the swirl patterns in the weld nugget, as suggested by Fuller et al. [14].

466

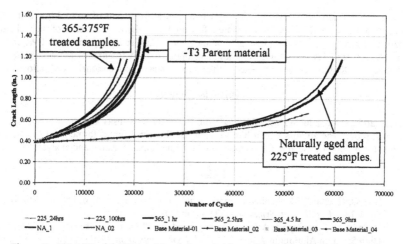

Figure 9. 2024-T3 with PWAA (Crack Length (a) vs. (N) Cycles) (R=0.5 at 1100lbs)

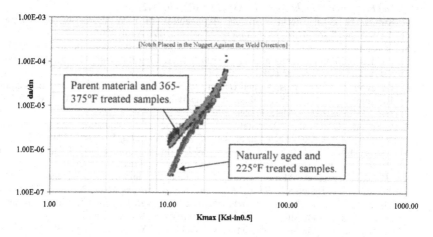

Figure 10. da/dN vs. K_{max} for 0.125-in 2024-T3 for Various Thermal Treatments

Conclusions

Through this study, it has been demonstrated that the exfoliation resistance of friction stir welded 2024 can be restored through post-weld artificial aging. Further, it was shown that when 2024-T3 is tested in the as-welded condition, it exhibits much lower crack growth rates than parent material. Friction stir welded 2024-T3, in which exfoliation corrosion resistance has been restored through PWAA, has moderately decreased tensile strength but a crack growth resistance comparable to (but slightly lower than) that of parent material values. Even better exfoliation resistance and higher tensile strengths can be achieved by friction stir welding 2024 in the T81 condition. It was also found that for the −T81 temper heat treatments of up to 4-hours at 375°F and up to 100-hours at 225°F had no notable effect on the exfoliation resistance of the weld zone or parent material, making it a potential candidate for dissimilar alloy joints with PWAA [16]. This research provides the groundwork for developing PWAA treatments for next generation over-aged 2XXX alloys designed to exceed the performance of conventional 2XXX alloys.

Acknowledgements

This research was conducted with funding from the State of Kansas and the Federal Aviation Administration. The authors would like to thank Dr. Brijesh Kumar, formerly of the National Institute for Aviation Research, for his assistance with mechanical testing and analysis.

References

1. Biallas, G. et al., "Mechanical Properties and Corrosion Behavior of Friction Stir Welded 2024-T3," *1st Intl. FSW Symposium*, Thousand Oaks, CA, 1999.
2. Hannour, F., Davenport, A.J., and Strangwood, M., "Corrosion of Friction Stir Welds in High Strength Aluminum Alloys," *2nd Intl. FSW Symposium*, Gothenburg, Sweden, 2000.
3. Davenport, A.J. et al., "Corrosion of Friction Stir Welds in Aerospace Alloys," *Corrosion and Protection of Light Metal Alloys*, 23 (2003), 403-412.
4. Kumar, B. et al. "Review of the Applicability of FSW Processing for Aircraft Applications," *46th AIAA SDM Conference*, Austin, TX, April, 2005.
5. Sankaran, K.K., Smith, H.L., and Jata, K., "Pitting Corrosion Behavior of Friction Stir Welded 7050-T74 Aluminum Alloy," *Trends in Welding Research*, Pine Mountain, GA, 15-19 April, 2002, pp. 284-286.
6. Genevois, C., Deschamps, A., and Denquin, A., "Characterization of the Microstructural Evolution during FSW of Al Alloys: A Comparative Study of 5251 and 2024 Alloys," *5th Intl. FSW Symposium*, Metz, France, 2004.
7. Genevois, C. et al., "Quantitative Investigation of Precipitation and Mechanical Behaviour for AA2024 Friction Stir Welds," *Acta Materialia*, 53 (2005), 2447-2458.
8. Nelson, T.W., Steel, R.J., and Arbegast, W.J., "Investigation of Heat Treatment on the Properties of Friction Stir Welds," Aeromat Conference Presentation, 2001.
9. Williams, S. et al., "Laser Treatment Method for Improvement of the Corrosion Resistance of Friction Stir Welds," *Materials Science Forum*, 426-432(2003), 2855-2860.
10. Hannour, F. et al., "Corrosion Behavior of Laser Treated Friction Stir Welds in High Strength Aluminum Alloys," *3rd Intl. FSW Symposium*, Kobe, Japan, 2001.
11. Corral, J. et al., "Corrosion of Friction-Stir Welded Al Alloys 2024 and 2195," *Journal of Materials Science Letters*, 19 (2000), 2117-2122.
12. Li, Z.X. et al., "Microstructure Characterization and Stress Corrosion Evaluation of Friction Stir Welded Al 2195 and Al 2219 Alloys," *Trends in Welding Research*, Pine Mountain, GA, 1-5 June, 1998, pp. 568-573.
13. Alfaro Mercado, U., Ghidini, T., and Dalle Donne, C., "Fatigue of Pre-corroded 2024-T3 Friction Stir Welds: Experiment and Prediction," *Friction Stir Welding and Processing III*, San Diego, CA, 2005, pp. 43-53.
14. Fuller, C. et al., "Unpredictable Stress Corrosion Crack Growth in Friction Stir Welded 7075 Al," *Friction Stir Welding and Processing III*, San Francisco, CA, 2005, pp. 11-18.
15. Bussu, G., and Irving, P.E., "The Role of Residual Stress and Heat Affected Zone Properties on Fatigue Crack Propagation in Friction Stir Welded 2024-T351 Aluminum Joints," *International Journal of Fatigue*, 25 (2003), 93-104.
16. Widener, C.A. et al., "Corrosion in Friction Stir Welded Dissimilar Aluminum Alloy Joints of 2024 and 7075," *Friction Stir Welding and Processing IV*, Orlando, FL, 2007.

AUTHOR INDEX

470

SUBJECT INDEX